Working with Stem Cells

Henning Ulrich • Priscilla Davidson Negraes

# Working with Stem Cells

## A Quick and Easy Approach of Methodologies and Applications

Springer

Henning Ulrich
Department of Biochemistry
Institute of Chemistry
University of Sao Paulo
São Paulo, SP, Brazil

Priscilla Davidson Negraes
Department of Biochemistry
Institute of Chemistry
University of São Paulo
São Paulo, SP, Brazil

ISBN 978-3-319-30580-6 ISBN 978-3-319-30582-0 (eBook)
DOI 10.1007/978-3-319-30582-0

Library of Congress Control Number: 2016945374



Printed on acid-free paper

This Springer imprint is published by Springer Nature
The registered company is Springer International Publishing AG Switzerland

# Preface

We are in a unique moment in human history where stem cells, previously viewed as a great promise in medicine, are now slowly moving into clinical trials. Scientists are now confident to start testing different stem cell-based protocols in humans. For the first time, we will be the pioneers, having the opportunity to do it right, setting the example for the generations to come. Contrary to conventional medicine, cellular therapies will replace damaged tissues and stay there for the lifespan of a person. Alternatively, stem cells can be used as a tool to recapitulate human development and inform us about the disease pathology, proving opportunities for novel diagnostics and therapies. This is an extraordinary example of plasticity, never seen before in medical history. The fundamental aspects of stem cells come back to the methods that were developed over the years to reach to this point. We have learnt how to isolate, grow, enrich, and differentiate stem cells. These methods are still evolving in the scientific laboratories all over the world and will be optimized for the different medical strategies. This book summarizes the most current and important methods, and state-of-the art protocols to manipulate stem cells. It should not be seen as a fixed textbook, but a snapshot of a new and dynamic era in medicine.

San Diego, CA, USA — Alysson Renato Muotri, Ph.D.

# Contents

# About the Editors

**Henning Ulrich** studied Biology at the Universities of Hamburg and Kiel (Germany) and performed his master thesis in Biochemical Parasitology, followed by a Ph.D. in Biochemistry and Neuroscience at the University of Hamburg, Germany. He completed his training by postdoctoral research at the Center of Molecular Neurobiology at the University of Hamburg and Cornell University, NY, the latter financed by a fellowship of the American Heart Association. He came to the Institute of Chemistry at the Sao Paulo University as a visiting scientist, where he was then appointed as faculty member. Here he habilitated in Biochemistry and is currently Associate Professor and head of the laboratory of neuroscience, working on stem cells and mechanisms of neurogenic differentiation in diverse types of stem cells and stem cell techniques for possible therapeutic use of neurodegenerative disease. Dr. Ulrich has published more than 100 papers in peer-reviewed journals (H-factor of 30, more than 2600 citations), four books (three of them published with Springer), 17 book chapters, and participated in three patents. His research results were commented in major Brazilian newspapers. His is acting as editorial board member of renowned journals, Associate Editor of *Cytometry Part A* and Academic Editor of *PLoS ONE*.

**Priscilla Davidson Negraes** has a bachelor's degree in Biological Sciences from the State University of Londrina (Brazil), where she also got her master's degree in Genetics and Breeding after evaluating the antimutagenic effect of natural compounds in mammalian cells. She completed her Ph.D. in Genetics at São Paulo State University (Brazil), studying the relationship and relevance of DNA methylation for the prognosis of human bladder cancer. Dr. Negraes joined the Neuroscience and Stem Cell fields during her postdoctoral training at the University of São Paulo (Brazil), working on neural differentiation of stem cells, neurotransmitter systems and cell fate determination. Next, she became a Visiting Scholar at the University of California San Diego (USA) and started to model human neurological diseases using induced pluripotent stem cells. As a result of these experiences, Dr. Negraes developed an extensive knowledge in cell culture, cellular and molecular biology, and neural differentiation of stem cells. Currently, she is a reviewer for several scientific journals and works on modeling neurodevelopmental disorders to ultimately disclose new therapies and improve patients' quality of life.

# Contributors

**Jaime E. Anderson** Department of Neurosurgery, Harvard Medical School and Brigham and Women's Hospital, Boston, MA, USA

Division of SCI Research, VA Boston Healthcare System, Boston, MA, USA

**Treena L. Arinzeh** Department of Biomedical Engineering, New Jersey Institute of Technology, Newark, NJ, USA

**Amanda Faria Assoni** Department of Genetics and Evolutionary Biology, Human Genome and Stem Cell Research Center, Biosciences Institute, University of São Paulo (USP), São Paulo, SP, Brazil

**Patricia Cristina Baleeiro Beltrão-Braga** Faculdade de Medicina Veterinária e Zootecnia (FMVZ), Universidade de São Paulo, São Paulo, SP, Brazil

**David Brafman** School of Biological and Health Systems Engineering, Arizona State University, Tempe, AZ, USA

**Raul Bardini Bressan** MRC Centre for Regenerative Medicine, The University of Edinburgh, Edinburgh, UK

**Amanda-Jayne F. Carr** Division for Ocular Biology and Therapeutics, UCL Institute of Ophthalmology, London, UK

**Cassiano Carromeu** University of California San Diego, La Jolla, CA, USA

**David A. Carter** Division for Ocular Biology and Therapeutics, UCL Institute of Ophthalmology, London, UK

**Giuliana Castello Coatti** Department of Genetics and Evolutionary Biology, Human Genome and Stem Cell Research Center, Biosciences Institute, University of São Paulo (USP), São Paulo, SP, Brazil

**Pete J. Coffey** Division for Ocular Biology and Therapeutics, UCL Institute of Ophthalmology, London, UK

BMRC (NIH), Moorfields Eye Hospital, London, UK

Center for Stem Cell Biology and Engineering, Neuroscience Research Institute, University of California, Santa Barbara, CA, USA

**Juliana da Cruz Corrêa-Velloso** Department of Biochemistry, Institute of Chemistry, University of São Paulo, São Paulo, SP, Brazil

**Roberta Montero Costas** National Laboratory of Embryonic Stem Cell (LaNCE), Department of Genetics and Evolutionary Biology, University of São Paulo, São Paulo, SP, Brazil

**Simone Aparecida Siqueira Fonseca** National Laboratory of Embryonic Stem Cell (LaNCE), Department of Genetics and Evolutionary Biology, University of São Paulo, São Paulo, SP, Brazil

**Ricardo Castilho Garcez** Department of Cell Biology, Embryology and Genetics, Federal University of Santa Catarina, Florianopolis, SC, Brazil

**Talita Glaser** Department of Biochemistry, Institute of Chemistry, University of São Paulo, São Paulo, SP, Brazil

**Juliana Plat Aguiar Gomes** Department of Genetics and Evolutionary Biology, Human Genome and Stem Cell Research Center, Biosciences Institute, University of São Paulo (USP), São Paulo, SP, Brazil

**Maxine Nicole Gonzalez** Department of Biochemistry, Universidad Central del Caribe, Bayamón, PR, USA

**Khadidiatou Guiro** Department of Medicine – Hematology/Oncology, Rutgers, New Jersey Medical School, Newark, NJ, USA

**Inbo Han** Department of Neurosurgery, Harvard Medical School and Brigham and Women's Hospital, Boston, MA, USA

Division of SCI Research, VA Boston Healthcare System, Boston, MA, USA

Department of Neurosurgery, Cha University, Seoul, South Korea

**Jaemin Kim** Department of Medicine, Lillehei Heart Institute, University of Minnesota, Minneapolis, MN, USA

**Nathan Kumar** Department of Bioengineering, UCSD, La Jolla, CA, USA

**Albert R. La Spada** Department of Pediatrics, University of California, San Diego, La Jolla, CA, USA

Sanford Consortium for Regenerative Medicine, University of California, San Diego, La Jolla, CA, USA

Cellular and Molecular Medicine, University of California, San Diego, La Jolla, CA, USA

Department of Neurosciences, University of California, San Diego, La Jolla, CA, USA

Division of Biological Sciences, University of California, San Diego, La Jolla, CA, USA

Institute for Genomic Medicine, University of California, San Diego, La Jolla, CA, USA

**Alessandro Magli** Department of Medicine, Lillehei Heart Institute, University of Minnesota, Minneapolis, MN, USA

**Antonio H. Martins** Pharmacology and Toxicology Department, School of Medicine, University of Puerto Rico, Medical Science Campus, San Juan, PR, USA

Department of Biochemistry, Universidad Central del Caribe, Bayamón, PR, USA

**Rebecca Vasconcellos Botelho de Medeiros** Cell Signaling and Nanobiotechnology Laboratory, Department of Biochemistry, Immunology, Universidade Federal de Minas Gerais and Instituto Nanocell, Belo Horizonte, MG, Brazil

**Ana Paula Diniz Mendes** Salk Institute for Biological Studies, La Jolla, CA, USA

**Helen Cristina Miranda** Department of Pediatrics, University of California, San Diego, La Jolla, CA, USA

Sanford Consortium for Regenerative Medicine, La Jolla, CA, USA

**Alysson R. Muotri** Department of Pediatrics, Stem Cell Program, School of Medicine, University of California San Diego, La Jolla, CA, USA

Department of Pediatrics and Cellular & Molecular Medicine, School of Medicine, University of California San Diego, La Jolla, CA, USA

Sanford Consortium for Regenerative Medicine, La Jolla, CA, USA

**Priscilla Davidson Negraes** Department of Biochemistry, Institute of Chemistry, University of São Paulo, São Paulo, SP, Brazil

Department of Pediatrics, Stem Cell Program, School of Medicine, University of California San Diego, La Jolla, CA, USA

Sanford Consortium for Regenerative Medicine, La Jolla, CA, USA

**Britta Nommiste** Division for Ocular Biology and Therapeutics, UCL Institute of Ophthalmology, London, UK

**Ágatha Oliveira** Department of Biochemistry, Institute of Chemistry, University of São Paulo, São Paulo, SP, Brazil

**Nicole de Cássia Oliveira Paiva** Cell Signaling and Nanobiotechnology Laboratory, Department of Biochemistry and Immunology, Universidade Federal de Minas Gerais and Instituto Nanocell, Belo Horizonte, MG, Brazil

**Mayra Vitor Pelatti** Department of Genetics and Evolutionary Biology, Human Genome and Stem Cell Research Center, Biosciences Institute, University of São Paulo (USP), São Paulo, SP, Brazil

**Lygia V. Pereira** National Laboratory of Embryonic Stem Cell (LaNCE), Department of Genetics and Evolutionary Biology, University of São Paulo, São Paulo, SP, Brazil

**Rita C.R. Perlingeiro** Department of Medicine, Lillehei Heart Institute, University of Minnesota, Minneapolis, MN, USA

**Denise Freitas Siqueira Petri** Department of Fundamental Chemistry, Institute of Chemistry, University of São Paulo, São Paulo, SP, Brazil

**Mauro Cunha Xavier Pinto** Cell Signaling and Nanobiotechnology Laboratory, Department of Biochemistry and Immunology, Universidade Federal de Minas Gerais and Instituto Nanocell, Belo Horizonte, MG, Brazil

**Pranela Rameshwar** Department of Medicine – Hematology/Oncology, Rutgers, New Jersey Medical School, Newark, NJ, USA

Rutgers Graduate School of Biomedical Health Sciences at New Jersey Medical School, Newark, NJ, USA

**Janina Ratajczak** Stem Cell Institute at James Graham Brown Cancer Center, University of Louisville, Louisville, KY, USA

**Mariusz Z. Ratajczak** Stem Cell Institute at James Graham Brown Cancer Center, University of Louisville, Louisville, KY, USA

Department of Regenerative Medicine, Warsaw Medical University, Warsaw, Poland

**Rodrigo Ribeiro Resende** Cell Signaling and Nanobiotechnology Laboratory, Department of Biochemistry and Immunology, Universidade Federal de Minas Gerais and Instituto Nanocell, Belo Horizonte, MG, Brazil

**Jose L. Roig-Lopez** School of Science and Technology, Universidad Del Este, AGMUS, Carolina, PR, USA

**Maike Sander** Departments of Pediatrics and Cellular and Molecular Medicine, Pediatric Diabetes Research Center, University of California San Diego, La Jolla, CA, USA

Sanford Consortium for Regenerative Medicine, La Jolla, CA, USA

**Oleta Sandiford** Department of Medicine – Hematology/Oncology, Rutgers, New Jersey Medical School, Newark, NJ, USA

**Laura Sardá-Arroyo** Department of Biochemistry, Institute of Chemistry, University of São Paulo, São Paulo, SP, Brazil

**Clarissa Schitine** Institute of Biophysics Carlos Chagas Filho, Federal University of Rio de Janeiro, Rio de Janeiro, RJ, Brazil

**Ana Teresa Semeano** Department of Biochemistry, Institute of Chemistry, University of São Paulo, São Paulo, SP, Brazil

**Atossa Shaltouki** Department of Neurosurgery, School of Medicine, Stanford University, Palo Alto, CA, USA

**Farah Sheikh** Cardiovascular Medicine Division, Department of Medicine, University of California-San Diego, La Jolla, CA, USA

**Garima Sinha** Department of Medicine – Hematology/Oncology, Rutgers, New Jersey Medical School, Newark, NJ, USA

Rutgers Graduate School of Biomedical Health Sciences at New Jersey Medical School, Newark, NJ, USA

**Malwina Suszynska** Stem Cell Institute at James Graham Brown Cancer Center, University of Louisville, Louisville, KY, USA

**Leon Tejwani** Department of Pediatrics, Stem Cell Program, School of Medicine, University of California San Diego, La Jolla, CA, USA

**Yang D. Teng** Department of Neurosurgery, Harvard Medical School and Brigham and Women's Hospital, Boston, MA, USA

Departments of Physical Medicine and Rehabilitation, Harvard Medical School and Spaulding Rehabilitation Hospital, Boston, MA, USA

Division of SCI Research, VA Boston Healthcare System, Boston, MA, USA

**Charles A. Thomas** Department of Pediatrics, Stem Cell Program, School of Medicine, University of California San Diego, La Jolla, CA, USA

**Fernanda Maria Policarpo Tonelli** Cell Signaling and Nanobiotechnology Laboratory, Department of Biochemistry and Immunology, Universidade Federal de Minas Gerais and Instituto Nanocell, Belo Horizonte, MG, Brazil

**Andréa Gonçalves Trentin** Department of Cell Biology, Embryology and Genetics, Federal University of Santa Catarina, Florianopolis, SC, Brazil

**Cleber A. Trujillo** Department of Pediatrics, Stem Cell Program, School of Medicine, University of California San Diego, La Jolla, CA, USA

Sanford Consortium for Regenerative Medicine , La Jolla , CA, USA

**Henning Ulrich** Department of Biochemistry, Institute of Chemistry, University of São Paulo, São Paulo, SP, Brazil

**Alexandre Vessoni** University of São Paulo, São Paulo, SP, Brazil

**Nicholas Vinckier** Departments of Pediatrics and Cellular and Molecular Medicine, Pediatric Diabetes Research Center, University of California San Diego, La Jolla, CA, USA

**Jinzhao Wang** Departments of Pediatrics and Cellular and Molecular Medicine, Pediatric Diabetes Research Center, University of California San Diego, La Jolla, CA, USA

Department of Cellular and Molecular Medicine, University of California San Diego, La Jolla, CA, USA

**Karl Willert** Department of Cellular and Molecular Medicine, UCSD, La Jolla, CA, USA

Sanford Consortium for Regenerative Medicine, La Jolla, CA, USA

**Sara Alves Xapelli** Instituto de Farmacologia e Neurociências and Instituto de Medicina Molecular, Faculdade de Medicina da Universidade de Lisboa, Lisbon, Portugal

**Fabian Zanella** Cardiovascular Medicine Division, Department of Medicine, University of California-San Diego, La Jolla, CA, USA

**Mayana Zatz** Department of Genetics and Evolutionary Biology, Human Genome and Stem Cell Research Center, Biosciences Institute, University of São Paulo (USP), São Paulo, SP, Brazil

**Xiang Zeng** Department of Neurosurgery, Harvard Medical School and Brigham and Women's Hospital, Boston, MA, USA

Division of SCI Research, VA Boston Healthcare System, Boston, MA, USA

# Chapter 1
# Stem Cells: Principles and Applications

**Ágatha Oliveira, Juliana da Cruz Corrêa-Velloso, Talita Glaser, and Henning Ulrich**

**Abstract** Stem cell research is a promising and markedly emerging area of investigation concerning basic and clinical research. Since the 50s, the understanding that undifferentiated cells are able to originate different cell types held great promise for regenerative medicine, making until today this field to one of intense and growing research. The possibility to artificially replace damaged tissue unlocked new possibilities for clinical treatment of so far incurable diseases. This chapter highlights basic concepts about stem cells, as well as their current and potential future applications. Moreover, it brings an overview of important historical facts of the path taken by science to get to the current status of this research field.

**Keywords** Stem cells • Differentiation • Therapeutic use

## 1.1 Historical Remarks

Stem cell history began far ago in the 1950s, when researchers first isolated embryonal carcinoma cells (ECCs) from teratocarcinomas (Yu and Thomson 2008; Stevens and Little 1954). These cells could differentiated into all thee germ layers and, in 1964, Kleinsmith and Pierce (1964) showed that a single ECC could undergo unlimited self-renewal and multi-lineage differentiation, defining the existence of a pluripotent stem cell and thus providing the intellectual framework for mouse and human embryonic stem cells (ESCs). In the earlies 1970s, ECCs were stably propagated *in vitro* and studied as "an *in vitro* model of development (Kahan and Ephussi 1970)" due to their properties, many research groups started to search for an *in vivo* counterpart of these cells.

During the embryonic development, as the zygote embryo divides, it forms a morula and the first differentiation occurs: cells from the outer layer differentiate to originate the trophectoderm and to form the blastocyst. The inner cell mass of the blastocyst (ICM) gives rise to all cells of the adult body, while the trophectoderm

Á. Oliveira • J.C. Corrêa-Velloso • T. Glaser • H. Ulrich (✉)
Department of Biochemistry, Institute of Chemistry, University of São Paulo,
Av. Prof. Lineu Prestes 748, São Paulo, SP 05508-000, Brazil
e-mail: henning@iq.usp.br


H. Ulrich, P. Davidson Negraes, *Working with Stem Cells*,
DOI 10.1007/978-3-319-30582-0_1

differentiates into the placenta. In 1980, it was found that the cells from the ICM are the counterpart of ECCs (Martin 1980). Differently from cells from the ICM, most ECC lines have limited potential of differentiation, are highly aneuploid and poorly contribute to chimeric mice (Atkin et al. 1974), which limits their utility as an *in vitro* model for development, favoring the use of ICM cells.

The first mouse ESC lines were derived from the ICM of mouse blastocysts and maintained in culture in the presence of fibroblast feeder layers and serum, as previously used for mouse ECCs (Martin 1981; Evans and Kaufman 1981). In 1988, it was found that a cytokine, the leukemia inhibitory factor (LIF), was the element secreted by the feeder layer responsible for sustaining ESCs in an undifferentiated state (Smith et al. 1988; Williams et al. 1988).

Human (hESC) derivation was achieved in 1998 (Thomson et al. 1998). These cells are karyotypically normal and differentiate into all three germ layers (Amit et al. 2000). In contrast to mouse ESCs, hESCs or nonhuman primate ESCs do not maintain pluripotency in the presence of LIF and its related cytokines in serum-containing media (Dahéron et al. 2004; Thomson et al. 1998; Humphey et al. 2004).

Due to many ethical issues related to the use of human embryos for obtaining hESC, a new model with similar characteristics was necessary. In this context, the reprogramming of mouse somatic cells into a pluripotent state by transfection with specific pluripotency-coding vectors was successfully conducted by Yamanaka's group (Takahashi and Yamanaka 2006), giving rise to induced pluripotent stem cells (iPSCs). Shortly after, this technique was applied to human cells (Takahashi et al. 2007; Yu et al. 2007; Lowry et al. 2008).

In 1976, during the same period when ICM cells had been shown to be pluripotent, Friedenstein and colleagues placed a whole bone marrow in plastic dishes and, after removal of the non-adherent hematopoietic cells, they found that the adherent cells could differentiate into all bone cell subtypes, such as osteoblasts, chondrocytes, adipocytes, and even myoblasts, defining the multipotency (Friedenstein et al. 1976; reviewed by Chamberlain et al. 2007). These cells were referred to as mesenchymal stem cells (MSCs), once they differentiate into mesenchymal-type cells, or as marrow stromal cells (Prockop 1997) due to the complex array found in the marrow from which they derive (Ashton et al. 1980; Bab et al. 1986; Castro-Malaspina et al. 1980). In summary, along the last six decades, stem cells have become an expanding research field that promises to strongly contribute to the advancement of basic and clinical sciences (Fig. 1.1).

## 1.2 Stem Cell Characteristics and Potency Concepts

Stem cells have the remarkable potential to differentiate into more than 200 cell types found in an adult body. Throughout life, they give rise to cells that can become highly specialized and replace injured tissues, or participate in normal tissue regeneration. The classical definition of stem cells, which distinguishes them from other cell types, is determined by two key properties: first, stem cells have the ability to

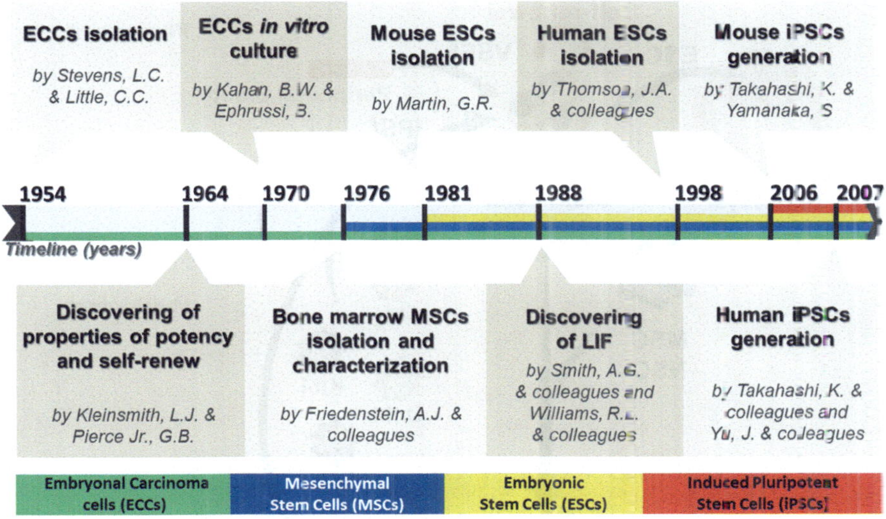


**Fig. 1.1** Schematic timeline showing the most important historical milestones in stem cell research since the isolation of ECC in 1954, until the development of human iPSC in 2007. *ECCs* embryonal carcinoma cells, *MSCs* mesenchymal stem cells, *ESCs* embryonic stem cells, *iPSCs* induced pluripotent stem cells

self-renew, dividing in a way that generates copies of themselves; second, under specific physiologic or experimental conditions, they are able to differentiate, giving rise to mature types of cells that constitute distinct organs and tissues (Potten and Loeffler 1990).

The developmental stage of a stem cell defines its potential of differentiation. At the beginning of development, just after the fertilization, cells within the first few rounds of cell division are the only ones defined as totipotent. Under the right conditions, totipotent cells can generate not only a whole viable embryo, but also temporary support tissues and structures, including the placenta and the umbilical cord (Brook and Gardner 1997). The totipotency of these cells lasts until the blastomeric stage, approximately 4 days after fertilization, when cells start to specialize and originate pluripotent cells, as the inner cell mass within the blastocyst (Thomson et al. 1998; Reubinoff et al. 2000). Pluripotent stem cells can differentiate into cells derived from the three germ layers, generating any tissue type present in the organism, but they lose the ability to form the placenta or other extraembryonic tissues (Smith 2012). During embryonic maturation and tissue formation, when stimulated by transcriptional and epigenetic signals affecting gene expression, pluripotent stem cells can also give rise to multipotent stem cells. These cells are capable to differentiate into only a few different cell types originating or repairing a given tissue (Spangrude et al. 1988; Slack 2000). When the organism is completely formed and progenitor cells are committed to their differentiation fate, these lose their potency and are no longer able to change their phenotype determination.

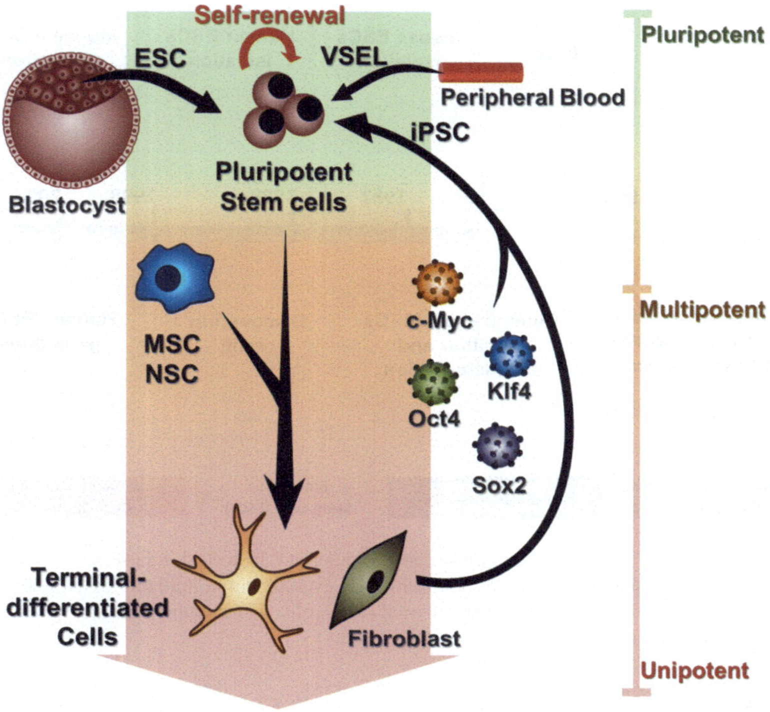


**Fig. 1.2** Differentiation potential of pluripotent stem cells and their origins. Pluripotent stem cells can be extracted from a blastocyst under development and as Very Small Embryonic Like (VSEL) stem cells from adult tissues. Throughout development, pluripotent stem cells originate different cell lines, and their potency is lost by the time each cell line becomes committed to its own phenotype/fate. In this process, intermediate multipotent cells like MSCs and NSCs are also generated. *In vitro*, the overexpression of specific reprogramming factors [c-Myc, Oct4 (octamer binding transcription factor-4), Klf4 (Krüppel-Like Factor 4) and Sox2 (sex determining region Y, box 2)] induces pluripotency in specialized cells, such as fibroblasts, originated iPSCs. *ESC* embryonic stem cell, *VSEL* very small embryonic/epiblast-like stem cell, *MSC* mesenchymal stem cell, *NSC* neural stem cell, *iPSC* induced pluripotent stem cell

## 1.3 Stem Cell Origins

Pluri- and multi-potent stem cells can be obtained of embryos, some tissues from an adult individual, and also be generated through *in vitro* interventions (Fig. 1.2). The following items focus on the most prominent sources of stem cells.

### *1.3.1 Embryonic Stem Cells (ESCs)*

ESCs are pluripotent stem cells obtained from the inner cell mass of the blastocyst, as mentioned before. Due to their capability to generate every adult tissue type, they provide a renewable resource for studying normal and disease development, besides their potential therapeutic applications (Lerou and Daley 2005). As a matter of fact, the establishment and optimization of embryonic cell lineage protocols are crucial for improving knowledge about both physiological and pathological states.

Since mouse ESC isolation (Evans and Kaufman 1981), molecular mechanisms involved in the maintenance of self-renewing and pluripotency have been extensively studied. Among these molecular mechanisms are induction of conformational changes in chromatin by the epigenetic machinery, transcription factor networks and specific signaling pathways, which are able to orchestrate the pluripotency of ESCs (Marks and Stunnenberg 2014; Welling and Geijsen 2013). Several transcription factors have been shown to be indispensable in regulating the pluripotent state of ESCs *in vivo* and *in vitro* (Dunn et al. 2014; Takashima et al. 2014), including Oct4 and Nanog, being part of well-characterized core network factors with crucial roles in maintaining pluripotency (Boyer et al. 2014; Loh et al. 2006).

### *1.3.2 Induced Pluripotent Stem Cells (iPSCs)*

The somatic cell nuclear transplantation (SCNT) technique was developed aiming to engineer cells with pluripotency properties. In this method, the nucleus of a differentiated cell is transferred to an enucleated oocyte, reaching nearly 100% of transfection efficiency in mice (Wakayama et al. 1998). However, the method involves a range of ethical issues regarding human cells, once the resulting oocyte development, despite countless obstacles to bypass, could result in a cloned individual. Furthermore, these cells are inapt for cell transplantation due to the fact that they are triploid.

In view of that, iPSCs were developed by reprogrammation of somatic cells into a pluripotent state, originating cells with morphology, self-renewal and pluripotency properties similar to ESCs. Since the pioneering work of Takahashi and Yamanaka in reprogramming mouse fibroblasts (Takahashi and Yamanaka 2006) and introducing the concept of iPSCs, many studies were conducted to refine the reprogramming procedure for increasing effectiveness and eliminating traces of the viral genome that could have been incorporated into the genome of the resulting iPSCs. That is because the initial reprogramming technique occurred by retroviral transduction of factors including Oct4 (octamer binding transcription factor-4), Sox2 (sex determining region Y, box 2), Klf4 (Krüppel-Like Factor 4) and c-Myc. Reprogramming of human cells was done by the same group (Takahashi et al. 2007) and, simultaneously, by Yu and colleagues (2007), with the latter research group introducing a reprogramming method based on the use of Nanog and Lin28 instead of Klf4 and c-Myc. In fact, the combination of these six transcription factors resulted in increased efficiency in reprogramming human fibroblast cells (Liao et al. 2008). Next, numerous reprogramming factors were found to interfere with the efficacy of pluripotency induction, including c-Myc, which seems to be dispensable (Wernig et al. 2008). As further improvement, combinations of these factors with proteins, peptides and RNA interference, among other mechanisms, gave rise to different protocols that do not necessarily involve viral infection, but transposon and nucleofection with plasmids (O'Malley et al. 2009; Malik and Rao 2013).

### 1.3.3 Very Small Embryonic/Epiblast-Like Stem Cells (VSELs)

Very small embryonic/epiblast-like stem cells (VSELs) are a developmentally early stem cell population that remains in an undifferentiated state and resides in adult tissues. They are rare and slightly smaller than red blood cells, and were first described in 2006 (Kucia et al. 2006).

VSELs keep circulating in the adult body during stress situations through peripheral blood and express markers of pluripotency, including Oct4, Nanog, and SSEA, and are able to differentiate into all three germ layers. These cells are Sca1+Lin−CD45− in mice and CD133+Lin−CD45− in humans, and their morphology is characterized by a high nuclear/cytoplasmic ratio and euchromatin content, which are typical for ESCs. VSELs are a promising source for future cell therapies (Ratajczak et al. 2012).

### 1.3.4 Mesenchymal Stem Cells (MSCs)

MSCs are non-hematopoietic stromal cells capable of differentiating into mesenchymal tissues, such as bone, cartilage, muscle, ligament, tendon, and adipose, contributing to the regeneration of these tissues (Chamberlain et al. 2007). They can be isolated from different sources including adipose tissue, bone marrow, amniotic fluid, umbilical cord, placenta, menstrual blood and even dental pulps (Portmann-Lanz et al. 2006; Musina et al. 2008; Tirino et al. 2011; Ma et al. 2014). In addition, these cells have the ability to self-renew and are identified by their phenotype, being positive for CD29, CD44, CD73 and CD90 cell surface markers, while negative for the hematopoietic markers CD34, CD45 and CD14. Moreover, MSCs contribute to cellular homeostasis maintenance and many physiological and pathological processes such as aging, tissue damage and inflammatory diseases (Prockop 1997; Sordi et al. 2005; Le Blanc et al. 2003).

When transplanted, MSCs are able to migrate to injury sites. This trafficking into and through tissue is a process that involves adhesion molecules, chemokine receptors and their ligands. Several studies conducted to elucidate the mechanisms underlying this process reported the functional expression of various chemokine receptors and adhesion molecules on human MSCs (Chamberlain et al. 2007). The differentiation potential of MSCs is limited in comparison to ESCs and iPSCs, characterizing them as multipotent cells, even though they are a great promise for clinical applications especially due to their immunoregulatory functions.

### 1.3.5 Neural Stem Cells (NSCs)

NSCs are multipotent stem cells capable to differentiate into many neural cell types from the central nervous system (CNS). They are found in both the developing and the adult brain, with some distinct properties. Basically, during early embryo development the rearrangement of neuroepithelial cells leads to the neural tube formation. In the formed ventricular zone, these cells constantly proliferate to increase cell number and then migrate to form the CNS (Merkle and Alvarez-Buylla 2006). A niche of stem cells remains in the ventricular zone and gives rise to the radial glial cells, another NSC type that differentiate into distinct neural cell types. Moreover, NSCs seem to modify their morphology, gene expression profile and other properties throughout the embryonic development (reviewed by Götz et al. 2015), since they first originate a large amount of neurons and later start to produce more glial cells. In mammals, radial glial cells are no longer present in the after-birth brain, giving rise to multipotent adult NSCs (aNSCs) (Merkle et al. 2004).

The aNSCs expresses glial fibrillary acidic protein (GFAP), an astrocyte marker, and are located in the subventricular zone (SVZ) of the lateral ventricle's wall, and in the subgranular zone (SGZ) of the hippocampus' dentate gyrus in the adult brain (Doetsch et al. 1999; Gage et al. 1998). In the SVZ, aNSCs are known as type B cells and their derived neural progenitors are type C cells, which can be identified by Mash1 gene expression. The later give rise to neuroblasts, some expressing Olig2 that generate oligodendrocytes (Parras et al. 2004). In summary, prenatal neuroephitelial cells originate radial glial cells that disappear after birth and give rise to astrocyte-like cells (Merkle and Alvarez-Buylla 2006). Since these GFAP-expressing cells are able to replenish the SVZ after ablation and differentiate into neurons (Doetsch et al. 1997), aNSC seems to be a promising cell type for cell therapy use.

## 1.4 Stem Cell Applications

Once the possibility of differentiating stem cells into specific phenotypes *in vitro*, had been established, a variety of methods emerged aiming to increase cell fate specificity and differentiation efficiency. This book provides several advanced protocols regarding stem cell differentiation (Fig. 1.3). Such approach allows science to advance in different areas of basic and applied research, facilitating the understanding of physiological and pathological processes, and enabling the advancement of medicine to control and/or cure several diseases by unraveling cellular and molecular mechanisms involved in each process.

ESCs originate embryoid bodies *in vitro*, which tend to spontaneously differentiate into distinct tissues, mimicking the embryonic development (Ling and Neben 1997). In the presence of specific growth factors and small molecules, ESCs can differentiate into particular cell types mimicking mechanisms underlying development of different organs/tissues, such as pancreas and liver (Zaret and Grompe 2008), muscles (Xie et al. 2011), the cardiovascular (Feraud and Vittet 2003; Winkler et al. 2004) and the nervous systems (Lupo et al. 2014). Furthermore, ESCs provide a suitable model to study the impact of genetic mutations and toxicity of diverse substances during early development. In combination with adult stem cells as further model systems, these can be employed for unraveling bases of differentiation, physiology, biochemistry and potential pathologic processes during embryonic development and adult cell differentiation.

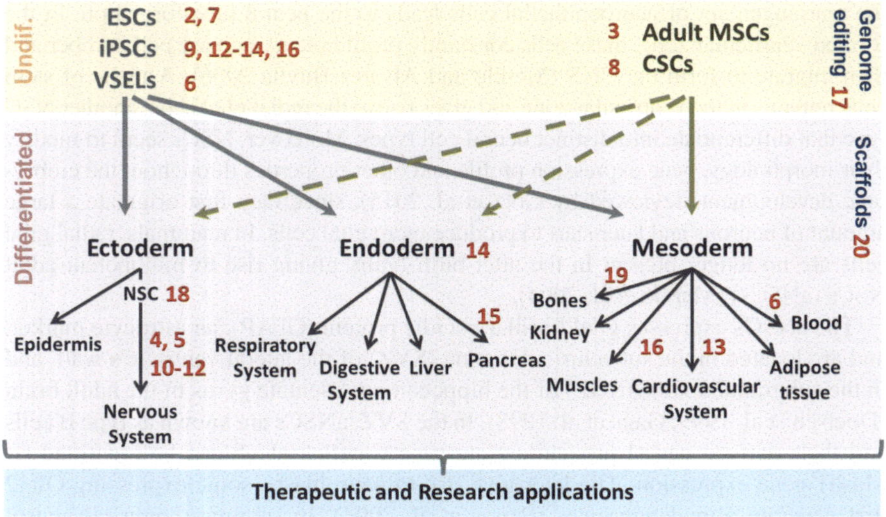


**Fig. 1.3** Stem cell topics covered by this book. *ESCs* embryonic stem cells, *iPSCs* induced pluripotent stem cells, *VSELs* very small embryonic-like stem cells, *MSCs* mesenchymal stem cells, *CSCs* cancer stem cells, *NSC* neural stem cell

Based on this principle, stem cells, and particularly iPSCs, have gained importance as simplified disease models. The use of patients' cells to generate iPSCs provides a more accurate understanding of how the genome contributes to erroneous or defective differentiation processes underlying neurodegenerative diseases. For instance, in Parkinson's disease, death of dopaminergic neurons in a specific stage of life may result from compromised development, whose etiology may involve genetic inheritance (reviewed by Badger et al. 2014). Thus, dopaminergic neurons differentiated from reprogrammed patients' fibroblasts may help to elucidate the participation of genetic preprogrammed mechanisms in the disease. Furthermore, the use of patients' cells enables the refinement of individualized treatment, since some interventions might lead to different effects depending on the individual background.

iPSC-based approaches may deflect issues involved in the development and application of human cell therapy in curing diseases, since sources of pluripotent human stem cells are scarce and comprises several ethical issues. Cell therapy is based on the transplantation of cells, whether differentiated or undifferentiated, to reverse the injury in the subject. For this purpose, stem cells from different sources are cultured *in vitro* and transplanted into animal models to assess their effectiveness. Many studies are being conducted in the hope for developing efficient therapies for so far irreversible conditions. Therefore, stem cells can be manipulated *in vitro* for effective differentiation and integration and survival in a living tissue without immunological rejection after transplantation.

One way to manipulate stem cells is by means of genome editing, as further discussed in Chap. 17. DNA modifications can be incorporated or excluded from the genome in knock-in and -out models, respectively. Moreover, gene expression can also be modulated by using RNA interference (Martin and Caplen 2007), the Cre/loxP system (Van Duyne 2015) and other recombinase systems (Kilby et al. 1993; Gaj and Barbas 2014). Despite the available technologies, genetic modifications still need refinement and this intervention has opened a new range of research possibilities to analyze gene function, genetic diseases, mutation studies and pharmacological applications.

## 1.5 Stem Cell Research's Ethical Issues

Despite of its huge potential regarding regenerative medicine and contribution to basic science, stem cell research faces ethical and political challenges that delay the advancement of this field. Some sources of stem cells such as adult MSCs and VSELs do not bring up strong ethical concerns, while hESCs and iPSCs are constant subjects of discussions.

Currently, the extraction of hESCs requires the destruction of a human embryo, being the main reason why ESCs raise ethical discussions. There are two main positions in relation to embryos' use in research: (1) those who are strictly against embryo utilization for research purposes, because they consider the embryo morally equivalent to an adult human being; and (2) those who defend the utilization of embryos for therapeutic purposes in research, and also consider this as an obligation in view of the benefits for patients (Devolder and Savulescu 2006).

From a legal standpoint, these distinct ethical perspectives translate into stem cell legislations that greatly vary from country to country (for a review of different policies around the world, see Dhar and Ho 2009). Brazil was the pioneer country to develop a law regulating the use of ESCs in 2005 (Dhar and Ho 2009). In the United States, for example, embryos produced during *in vitro* fertilization, which eventually need to be discarded, can be used for scientific experiments, since they are not produced for research purposes (Green 2002). Stem cells produced by SCNT share the same ethical concerns of hESC, once this technique enables the development of a cloned embryo. Moreover, obtaining human oocytes involves trade, which in turn results in additional ethical problems (Alpers and Lo 1995).

Developed as a promising alternative for ESCs and aiming to bypass the ethical concerns that accompany these cells' use, iPSCs rapidly are controversely discussed. The central topic relies on the possibility that iPSCs could originate, accidentally or on purpose, a totipotent cell similar to oocyte, which raises ethical concerns similar to those of SCNT cells, such as cloning possibilities (de Miguel-Beriain 2015). Additionally, there are further questions regarding iPSCs research hovering over bioethics, since the manipulation of these cells could culminate in a range of unthinkable possibilities, including the generation of gametes *in vitro* and the creation of a human chimaera (Carvalho and Ramalho-Santos 2013).

At this moment, the scientific community has not yet reached a consensus on the use of stem cells and its ethical implications. Thus, highlighting the importance of the topic is each researcher's responsibility, who must be aware of the boundaries set by the legislation in the country where the experiments are being conducted. In addition to legal considerations, it is noteworthy mentioning the need for a personal judgment in developing any stem cell research.

**Acknowledgments** This work was supported by research grants from Brazilian funding agencies Sao Paulo Research Foundation (FAPESP; Proc. No. 2012/50880-4, 2015/13345-1), National Council for Scientific and Technological Development (CNPq; Proc. No. 467465/2014-2, 141979/2014-3, 403745/2014-4), and Provost's Office for Research of the University of Sao Paulo, Grant number: 2011.1.9333.1.3 (NAPNA-USP), Brazil.

## References

Alpers A, Lo B (1995) Commodification and commercialization in human embryo research. Stanf Law Policy Rev 6:39–46

Amit M, Carpenter MK, Inokuma MS, Chiu CP, Harris CP, Waknitz MA, Itskovitz-Eldor J, Thomson JA (2000) Clonally derived human embryonic stem cell lines maintain pluripotency and proliferative potential for prolonged periods of culture. Dev Biol 227:271–278

Ashton BA, Allen TD, Howlett CR, Eaglesom CC, Hattori A, Owen M (1980) Formation of bone and cartilage by marrow stromal cells in diffusion chambers *in vivo*. Clin Orthop Relat Res 151:294–307

Atkin NB, Baker MC, Robinson R, Gaze SE (1974) Chomosome studies on 14 near-diploid carcinomas of the ovary. Eur J Cancer 10:144–146

Bab I, Ashton BA, Gazit D, Marx G, Williamson MC, Owen ME (1986) Kinetics and differentiation of marrow stromal cells in diffusion chambers *in vivo*. J Cell Sci 84:139–151

Badger JL, Cordero-Llana O, Hartfield EM, Wade-Martins R (2014) Parkinson's disease in a dish—using stem cells as a molecular tool. Neuropharmacology 76A:88–96

Boyer LA, Lee TI, Cole MF, Johnstone SE, Levine SS, Zucker JP, Guenther MG, Kumar RM, Murray HL, Jenner RG, Gifford DK, Melton DA, Jaenisch R, Dunn SJ, Martello G, Yordanov B, Emmott S, Smith AG (2014) Defining an essential transcription factor program for naive pluripotency. Science 344:1156–1160

Brook FA, Gardner RL (1997) The origin and efficient derivation of embryonic stem cells in the mouse. Proc Natl Acad Sci U S A 94:5709–5712

Carvalho AS, Ramalho-Santos J (2013) How can ethics relate to science? The case of stem cell research. Eur J Hum Genet 21(6):591–595

Castro-Malaspina H, Gay RE, Resnick G, Kapoor N, Meyers P, Chiarieri D, McKenzie S, Broxmeyer HE, Moore MA (1980) Characterization of human bone marrow fibroblast colony-forming cells (CFU-F) and their progeny. Blood 56:289–301

Chamberlain G, Fox J, Ashton B, Middleton J (2007) Concise review: mesenchymal stem cells: their phenotype, differentiation capacity, immunological features, and potential for homing. Stem Cells 25:2739–2749

Dahéron L, Opitz SL, Zaehes H, Lensch MW, Andrews PW, Itskovitz-Eldor J, Daley GQ (2004) LIF/STAT3 signaling fails to maintain self-renewal of human embryonic stem cells. Stem Cells 22:770–778

de Miguel-Beriain I (2015) The ethics of stem cells revisited. Adv Drug Deliv Rev 82–83:176–180

Devolder K, Savulescu J (2006) The moral imperative to conduct embryonic stem cell and cloning research. Ethics 15:7–21

Dhar D, Ho JH (2009) Stem cell research policies around the world. Yale J Biol Med 82(3):113–115

Doetsch F, Garcia-Verdugo JM, Alvarez-Buylla A (1997) Cellular composition and thee-dimensional organization of the subventricular germinal zone in the adult mammalian brain. J Neurosci 17:5046–5061
Doetsch F, Caille I, Lim DA, Garcia-Verdugo JM, Alvarez-Buylla A (1999) Subventricular zone astrocytes are neural stem cells in the adult mammalian brain. Cell 97:703–716
Dunn SJ, Martello G, Yordanov B, Emmott S, Smith AG (2014) Defining an essential transcription factor program for naïve pluripotency. Science 344:1156–1160
Evans MJ, Kaufman MH (1981) Establishment in culture of pluripotential cells from mouse embryos. Nature 292:154–156
Feraud O, Vittet D (2003) Murine embryonic stem cell *in vitro* differentiation: applications to the study of vascular development. Histol Histopathol 18:191–199
Friedenstein AJ, Gorskaja JF, Kulagina NN (1976) Fibroblast precursors in normal and irradiated mouse hematopoietic organs. Exp Hematol 4:267–274
Gage FH, Kempermann G, Palmer TD, Peterson DA, Ray J (1998) Multipotent progenitor cells in the adult dentate gyrus. J Neurobiol 36:249–266
Gaj T, Barbas CF 3rd (2014) Genome engineering with custom recombinases. Methods Enzymol 546:79–91
Götz M, Sirko S, Beckers J, Irmler M (2015) Reactive astrocytes as neural stem or progenitor cells *in vivo* lineage, *In vitro* potential, and Genome-wide expression analysis. Glia 63:1452–1468
Green R (2002) Benefiting from "evil": an incipient moral problem in human stem cell research. Bioethics 16:544–556
Humphey RK, Beattie GM, Lopez AD, Bucay N, King CC, Firpo MT, Rose-John S, Hayek A (2004) Maintenance of pluripotency in human embryonic stem cells is STAT3 independent. Stem Cells 22:522–530
Kahan BW, Ephussi B (1970) Developmental potentialities of clonal *in vitro* cultures of mouse testicular teratoma. J Natl Cancer Inst 44:1015–1036
Kilby NJ, Snaith MR, Murray JA (1993) Site-specific recombinases: tools for genome engineering. Trends Genet 9:413–421
Kleinsmith LJ, Pierce GB Jr (1964) Multipotentiality of single embryonal carcinoma cells. Cancer Res 24:1544–1551
Kucia M, Reca R, Campbell FR, Zuba-Surma E, Majka M, Ratajczak J, Ratajczak MZ (2006) A population of very small embryonic-like (VSEL) CXCR4(+)SSEA-1(+)Oct-4+ stem cells identified in adult bone marrow. Leukemia 20:857–869
Le Blanc K, Tammik C, Rosendahl K, Zetterberg E, Ringdén O (2003) HLA expression and immunologic properties of differentiated and undifferentiated mesenchymal stem cells. Exp Hematol 31:890–896
Lerou PH, Daley GQ (2005) Therapeutic potential of embryonic stem cells. Blood Rev 19:321–331
Liao J, Wu Z, Wang Y, Cheng L, Cui C, Gao Y, Chen T, Rao L, Chen S, Jia N, Dai H, Xin S, Kang J, Pei G, Xiao L (2008) Enhanced efficiency of generating induced pluripotent stem (iPS) cells from human somatic cells by a combination of six transcription factors. Cell Res 18:600–603
Ling V, Neben S (1997) *In vitro* differentiation of embryonic stem cells: immunophenotypic analysis of cultured embryoid bodies. J Cell Physiol 171:104–115
Loh YH, Wu Q, Chew JL, Vega VB, Zhang W, Chen X, Bourque G, George J, Leong B, Liu J, Wong KY, Sung KW, Lee CW, Zhao XD, Chiu KP, Lipovich L, Kuznetsov VA, Robson P, Stanton LW, Wei CL, Ruan Y, Lim B, Ng HH (2006) The Oct4 and Nanog transcription network regulates pluripotency in mouse embryonic stem cells. Nat Genet 38:431–440
Lowry WE, Richter L, Yachechko R, Pyle AD, Tchieu J, Sridharan R, Clark AT, Plath K (2008) Generation of human induced pluripotent stem cells from dermal fibroblasts. Proc Natl Acad Sci U S A 105:2883–2888
Lupo G, Bertacchi M, Carucci N, Augusti-Tocco G, Biagioni S, Cremisi F (2014) From pluripotency to forebrain patterning: an *in vitro* journey astride embryonic stem cells. Cell Mol Life Sci 71:2917–2930
Ma S, Xie N, Li W, Yuan B, Shi Y, Wang Y (2014) Immunobiology of mesenchymal stem cells. Cell Death Differ 21:216–225

Malik N, Rao MS (2013) A review of the methods for human iPSC derivation. Methods Mol Biol 997:23–33

Marks H, Stunnenberg HG (2014) Transcription regulation and chomatin structure in the pluripotent ground state. Biochim Biophys Acta 1839:129–137

Martin GR (1980) Teratocarcinomas and mammalian embryogenesis. Science 209:768–776

Martin GR (1981) Isolation of a pluripotent cell line from early mouse embryos cultured in medium conditioned by teratocarcinoma stem cells. Proc Natl Acad Sci U S A 78:7634–7638

Martin SE, Caplen NJ (2007) Applications of RNA interference in mammalian systems. Annu Rev Genomics Hum Genet 8:81–108

Merkle FT, Alvarez-Buylla A (2006) Neural stem cells in mammalian development. Curr Opin Cell Biol 18:704–709

Merkle FT, Tramontin AD, Garcia-Verdugo JM, Alvarez-Buylla A (2004) Radial glia give rise to adult neural stem cells in the subventricular zone. Proc Natl Acad Sci U S A 101:17528–17532

Musina RA, Belyavski AV, Tarusova OV, Solovyova EV, Sukhikh GT (2008) Endometrial mesenchymal stem cells isolated from the menstrual blood. Bull Exp Biol Med 145:539–543

O'Malley J, Woltjen K, Kaji K (2009) New strategies to generate induced pluripotent stem cells. Curr Opin Biotechnol 20:516–521

Parras CM, Galli R, Britz O, Soares S, Galichet C, Battiste J, Johnson SE, Nakafuku M, Vescovi A, Guillemot F (2004) Mash1 specifies neurons and oligodendrocytes in the postnatal brain. EMBO J 23:4495–4505

Portmann-Lanz CB, Schoeberlein A, Huber A, Sager R, Malek A, Holzgreve W, Surbek DV (2006) Placental mesenchymal stem cells as potential autologous graft for pre- and perinatal neuroregeneration. Am J Obstet Gynecol 194:664–673

Potten CS, Loeffler M (1990) Stem cells: attributes, cycles, spirals, pitfalls and uncertainties: Lessons for and from the crypt. Development 110:1001–1020

Prockop DJ (1997) Marrow stromal cells as stem cells for nonhematopoietic tissues. Science 276:71–74

Ratajczak MZ, Shin DM, Liu R, Mierzejewska K, Ratajczak J, Kucia M, Zuba-Surma EK (2012) Very small embryonic/epiblast-like stem cells (VSELs) and their potential role in aging and organ rejuvenation—an update and comparison to other primitive small stem cells isolated from adult tissues. Aging (Albany NY) 4:235–246

Reubinoff BE, Pera MF, Fong CY, Trounson A, Bongso A (2000) Embryonic stem cell lines from human blastocysts: somatic differentiation *in vitro*. Nat Biotechnol 18:399–404

Slack JM (2000) Stem cells in epithelial tissues. Science 287:1431–1433

Smith AG (2012) Embryo-derived stem cells: of mice and men. Annu Rev Cell Dev Biol 17:435–462

Smith AG, Heath JK, Donaldson DD, Wong GG, Moreau J, Stahl M, Rogers D (1988) Inhibition of pluripotential embryonic stem cell differentiation by purified polypeptides. Nature 336:688–690

Sordi V, Malosio ML, Marchesi F, Mercalli A, Melzi R, Giordano T, Belmonte N, Ferrari G, Leone BE, Bertuzzi F, Zerbini G, Allavena P, Bonifacio E, Piemonti L (2005) Bone marrow mesenchymal stem cells express a restricted set of functionally active chemokine receptors capable of promoting migration to pancreatic islets. Blood 106:419–427

Spangrude GJ, Heimfeld S, Weissman IL (1988) Purification and characterization of mouse hematopoietic stem cells. Science 241:58–62

Stevens LC, Little CC (1954) Spontaneous testicular teratomas in an inbred strain of mice. Proc Natl Acad Sci U S A 40:1080–1087

Takahashi K, Yamanaka S (2006) Induction of pluripotent stem cells from mouse embryonic and adult fibroblast cultures by defined factors. Cell 126:663–676

Takahashi K, Tanabe K, Ohnuki M, Narita M, Ichisaka T, Tomoda K, Yamanaka S (2007) Induction of pluripotent stem cells from adult human fibroblasts by defined factors. Cell 131:861–872

Takashima Y, Guo G, Loos R, Nichols J, Ficz G, Krueger F, Oxley D, Santos F, Clarke J, Mansfield W, Reik W, Bertone P, Smith A (2014) Resetting transcription factor control circuitry toward ground-state pluripotency in human. Cell 158:1254–1269

Thomson JA, Itskovitz-Eldor J, Shapiro SS, Waknitz MA, Swiergiel JJ, Marshall VS, Jones JM (1998) Embryonic stem cell lines derived from human blastocysts. Science 282:1145–1147

Tirino V, Paino F, d'Aquino R, Desiderio V, De Rosa A, Papaccio G (2011) Methods for the identification, characterization and banking of human DPSCs: current strategies and perspectives. Stem Cell Rev 7:608–615

Van Duyne GD (2015) Cre recombinase. Microbiol Spectr 3(1):119–138

Wakayama T, Perry AC, Zuccotti M, Johnson KR, Yanagimachi R (1998) Full-term development of mice from enucleated oocytes injected with cumulus. Nature 394:369–374

Welling M, Geijsen N (2013) Uncovering the true identity of naive pluripotent stemcells. Trends Cell Biol 23:442–448

Wernig M, Meissner A, Cassady JP, Jaenisch R (2008) c-Myc is dispensable for direct reprogramming of mouse fibroblasts. Cell Stem Cell 2:10–12

Williams RL, Hilton DJ, Pease S, Willson TA, Stewart CL, Gearing DP, Wagner EF, Metcalf D, Nicola NA, Gough NM (1988) Myeloid leukaemia inhibitory factor maintains the developmental potential of embryonic stem cells. Nature 336:684–687

Winkler J, Hescheler J, Sachinidis A (2004) Embryonic stem cells for basic research and potential clinical applications in cardiology. Biochim Biophys Acta 1740:240–248

Xie C, Ritchie RP, Huang H, Zhang J, Chen YE (2011) Smooth muscle cell differentiation *in vitro*: models and underlying molecular mechanisms. Arterioscler Thromb Vasc Biol 31:1485–1494

Yu J, Thomson JA (2008) Pluripotent stem cell lines. Genes Dev 22:1987–1997

Yu J, Vodyanik MA, Smuga-Otto K, Antosiewicz-Bourget J, Frane JL, Tian S, Nie J, Jonsdottir GA, Ruotti V, Stewart R, Slukvin II, Thomson JA (2007) Induced pluripotent stem cell lines derived from human somatic cells. Science 318:1917–1920

Zaret KS, Grompe M (2008) Generation and regeneration of cells of the liver and pancreas. Science 322:1490–1494

# Chapter 2
# Human Embryonic Stem Cell Line Derivation

**Simone Aparecida Siqueira Fonseca, Roberta Montero Costas, and Lygia V. Pereira**

**Abstract** Human embryonic stem cells are commonly derived from the inner cell mass of developing blastocyst-stage embryos. They are capable of unlimited expansion *in vitro* and can be maintained in culture indefinitely in their undifferentiated state. These cells can also spontaneously differentiate into different cell types that are representative of the three germ layers (ectoderm, mesoderm and endoderm) both *in vitro* and *in vivo*, by generating teratomas after injection into immunocompromised mice. The capacity to differentiate into a variety of cells gives them a promising applicability in cell replacement therapies, and makes hESCs powerful tools for studying the molecular mechanisms underlying cellular differentiation. Here, we will describe protocols to derive hESC lines using conventional procedures and in a xeno-free culture condition, as well as discuss the potential therapeutic use of these cells in regenerative medicine and in pharmaceutical drug screening.

**Keywords** hESCs • Embryonic stem cells • ES cell derivation

## 2.1 Introduction

Human embryonic stem cells (hESCs) are pluripotent cells derived from the inner cell mass (ICM) of blastocysts (Thomson et al. 1998). They have the capacity of maintaining an undifferentiated state (self-renewal) during proliferation, and of differentiating into tissues from all three germ layers (pluripotency). Therefore, hESCs have been considered as a potential source of tissue for cell therapy as well as for basic research on different aspects of human development (Doss et al. 2004). Usually, lineages of hESCs are derived from surplus embryos produced for reproductive reasons. The first hESC lines were established similarly to mouse embryonic stem cells (Thomson et al. 1998; Reubinoff et al. 2000), using murine

S.A.S. Fonseca • R.M. Costas • L.V. Pereira (✉)
National Laboratory of Embryonic Stem Cell (LaNCE), Department of Genetics and Evolutionary Biology, University of São Paulo,
Rua do Matão, 277 – sala 350, São Paulo, SP 05508-090, Brazil
e-mail: lpereira@usp.br



H. Ulrich, P. Davidson Negraes, *Working with Stem Cells*,
DOI 10.1007/978-3-319-30582-0_2

**Table 2.1** Differences between mouse and human ES cells

| | Morphology | Clonal passage | Lif | Fgf | Xci | SSEA-1 |
|---|---|---|---|---|---|---|
| mESCs | Dome | Yes | Yes | No | XaXa | Express |
| hESCs | Flat | More difficult | No | Yes | XaXi (unstable) | Low or absent |

embryonic fibroblasts (MEFs) as feeder cells, and a culture medium containing fetal bovine serum (Evans and Kaufman 1981; Martin 1981). Next, some groups have invested in studying animal substrate-free feeder cells, as a xeno-free environment (Rodin et al. 2010; Melkoumian et al. 2010; Villa-Diaz et al. 2010); these matrices provide a better system culture to hESCs, expanding the use of hES cells in biotechnologies and allowing potential clinical applications.

Human ESCs differ from their murine counterparts in many ways, including morphology, passage in culture, growth factors dependence, genomic stability and the epigenetic state (Table 2.1). Human ESCs, although also pluripotent, seem to be in a more advanced developmental stage than murine cells. Because of the differences between mouse embryonic stem cells (ESCs) and hESCs, culture conditions were established in order to support mouse ESC-like human ESCs, the so-called naïve human ESCs (reviewed in Fonseca et al. 2015a).

Although most hESC lines have been established from normal/good quality embryos, hESC derivation has also been achieved from morphologically abnormal embryos (Chen et al. 2005; Lerou et al. 2008; Shetty and Inamdar 2012), and from embryos scored as aneuploid by preimplatational analysis of cleavage-stage blastomeres (Mateizel et al. 2006; Peura et al. 2008; Tropel et al. 2010; Fonseca et al. 2015b).

## 2.2 Protocos for Estabhisment of New Linages of HESC

### *2.2.1 Feeder Cells*

Mouse embryonic fibroblasts (MEFs) are the most commonly used support cells for hESC derivation. However, in case of xeno-free protocols one can use human foreskin fibroblasts (HFF) or human recombinant laminin. Here, we describe the general protocol of MEF isolation and cultivation, and how these can be used as feeder cells during the establishment and cultivation of hESC.

#### 2.2.1.1 Isolation, Preparing and Initial Cultivation of MEFs

**Material**

- 1× DPBS free of $Ca^{2+}$ e $Mg^{2+}$ (Dulbecco's Phosphate-Buffered Saline, no calcium, no magnesium, Life Technologies, Gibco, Cat. No. 14190-144).
- 0.05 % Trypsin/EDTA (Life Technologies, Gibco, Cat. No. 25300-054).
- 0.25 % Trypsin/EDTA (Life Technologies, Gibco, Cat. No. 25200-056).

- Hydrated Alcohol 70 %.
- 0,1 % Gelatin in water (StemCell Technologies, Cat. No. 07903).
- Centrifuge tubes (50 ml).
- Sterile Petri Dishes 100 mm.
- Sterile Petri Dishes 60 mm.
- Sterile Adhere Culture flasks (T75).
- Sterile Surgical Scissors and forceps.
- MEF medium (below).
- DNAse I (optional).

**MEF Medium**

- 1× DMEM High Glucose (Life Technologies, Gibco, Cat. No. 11960-044).
- 10 % FBS ES-Qualified (Life Technologies, Gibco, Cat. No. 16141-079).
- 2 mM GlutaMAX™ Supplement (Life Technologies, Gibco, Cat. No. 35050-061).
- 100 U/ml Penicillin/100 Ug/ml Streptomicin (Life Technologies, Gibco, Cat. No. 15140-122).
- 1× (0.1 mM) Non-Essential Amino Acids (Life Technologies, Gibco, Cat. No. 11140-050).
- 1 mM Sodium Pyruvate (Life Technologies, Gibco, Cat. No. 11360-070).
- 10 mM HEPES (Life Technologies, Gibco, Cat. No. 15630-080).
- 0.055 mM 2-Mercaptoethanol (Life Technologies, Gibco, Cat. No. 21985-023).
- After preparing and filtering the medium in a 22 mm filter, store at 4 °C for up to 2 weeks.

**Protocol**

1. To isolate fetal murine fibroblasts, the best age is 12.5–14.5 d.p.c. Pregnant mice should be euthanized by cervical dislocation according to animal care regulation;
2. Clean the abdomen of the animal using 70 % ethanol before proceeding to the surgery;
3. Open the abdominal cavity with a sterile scissor and forceps; this procedure can be performed in a non-sterile environment;
4. Dissect the uterine horns and briefly wash then in a petri dish containing DPBS; with a scissor, quickly separate the embryos;

   The following steps should be carried out under aseptic conditions in a tissue culture hood and using sterile instruments;
5. Transfer the embryos to a new 100 mm petri dish containing DPBS;
6. Wash the embryos at least three times until the DPBS remains clear (without blood);
7. Under a stereomicroscope, dissect the head and red organs, and wash carcass in DPBS;
8. Wash the carcass in a new petri dish (60 mm diameter) with 2 ml of DPBS;
9. To isolate the fibroblasts, transfer the carcass to a new petri dish (60 mm) containing around 1 ml of trypsin/EDTA 0.05 % per two embryo carcasses (and no more than five per petri dish);

10. While the embryos are still in the trypsin/EDTA, quickly cut the fragments in small pieces (1–2 mm) using scissors or blades;
11. Transfer the fragments in trypsin to a centrifuge tube and incubate for 10 min at 37 °C, 5 % $CO_2$, 95 % humidity. *Attention*: tissues should not be left in trypsin for long periods to prevent cell death;
12. After 10 min, neutralize the trypsin by adding 2 volumes of MEF medium to the tube;
13. Dissociate the cells by thoroughly pipetting up and down; at this stage, it is possible to get a gel solution (DNA) resulting from cell break. To avoid this, add DNAse I (1 Ug/ml) to the solution and incubate for an extra time (5 min) at 37 °C, 5 % $CO_2$, 95 % humidity;
14. Centrifuge at 230 × *g* for 5 min at room temperature; carefully remove the supernatant and resuspend cell pellet using warm MEF medium;
15. Pre-treat T75 $cm^2$ culture flask with gelatin (0,1 %) by adding 10 ml of gelatin solution and leave the flasks at room temperature for 20 min. Before plating the cells, remove the gelatin solution from the flasks. Plate approximately a number of cells equivalent to two embryos in each flask and incubate overnight (ON);
16. 24 h after plating, ideal cells would reach 80–90 % of confluency and the media needs to be changed every other day;
17. Cells must be split each 3–5 days after plating by trypsin/EDTA (0.05 %) treatment—when cells are 90 % confluent, split at proportion of 1:3;
18. Freeze cells in lower passages (P0 or P1), allowing these cells to be expanded for two more passages, inactivated (by Mitomicin treatment or radiation) and used as feeder cells in the derivation and maintenance of hESC. Freeze cells using freezing medium (70 % DMEM, 20 % FBS, 10 % DMSO) by slow freezing process.

#### 2.2.1.2 Passage of MEFs

**Material**

- 1× DPBS $Ca^{2+}$ e $Mg^{2+}$ free (Life Technologies, Gibco, Cat. No. 14190-144).
- Sterile gelatin solution in water 0,1 % (StemCell Technologies, Cat. No. 07903).
- 0.25 % Trypsin/EDTA (Life Technologies, Gibco, Cat. No. 25200-056).
- Sterile centrifuge tubes (15 or 50 ml).
- Sterile adherent culture flasks (T75).
- MEF medium (as described above).

**Protocol**

1. Remove the media from culture flasks by aspiration and gentle wash (twice) the flasks with 10 ml of DPBS;
2. Remove the DPBS and add to the culture flasks around 5 ml of trypsin/EDTA 0.25 %; gently shake the flasks and verify if all cells are covered by the trypsin solution;

3. Incubate at 37 °C, 5 % $CO_2$, 95 % humidity for 3–5 min, or until the cells are completely detached (observed under microscope); if the cells still adherent, leave them in the incubator for an extra time (3 min);
4. Block the trypsin action by adding 10–15 ml of MEF medium to the culture flasks;
5. Transfer the cell suspension to centrifuge tubes and centrifuge for 5 min at 230 × *g* (room temperature);
6. Remove supernatant and resuspend the cells in 15 ml of MEF medium;
7. Split the cells in a proportion of 1:3, plating into T75 culture flask dishes pre-treated with gelatin. Alternatively, MEFs can be frozen.

#### 2.2.1.3 MEF Inactivation with Mitomicin-C

The main goal of inactivation is to block the proliferation of MEFs, avoiding these cells to inhibit the growth of hESC colonies. This inactivation can be performed by the use of gamma-radiation or by using the antibiotic antitumural drug Mitomicin-C, which acts by preventing the separation of DNA helices, resulting in the inhibition of cell division. Inactivated MEFs (iMEF) can be cryopreserved to be used in the future or can be immediately used after the inactivation treatment. Here, we describe the protocol for MEF inactivation by Mitomicin-C treatment. It is important to remember that Mitomicin-C is a toxic reagent, which requires attention during the manipulation; all materials that have contact with the reagent (including gloves) need to be disposed in special recipients, according to the rules for toxic and chemical refuse.

**Material**

- Flasks with active MEF cell culture (T75).
- 1× DPBS de $Ca^{2+}$ e $Mg^{2+}$free (Life Technologies, Gibco, Cat. No. 14190-144).
- Sterile gelatin solution in water 0,1 % (StemCell Technologies, Cat. No. 07903).
- 0.25 % Trypsin/EDTA (Life Technologies, Gibco, Cat. No. 25200-056).
- Mitomicin-C (Sigma-Aldrich, Cat. No. M4287).
- Sterile Centrifuge tubes (50 ml).

Cells are counted using a Neubauer or hemacytometer chamber, or any counter equipment (e.g. Countess, Invitrogen, Life Technologies). For manual counting, we will need sterile microtubes (1.5 ml), Trypan Blue solution, and the chambers

**Protocol**

1. The solution for Mitomycin-C treatment is prepared by adding 180 μl of Mitomycin-C stock solution (0.5 μg/μl in PBS) in 9 ml of MEF medium (final concentration 10 μg/ml);
2. Remove the medium from T75 culture flasks with active MEFs and add the Mitomycin solution; incubate for 3 h at 37 °C, 5 % $CO_2$, 95 % humidity;
3. Gently remove the Mitomycin-C medium, wash the cells twice with DPBS (25 ml of DPBS). The medium and DPBS removed during this procedure should be disposed in special recipients for toxic and chemical waste;

4. At this point, you can either freeze or plate the iMEFs; to split the cells, proceed as described above;
5. To count the cells, add 10 μl of cell suspension into a microtube with around 10 μl of Trypan Blue solution. After 2 min, place 10 μl of this mixture into a Neubauer chamber and estimate the cell number and viability under the microscope;
6. MEFs treated with Mitomicin can be used immediately, by plating the cells in a petri dish pretreated with gelatin, and incubated at 37 °C, 5 % $CO_2$, 95 % humidity. Alternatively, centrifuged the cells at 230 × *g* for 5 min, resuspend in freezing MEF medium, and freeze to use in the future.

#### 2.2.1.4 Plating Inactivated MEFs for Derivation and Cultivation of hESC

**Material**

- Inactivated MEFs.
- Sterile gelatin solution in water 0.1 % (StemCell Technologies, Cat. No. 07903).
- Sterile adherent plates with central well (Falcon 60 mm Center Well Organ Culture Dish, Corning, Cat. No. 353037, for derivation of hESC) or Sterile adherent plates 35 mm (for derivation or maintenance of hESC).
- MEF medium (the same previously described).

**Protocol**

1. Pre-treate plates with gelatin 0.1 % (500 μl for plates with central well or 1 ml for 35 mm plates) for at least 20 min at room temperature. Remove gelatin immediately before plating the MEF cells;
2. For each well, plate $1 \times 10^5$ inactivated MEF cells/$cm^2$ in MEF medium (800 μl of MEF medium for central well plates or 2 ml for 35 mm plates);
3. Incubate at 37 °C, 5 % $CO_2$, 95 % humidity;
4. Around 24 h after plating, verify if the cells are well distributed in the plate; the confluency must be 90 %). The plates containing inactivated MEFs are ready to be used in the process of derivation and maintenance of hESC. Once inactivated, MEFs have a short period of life; if the cells are plated for more than 3 days, they should be discarded and fresh plates must be prepared.

### *2.2.2 Derivation and Maintenance of New ESC Lines*

#### 2.2.2.1 Derivation of New ESC Lines in MEF or Matrix

Embryos used for the establishment of hESC lineages need to be in the blastocyst stage of development (days 5–6) and, ideally, with good morphology and good ICM, and with trophectoderm (TE) visualization (Fig. 2.1a). If the embryos were frozen, they

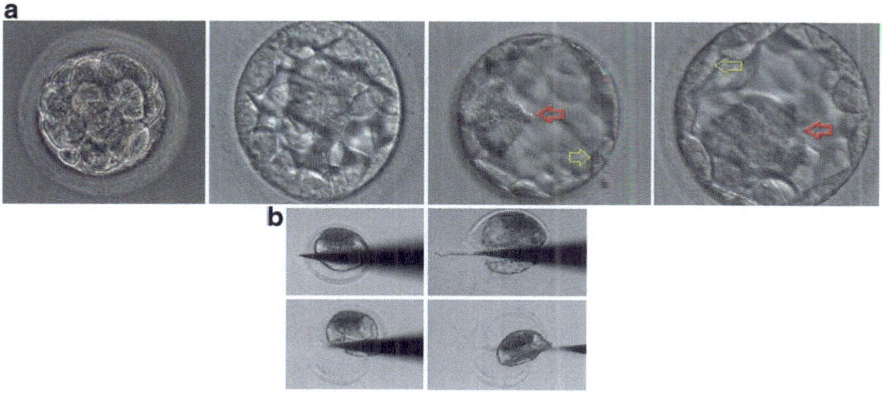


**Fig. 2.1** Derivation of hESC. (**a**) Embryonic development, morula, cavitating morulae, early blastocyst and expanded blastocyst, respectively from left to right, showing throphoblasts (*yellow arrow*) and inner cell mass (ICM) (*red arrow*); (**b**) mechanical removal of *zona pellucida* and isolation of ICM

need to be thawed and culture until the blastocyst stage before proceeding with the plating. Before plating, it is necessary to remove the *zona pellucida* (ZP) of the embryo, which can be made in different ways, according to the laboratory availability/routine: (1) chemically, using Tyrode's Solution Acidified (Irvine Scientific, Cat. No. 99252); (2) enzymatically, using *Pronase* (Sigma-Aldrich, Cat. No. P5147); (3) mechanically, with needles (BD Ultra-Fine Insulin Syringes, Cat. No. 14502) or micro blades (Fig. 2.1b) or, by using an especial equipment such as micromanipulator and laser (Nikon Instruments). In the last case, you can remove and plate only the ICM. After removing the ZP, the whole embryo or isolated ICM can be plated onto plates pretreated with inactivated feeder cells or, alternatively, in other types of matrices, such as Matrigel (Corning Matrigel Growth Factor Reduced (GFR) Basement Membrane Matrix, LDEV-Free, Cat. No. 354230) or Geltrex (Geltrex hESC-Qualified Reduced Growth Factor Basement Membrane Matrix, Gibco, Life Technologies, Cat. No. A15696), according the manufacture instructions. Different types of media can be used for the establishment of new lines; homemade media or commercial defined medium (e.g. mTeSR1, Stem Cell Technologies, Cat. No. 05850 or Essential 8, Life Technologies, Gibco, Cat. No. A1517001). Here, we will describe the derivation procedure using manual removal of ZP and ICM, and culture medium based KO-DMEM, KSR and bFGF. We will also describe a xeno-free protocol for derivation and cultivation of hESC.

## Material

- Surplus human embryos donate for research with signed informed consent from parents.
- Pretreated plates with inactivated feeder cells.
- Sterile non-adherent plates (35 mm) to wash, remove the ZP and isolate the ICM.
- Media for embryos.

*Note: Different types of media can be used, including* Modified HTF Medium with Gentamicine—HEPES, Irvine Scientific, Cat. No. 90126)*, that is usually used for washing and transportation of embryos from the clinic to the laboratory. It is better to keep the same medium that was used for the embryos.*

- 1× DPBS $Ca^{2+}$ and $Mg^{2+}$ Free (Life Technologies, Gibco, Cat. No. 14190-144).
- Stripper Micropipetter (Mid Atlantic, Cat. No. MXL3-STR).
- Stripper tips (150, 175 and 200 μm, MidAtlantic Inc., Cat. No. MXL3-150, MXL3-175, MXL3-200).
- Blades—Ultra Sharp Splitting Blades (Bioniche Animal Health USA, Cat. No. ESE020).
- hESC medium (describe below).

**hESC Medium**

- 1× KO-DMEM (Knockout-DMEM, Life Technologies, Gibco, Cat. No. 10829-018).
- 20 % KnockOut Serum Replacement (Life Technologies, Gibco, Cat. No. 10828-028).
- 1× Non-essential Aminoacids (0.1 mM, Life Technologies, Gibco, Cat. No. 11140-050).
- 50 U/ml Penicilin/50 μg/ml Streptomicin (Life Technologies, Gibco, Cat. No. 15140-122).
- 2 mM GlutaMAX™ Supplement (Life Technologies, Gibco, Cat. No. 35050-061).
- 0.055 mM 2-Mercaptoetanol (Life Technologies, Gibco, Cat. No. 21985-023).
- 8 ng/ml bFGF (EMD Millipore, Cat. No. GF003).

*Note: The bFGF factor should be added into the medium immediately before use to avoid degradation.*

Mix all reagents, except bFGF, and filter (22 μm filter). Store the medium at 4 °C for up to 10 days. Before use, aliquot the exact volume that will be used and pre-warm at 37 °C. Add the bFGF immediately before adding the medium into the cells.

**Protocol**

1. Prepare plates with inactivated MEFs the day before plating the embryo (*Note: maximum of 3 days before*), according to the previously described protocol;
2. In the day of plating embryos, remove the MEF medium from the plates and wash the cells twice with DPBS. This process is important to eliminate all FBS-based MEF medium. Add hESC medium to the plate and incubate at 37 °C until embryo/ICM plating;

3. Verify and make notes on the quality of the embryo. If the blastocysts have a good quality, with an easy visualization of ICM and TE (Fig. 2.1a), the ICM can be removed. Normally, embryos donated for research are considered unviable for transfer, which reflects their poor quality. Whole, poor quality embryos can be plated after ZP removal;
4. If the blastocysts are in spontaneous *hatching* (ZP free), transfer them directly to a 35 mm petri dish containing embryo medium (2 ml) to isolate de ICM. If the blastocyst embryo is not in spontaneous *hatching*, the ZP must be removed. The mechanical removal of ZP should be made with blades (Ultra Sharp Splitting Blades) previously treated with 70 % ethanol, air dried and washed in embryo's medium. Observe the embryo under the stereomicroscope to verify the correct localization of ZP and ICM, and cut with the blade from opposite side of ICM (Fig. 2.1b). The ZP can show some resistance being elastic. Make a small and gentle pressure in the embryo to facilitate ZP opening using the blade or fine needles. After making a hole in the ZP, the embryo can be isolate from ZP with the blade (by pressure) or using a Stripper micropipetter (175 μm or 200 μm) (Fig. 2.1b). In this step, the blastocyst free of ZP can collapse and isolation of ICM is difficult. If this is the case, proceed with whole embryo plating;
5. If it is possible to visualize the ICM after ZP was removed, the ICM can be isolate mechanically from the TE with blades or Stripper micropipette (150 μm or 175 μm), and plating free of TE. This blastocyst free of ZP is less resistant and can be easily cut by blades or fine needles. After localizing the ICM, cut very close to the ICM maintaining a small number of flanking TE cells. Important: protect the ICM as much as possible to avoid damages, if it was not possible, prefer plating whole embryos;
6. With a Stripper micropipetter, transfer the ZP free embryo or ICM into new plates with prewarmed hESC medium in order to wash the embryos. Repeat the washing by transferring the embryos or ICM to new dishes (you can use drops of medium ~50 μl for each wash);
7. Transfer whole embryos or ICM to the petri dishes containing iMEFs and hESC medium, and incubate at 37 °C, 5 % $CO_2$, 95 % humidity. If possible, use incubator with oxygen control, at 5 % $O_2$. Avoid moving and/or manipulating the dishes for 48 h to allow the ICM to adhere to the plate;
8. After this 48 h, performe daily media changes and visually follow cell growth. The criteria to choose the best moment for the first passage will depend on the quality of the embryo or ICM plated, and the quantity of TE present. If whole embryos with much TE were plated, you will need to follow the growing of epiblastic-like cells (ICM) or TE that show different morphology. Cells derived from ICM are very small, rounded and grow quite attached. Cells derived from TE are large and have a more isolated and flatter morphology. If the cells from ICM are isolated from TE cells, the first passage can be performed around 10 days after plating. However, TE cells can grow more quickly than ICM and inhibit the ICM growth. In this case, you should remove the excess of TE cells with fine tips and, sometimes, the

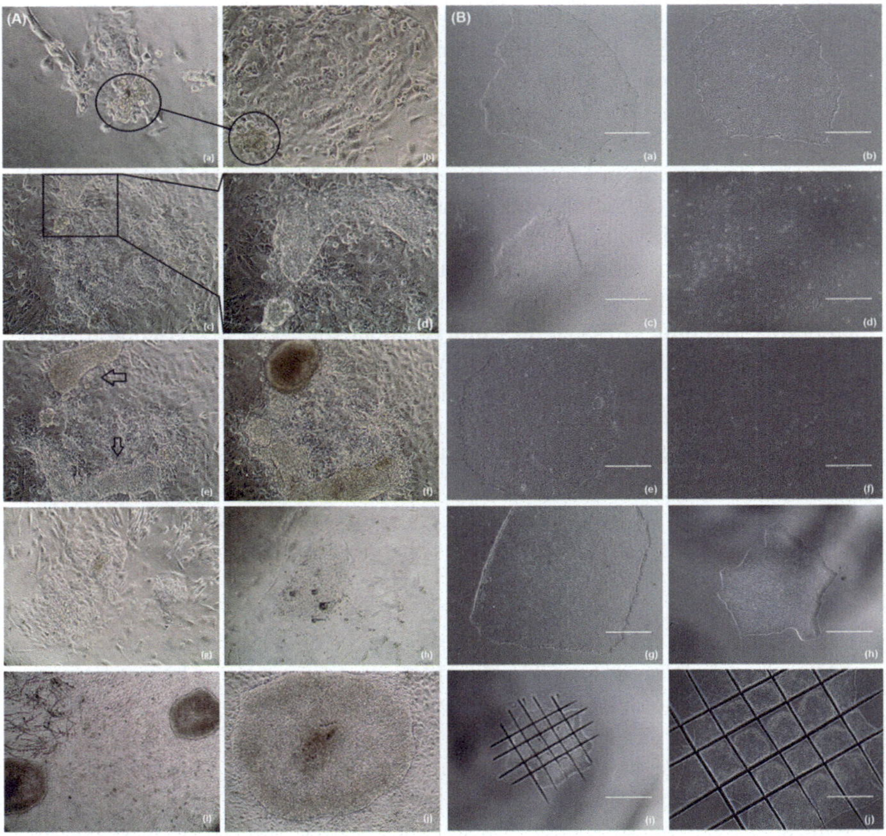

**Fig. 2.2** Derivation of hESC after ICM isolation and transfer to culture plates. (**A**) Different days of culture: (*a*) 6 days after plating of ICM in Matrigel treated plates (10× magnification); (*b*) 9 days of culturing, showing cellular growth (4× magnification); c) and d) 10 days of culture (10× and 20× magnification, respectively); cell proliferation at day 11 (*e*), *arrows* indicate the cell proliferation that represent the ICM and, at day 12 (*f*) (10× magnification); in (*g-h*) first passage of cells after 14 days of culture (10× magnification); (*i-j*) colonies of hESC after the second passage around the 20° day of culture (10× and 20× magnification, respectively). (**B**) Representative images hESC colony: (*a-d*) the *good* quality/morphology of a hESC colony suitable for a mechanical/manual passaging, (*e-f*) spontaneous cell differentiation; (*g-h*) hESC colony after cleaning by removing the potencial differentiated cells from the border; (*i*) scratches layout performed in a colony using a needle to generate small pieces (4× magnification, scale bar 1000 μm) and (*j*) fragments of hESC colony suitable to be transfered to a new petri dish, (10× magnification, scale bar 400 μm)

passage should be made more precocious (4–5 days after plating). Usually, the first passage is made between 5–8 days after plating and no more than 15 days, and should be made by manually fragmentation of cells derived only from the ICM. After the first passage, colonies of hESC have a homogeneous surface and are flattened with clear borders (Fig. 2.2A, B). The protocol for cell passages is described below.

#### 2.2.2.2 Maintenance, Manual Passage and Cultivation of hESCs

Long-term culture of hESC is challenging and some parameters should be considered to maintain these cells in an undifferentiated pluripotent stage. Several methods of passage can be used for culturing hESC, including enzymatic (colagenase, dispase, Accutase), chemical (based in EDTA, Versene, Life technologies or Gentle Cell Dissociation, Stem Cell Technologies), or manual passage (Fig. 2.2Bi–j).

Here, we will describe the manual method to pass cells cultivate in feeder cell layers. The same procedure can be adapted for hESC cultivate in Matrigel, geltrex or other type of matrix. The period between the passages should be determined by colony size and it is usually every 4–6 days. Some lineages have faster growth rate than others. At the moment of manual passage, colony size should be between 300 and 500 cells, independent of the number of colonies in the dish. However, the colony density in the plate should not be too high and should not form a monolayer. Very large colonies (occupying the entire microscope field when viewed under a 10× objective) or very high densities of colonies may result in spontaneous cell differentiation.

**Material**

- Sterile adherent plates (35 mm) covered with inactivated MEFs or other types of matrices.
- hESC medium with bFGF (as described above) or defined hESC medium.
- Sterile Fine Needle Syringe (Insulin or Tubulin).
- Sterile Tips for micropipets P200 or Cell Scraper.

**Protocol**

1. Prepare the plates with inactivated MEFs in the day before passing the cells;
2. Before the manual passage, verify if the culture and or the colonies have any type of spontaneous differentiation (Fig. 2.2Bg). In this case, you should remove the differentiation from the plate/colonies with micro tips under the stereomicroscope under sterile conditions (Laminar Flow cabinets);
3. Remove the medium containing the differentiated fragments in suspension and add fresh media into the plate (2 ml of hESC medium for each 35 mm plate);
4. Under an inverted microscope or stereomicroscope, using fine needles make small scratches in the colony as a grid (Fig. 2.2Bi–j). The size of the fragments is important and should be around 50–100 cells. For example, one colony with 300 cells should be fragmented into six pieces with around 50 cells each;
5. With a cell scraper or a tip for micropipettes (P200) gently lift each fragment of the colony;
6. After every colony fragments are detached from the plate, transfer the fragments to new plates (with MEFs) using serological pipettes (2 ml). Split the fragments in the proportion of 1:2 or 1:3 plates. If a plate has a high density and/or large colonies, you can increase the split ratio;

   *Note: very high or very low plating densities can result in cell differentiation.*

7. Incubate the plates at 37 °C, 5 % $CO_2$, 95 % humidity and leave them for 48 h before replacing the medium, avoiding any movement. This period is necessary for the fragments to adhere to the new plate and initiate growth;
8. After 48 h, remove the medium from the plates and gently add new fresh pre-warmed medium;
9. Replace the media daily until the moment of the passage.

#### 2.2.2.3 Freezing and Thawing of hESC

The hESC are extremely fragile to the freezing and thawing process and, therefore, must be carefully handled. For freezing, the size of fragments needs to be twice bigger than the size for passage for subculturing.

Freezing hESC

**Material**

- hESC medium.
- hESC freezing medium (below).
- Sterile fine needle (insulin fine needle, BD).
- Plastic sterile tips for micropipettes P200 or Cell Scraper.
- Sterile Centrifuge tubes (15 ml).
- Cryotube (2 ml).
- Mr. Frosty Freezing Container (Nalgene, Thermo-Scientific, Cat. No. 5100-001).
- Isopropanol.

**hESC Freezing Medium**

- 50 % hESC medium.
- 40 % FBS ES-Qualified (Life Technologies, Gibco, Cat. No. 16141-079).
- 10 % DMSO (Dimethyl Sulfoxide, Sigma-Aldrich, Cat. No. D2650).

**Protocol**

1. Remove the medium from the plates containing the colonies and add 2 ml of fresh prewarmed hESC medium;
2. Under the stereomicroscope, using fine needles, cut the colonies in small fragments with around 100–200 cells, similar to hESC manual passage;
3. As described before, use cell scraper or micropipettes tips to manually lift the fragments;
4. Transfer the medium containing the fragments to centrifuge tubes, centrifuge $300 \times g$ for 5 min at room temperature;
5. Remove supernatant and gently ressuspend the cell fragments in 1 ml of iced hESC freezing medium;

6. Transfer the cells to cryotubes, put them in an appropriate container (Mr. Frosty) and deep at −80 °C;
7. After 24–48 h, transfer the cryotubes to liquid nitrogen to store the cells at −196 °C.

#### 2.2.2.4 Thawing of hESC

**Material**

- Sterile adherent plates with inactivated MEFs.
- hESC medium supplemented with bFGF (as above).
- Sterile Centrifuge tube (15 ml).

**Protocol**

1. Prepare petri dishes with inactivated MEF as previously described in the day before thawing the cells; in MEF-free protocols, prepare the petri dishes with the appropriate matrix (Geltrex or Matrigel, 1 h before use)
2. Add 8 ml of prewarmed hESC medium to a centrifuge tube (or other defined hESC medium);
3. Quickly thaw the cryotubes in a waterbath at 37 °C for 2 min, or until the media is thawed;
4. Gentle and quickly transfer the cells to the pre-prepared centrifuge tube with medium;
5. Centrifuge 130 × *g* for 5 min;
6. Remove the supernatant and gently resuspend the cells in fresh prewarmed hESC medium by softly pipetting 2–3 times (avoid cell fragmentation);
7. Plate the cells in pretreated petri dishes and incubate at 37 °C, 5 % $CO_2$, 95 % humidity for 48 h;
8. After this time, change the medium daily until cells are ready to pass.

### *2.2.3 Derivation hESC on Xeno-Free Matrix*

For derivation in a xeno-free culture condition, you can use different types of matrices such as human Vitronectin (Vitronectin (VTN-N) Recombinant Human Protein, Truncated—Thermofisher, Cat. No. A14700), human recombinant Laminin (Biolamina Sundbyberg, Sweeden, Cat. No. Biolamin 521) or CloneStem kit (Biolamina). Here, we will describe a defined xeno-free culture condition using the CloneStem system to establish new ES cell lines. It contains recombinant Laminin-521 and E-cadherin as matrices, and we use E8 medium (Life Technologies, Thermo Fisher, Gibco, Cat. No. A1517001) as previously described (Rodin et al. 2014).

**Material**

- Sterile non-adherent plates (35 mm) for washing and isolating the ICM, and media for the embryos.
- 4-Well × 1 ml MultiDish Cell Culture Dish (Nunc—Thermo Scientific, Cat. No. 176740).
- Sterile centrifuge tubes (15 or 50 ml).
- Blades—Ultra Sharp Splitting Blades (Bioniche Animal Health USA, Cat. No. ESE020) for ZP removal or ICM separation in a mechanical way.
- 1× DPBS Ca2+ and Mg2+ (Life Technologies, Gibco, Cat. No.14040-133).
- Stripper Micropipetter (Mld Atlantic, Cat. No. MXL3-STR).
- Stripper tips (150, 175 and 200 μm, MidAtlantic Inc., Cat. No. MXL3-150, MXL3-175, MXL3-200, respectively).
- Sterile Petri dishes Universal GPS dish (LifeGlobal group, Cat. No. UGPS-010) with eight outer Wells (100 μl of medium) and two central (up to 150 μl).
- Sterile Oil for Embryo Culture (Irvine Scientific, Cat. No. 9305).
- CLONEstem kit (BioLamina) containing Laminin-521, E-cadherin and human albumin.
- LAMININ-521 Stem Cell Matrix (BioLamina).
- Essential 8 media (Life Technologies, Gibco, Cat. No. A1517001).
- Rock Inhibitor-Y27632 (StemGent, Inc, Cat. No. 04-0012).

**Solution for Passaging**

- 1× DPBS Ca2+ e Mg2+ free (Life Technologies, Gibco, Cat. No. 14190-144).
- Gentle Cell Dissociation Reagent (StemCell Technologies, Cat. No. 07174) or
- ReLeSR enzyme free (StemCell Technologies, Cat. No. 05872) or
- Versene solution (Life Technologies, Gibco, Cat. No.15040-066) or
- StemPro Accutase Cell Dissociation Reagent (Life Technologies, Gibco, Cat. No. A11105-01).

**Xeno-Free Cryopreservation Media**

- Synth-a-Freeze Cryopreservation Medium (Life Technologies, Gibco, Cat. No. A12542-01) or
- CryoStor CS10 (StemCell Technologies Cat. No. 07930).

#### 2.2.3.1 Embryo Plating

The preparation of embryo or ICM for culture is performed as described above.

**Protocol**

1. Prepare the E8 medium according to manufacturer's instructions and aliquot in small volumes;
2. Pre-coat the Universal GPS dish with Laminin-521 and E-cadherin, according to the manufacturer's instructions. Around 25 μl of laminin solution is enough

to each well. Leave the dishes at the incubator for 2 h before plating the embryos;

3. Supplement E8 media with 10 % human albumin serum and prewarm at 37 °C before use. You will need approximately 100 μl for each well and 500 μl to wash the embryo after ZP removal or ICM isolation before plating;
4. Immediately before plating the embryos, remove the laminin solution from the wells. Wash each well with prewarmed DPBS, add 75–100 μl of E8 media supplemented with albumin to each well and gently cover the dish with approximately 7 ml of prewarmed embryo oil;
5. Remove the ZP as described before wash the embryo once in embryo medium and wash again in E8 medium supplemented with albumin;
6. Using micropipettes, transfer the ZP-free embryos or ICM to Universal GPS dish, one in each well, and incubate the dishes at 37 °C, 5 % $CO_2$ (if possible, 5 % $O_2$) for 48 h. Avoid opening the incubator during this period;
7. After 48 h, the medium should be changed daily. Visually check if the embryos are attached to the matrix and are growing. At the first medium change, gently remove 50 % of the medium and replace with 50 % of prewarmed E8 supplemented with albumin. Avoid that the oil enters in contact with embryos;
8. If whole embryos were plated, the TE can overgrow and inhibit the ICM growth. Then, after 7–10 days, if you detect TE overgrowing, remove TE cells. Remove the medium containing the TE fragments and complete with fresh prewarmed E8 medium;
9. Around 10–15 days, the ICM should have formed a small colony and must be passaged;
10. Pretreat a universal GPS dish in the same way as described before;
11. The first passage is made in a mechanical/manual way. In this case, you can use Rho-associated kinase (ROCK) inhibitor (5 μM). ROCK inhibitor prevents apoptosis besides enhancing the survival and cloning efficiency from both the derivation of ES cell or during passages in small clumps or single cell passage;
12. After the passage, medium is changed every day by replacing around 50 % of the medium with fresh medium;
13. The next 3–4 passages are made mechanically; fragments are transferred to plates (you can use now 4-well dishes, NUNC) precoated with Laminin 521 (around 100 μl of laminin solution for each well; according to the manufacturer's instructions) in E8 medium without albumin supplementation, in the presence of ROCK Inhibitor, in a split ratio of 1:2;
14. After passage 5, passages can be performed using EDTA solution (Gentle Cell Dissociation or Versene) or Accutase, according to the manufacturer's instructions;
15. For cryopreservation of xeno-free hESC lines, you can proceed with the fragmentation in the same way for passing the cells (as described before) and use xeno-free cryopreservation media, such as Synth-a-freeze medium (Life Technologies, Gibco), CryoStor (Stem Cell Technologies) or similar, according to the manufacturer's instructions.

### 2.2.4 Characterization of New hESC Lines

#### 2.2.4.1 Immunocytochemistry

Immunofluorescence analysis is used to show pluripotency of the new lineages both in undifferentiated cells, using antibodies against pluripotency markers, such as OCT4, NANOG, SSEA4, TRA-1-60 and TRA-1-81 (Fig. 2.3A), and in differentiated cells (spontaneous differentiation) (Fig. 2.3B, C) to confirm that those cells are able to differentiate into cells representative from thee germ layers. In this case, we can use markers for differentiated cells (for example, NESTIN—ectoderm; MyoD—mesoderm and SOX17—endoderm). For immunofluorescence analysis, we need to cultivate the cells in slide chambers and immunostain according to standard protocols (Mitalipova and Palmarini 2006).

**Material**

- Slide chambers cover glass (Labtech, NUNC, Thermofisher) for 4-well (Cat. No. 177399) or 2-well (Cat. No. 177380), according to your preference.
- Coverslips.
- Sterile Centrifuge tubes (15 ml).
- VectaShield antifade mounting medium with DAPI (Vector Laboratories, Cat. No. H1200).

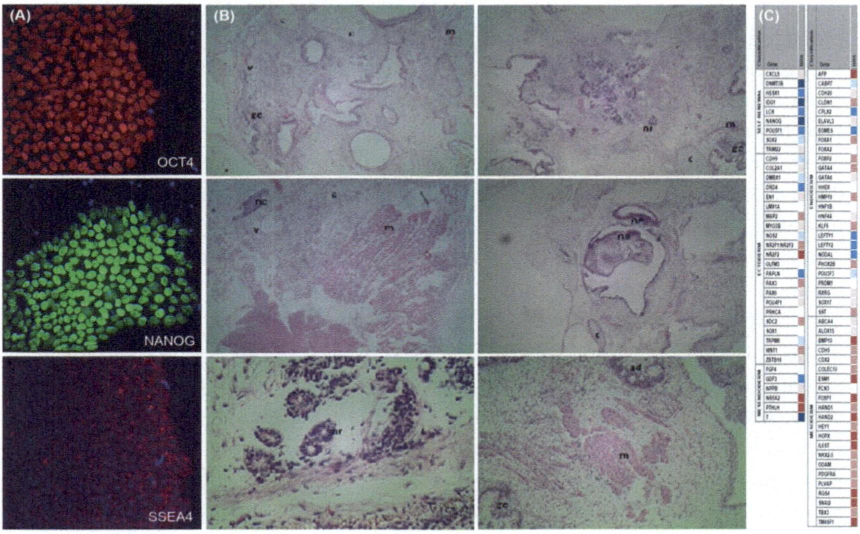


**Fig. 2.3** Characterization of hESC lines. (**A**) Imunoflurescence using pluripotency markers showing that cells are positive for OCT4, NANOG and SSEA4; (**B**) Differentiation in *vivo:* after the teratoma formation, tumors are sectioned and stained, showing representative cells from the three germ layers; (**C**) spontaneous differentiation *in vitro* and ScoreCard analysis showing that the cells are capable to differentiate into cells representative of three germ layers. In (**B**), *ne* neuro epithelium, *nr* neural rosettes, *m* muscle, c: cartilage, *ad* adipocyte, *v* vessel

- 1× DPBS $Ca^{2+}$ and $Mg^{2+}$ (Life Technologies, Gibco, Cat. No. 14040-133).
- Formaldehyde solution (Merck, Millipore, Cat. No. 104003) diluted to 4 % in DPBS.
- BSA (Sigma Aldrich, Cat. No. A3311) solution at 2 % in PBS.
- Triton solution (Sigma Aldrich, Cat. No. X100) diluted to 0.5 % in PBS.
- Primary antibodies for pluripotent markers or differentiated cells (conjugated or not).
- Secondary antibodies (Alexa Fluor, Molecular probes, ThermoFisher).
- ESC lines cultivated in slide chamber (cultivate the cells similar to plate in petri dishes).

**Protocol**

1. Wash the cells twice in DPBS and fix them using formaldehyde 4 % for 15 min. Next, cells are washed in DPBS and can be maintained in DPBS at 4 °C until the immunostaining;
2. To proceed with the immunostaining, remove DPBS from the slide chamber and add Triton solution for 10 min in an attempt to permeabilize the membrane;
3. Wash the cells with DPBS three times for 5 min each;
4. Proceed with the blocking step, using BSA solution for at least 1 h at room temperature;
5. Dilute the primary antibodies according to the manufacture suggestion, in BSA solution. For pluripotency markers, antibodies such as anti-OCT4 (1:200 to 1:100, Santa Cruz Biotechnologies, Cat. No. sc-8628), anti-NANOG (1:25, RD Systems, Cat. No. AF1997) and anti-SSEA4 (1:100, StemGent Inc., Cat. No. 09-0006,) can be used (Fig. 2.3A, respectively).
6. Incubate primary antibodies at 4 °C overnight in humidity chamber;
7. After washing thee times with DPBS, perform the incubation with secondary antibodies conjugated with Alexa-Fluor 488 (green) or 594 (red) (1:1000, Molecular Probes, Life Technologies) at room temperature;
8. Nuclei are counterstained with DAPI and the slides are mounted in VectaShield mounting medium. The immunostaining can be analyzed under an epifluorescence microscope.

#### 2.2.4.2 In Vitro Spontaneous Differentiation

Spontaneous differentiation of the hESCs is performed to verify if the cells are capable to differentiate into cells representative of the three germ layers. After differentiation, the cells can be analyzed by immunofluorescence using antibodies against markers representative of differentiated cells or, by the ScoreCard assay (From Applied Biosystem, ThermoFisher) (Fig. 2.3C). Spontaneous differentiation is performed as described (Silva et al. 2008), through embryoid body (EB) formation.

**Material**

- Sterile petri dishes (35 mm) or 6-wells dishes to culture cells or
- Slide chambers cover glass (Labtech, NUNC, Thermofisher) for 4-well (Cat. No. 177399) or 2-well (Cat. No. 177380), according to your preference, for immunostaining.
- E8 medium (Life Technologies, Thermo Fisher, Gibco, Cat. No. A1517001).
- Sterile agarose (Invitrogen, Life Technologies) 1 % solution to coated the petri dishes or you can use ultra low attachment petri dishes (NUNC).

**Differentiation Medium**

- 1× DMEM high glucose (Life Technologies, Gibco, Cat. No. 11965 or 11960).
- 15 % FBS ES-Qualified (Life Technologies, Gibco, Cat. No. 16141-079).
- 2 mM GlutaMAX™ Supplement (Life Technologies, Gibco, Cat. No. 35050-061).
- 100 U/ml Penicillin/100 µg/ml Streptomicin (Life Technologies, Gibco, Cat. No. 15140-122).
- 1× Non-Essential Amino Acids (0.1 mM, Life Technologies, Gibco, Cat. No. 11140-050).
- 1 mM Sodium Pyruvate (Life Technologies, Gibco, Cat. No. 11360-070).
- 10 mM HEPES (Life Technologies, Gibco, Cat. No. 15630-080).
- 0.055 mM 2-Mercaptoethanol (Life Technologies, Gibco, Cat. No. 21985-023).

**Protocol**

1. Culture the cells as the same to maintain cells in culture;
2. To prepare the cell fragments to form EBs, proceed as usual to mechanically pass the cells;
3. Treat non-adherent petri dishes with agarose 1 % (put 1 ml of agarose solution into the petri dish, remove the excess and leave the dishes to air dry for 20 min before use);
4. Transfer the media with colony fragments onto dishes precoated with agarose. Cell aggregates are grown in suspension in non-adherent plates with differentiation medium;
5. After 1 week in suspension, embryoid bodies (EBs) are transferred to adherent plates precoated with 0.1 % gelatin (Stem cell Technologies) or slide chambers, and grown for additional 8–10 days in differentiated medium; media changes are preformed every other day;
6. Proceed with immunofluoresce as described above using germlayer specific antibodies, for example: anti-Nestin (1:200–1:100, Millipore, Cat. No. AB5922), anti-MyoD (1:100, Millipore, Cat. No. MAB3878) and anti-SOX17 antibody (1:200 to 1:100, Millipore, Cat. No. 09-038);
7. To analyze differentiation using the ScoreCard assay, cells are harvested with Trizol (Life Technologies, Cat. No. 15596) and RNA extraction is performed using RNeasy kit (Qiagen, Hilden). One microgram of RNA is used for subsequent reverse transcriptase reactions using High Capacity cDNA kit (Life Technologies). Pluripotency and tri-lineage differentiation potential can be analyzed using the TaqMan hPSC Scorecard kit 96w following manufacturer's

instructions, and run on a StepOne Plus System (Applied Biosystems, Life Technologies). Data analysis is performed using the cloud based TaqMan hPSC Scorecard analysis software (Applied Biosystems, Life Technologies).

#### 2.2.4.3 Differentiation In Vivo: Teratoma Formation

Teratoma Assays are performed, *in vivo*, to show that the new ESC lines are capable to differentiate into a variety of tissues and cells when injected into immunecompromised mice (Fig. 2.3B).

**Protocol**

1. Cultivate the cells on MEFs. When colonies are ready to be passaged, mechanically dissect the colonies in small clumps;
2. Approximately $1 \times 10^5$ cells are ressuspended in 30 % BD Matrigel hESC-qualified Matrix (BD Biosciences) in DPBS and are subcutaneously injected in C57/B16 SCID mice, as already described (Prokhorova et al. 2009);
3. Ten weeks after injection, tumors can be observed. Mice are euthanized, and tumors are removed and subjected to standard histological analysis.

#### 2.2.4.4 Cytogenetic Analysis

To confirm the euploid constitution of the ES cells, we perform cytogenetic analysis by karyotyping the metaphases or by the CGH-Array assay. After cultivated, arrested and fixed, cells are subjected to standard G-banding karyotype analysis as described (Gosden et al. 1992). At least 50 metaphases need to be analyzed in order to detect the presence of mosaicism. Currently, CGH-array technique is more suitable to analyze the chomosomal balance in new hESC lines.

### 2.2.5 *Potential Use of hESC*

Human pluripotent stem cells (PSC) are a promising source for cell replacement therapy and regenerative medicine due to their capacity to differentiate into a variety of cells and tissues. Other types of stem cells, such as adult stem cells, have a limited potential of differentiation compared with PSC. The continuous interest in these cells is clearly evidenced by the fact that, since the derivation of the first line of hESC in 1998, the number of cell lines published and recorded in the databases has rapidly increased. In addition, pre-clinical and clinical studies performed to date, suggest that hESC display relevant biological activity, such as low risk of toxicity in short or long-term for the recipient.

However, to properly use these cells for cell replacement therapy, some strict culture conditions must be adopted during derivation and maintenance of hESC. In order to use these hESC or other types of PSC in regenerative medicine, it is preferable to have defined and free of any animal contaminants culture conditions. For this reason, there is an ongoing effort to establish cell supporting matrices free of animal origin, completely defined culture media and appropriate cultivation procedures for hESC according to good manufacturing pratice (Unger et al. 2008). However, most of hESC lines derived up to now were not available or maintained under these optimum culture conditions, since the most common methodology used for the establishment of these cell lines was based on protocols using MEFs as support cells or, if MEF-free, the matrices were derived from murine cells. Nevertheless, it is important to note that the first clinical study in which hESC were used for the treatment of spinal cord injuries (http://www.biotimeinc.com/) were conducted using a hESC line that has been derived in the presence of animal components and which was subsequently adapted to xeno-free culture conditions before being transplanted into patients (Geron Corporation 2009).

Possibly, the major concern in the use of hESC in cell therapy is related to the safety of the use of these cells, more specifically the intrinsic potential that pluripotent cells have to give rise to teratomas or teratocarcinomas after injected into the patient. This side effect of PSC could be avoided by *in vitro* pre-differentiation of these cells before their use in cell therapy procedures. However, the acquisition of a homogeneous population of differentiated cells in culture is not trivial, and even a small number of undifferentiated cells could lead to the formation of teratomas or teratocarcinomas afterwards.

In the last 5 years, several clinical trials using hESC-derived cells were initiated, including spinal cord injury, Stargardt's macular dystrophy and dry-age macular degeneration, and diabetes. The first clinical trial using hESC-derived oligodendrocytes cells was approved in the USA by the American Food and Drug Administration (FDA). Patients with recent neurologically complete traumatic spinal cord injuries (7–14 days of trauma) received a single administration of oligodendrocyte progenitor cells derived from a hESC line. This Phase I safety study was presented at the "2011 American Congress of Rehabilitation Medicine" and showed that the cells administered were well tolerated and did not cause any serious adverse effect (www.biotimeinc.com). Treated patients will be followed up for 15 years. Another example was conducted by the Advanced Cell Technology (Massachusetts, USA), which started a non-randomized, sequential, multi-center safety and tolerability Phase I/II trial clinical studies for dry age-related macular degeneration and Stargardt's macular dystrophy in 2011. These studies will serve to evaluate the effect of subretinal delivery of hESC-derived retinal pigmented epithelium cells in patients with macular degeneration (Schwartz et al. 2012). Finally, in 2014 Viacyte (www.viacyte.com) started a Phase I clinical trial for type I diabetes using hESC-derived pancreatic beta cell precursors. The company has already completed their first Phase I/II clinical trials (2005–2010, Clinical Trials ID: NCT00260234).

Human ESC and other PSCs can also be used as a model for drug discovery and testing for toxicology of new drugs directly in human cells, avoiding the use of other animal models for *in vivo* screening (reviewed in Maury et al. 2012; Song et al. 2012).

## References

Chen H, Qian K, Hu J, Liu D, Lu W, Yang Y, Wang D, Yan H, Zhang S, Zhu G (2005) The derivation of two additional human embryonic stem cell lines from day 3 embryos with low morphological scores. Hum Reprod 20:2201–2206

Doss MX, Koehler CI, Gissel C, Hescheler J, Sachinidis A (2004) Embryonic stem cells: a promising tool for cell replacement therapy. J Cell Mol Med 8:465–473

Evans MJ, Kaufman MH (1981) Establishment in culture of pluripotential cells from mouse embryos. Nature 292:154–156

Fonseca SAS, Costas RM, Pereira LV (2015a) Searching for naïve human pluripotent stem cells. World J Stem Cells 7(3):649–656

Fonseca SAS, Costas RM, Morato-Marques M, Costa S, Alegretti JR, Rosenberg C, da Motta EL, Serafini PC, Pereira LV (2015b) A euploid line of human embryonic stem cells derived from a 43,XX,dup(9q),+12,-14,-15,-18,-21 Embryo. PLoS One 10(11):e0140999.

Geron Corporation (2009) World's first clinical trial of human embryonic stem cell therapy cleared. Regen Med 4(2):161

Gosden C, Davdson C, Robertson M (1992) Lymphocyte culture. In: Rooney DE, Czepulkowsky BH (eds) Human cytogenetics. Oxford University Press, Oxford, pp 37–47

Lerou PH, Yabuuchi A, Huo H, Takeuchi A, Shea J, Cimini T, Ince TA, Ginsburg E, Racowsky C, Daley GQ (2008) Human embryonic stem cell derivation from poor-quality embryos. Nat Biotechnol 26:212–214

Martin GR (1981) Isolation of a pluripotent cell line from early mouse embryos cultured in medium conditioned by teratocarcinoma stem cells. Proc Natl Acad Sci U SA 78:7634–7638

Mateizel I, De Temmerman N, Ullmann U, Cauffman G, Sermon K, Van de Velde H, De Rycke M, Degreef E, Devroey P, Liebaers I, Van Steirteghem A (2006) Derivation of human embryonic stem cell lines from embryos obtained after IVF and after PGD for monogenic disorders. Hum Reprod 21:503–511

Maury Y, Gauthier M, Peschanski M, Martinat C (2012) Human pluripotent stem cells for disease modelling and drug screening. Bioessays 34(1):61–71

Melkoumian Z, Weber JL, Weber DM, Fadeev AG, Zhou Y, Dolley-Sonneville P, Yang J, Qiu L, Priest CA, Shogbon C, Martin AW, Nelson J, West P, Beltzer JP, Pal S, Brandenberger R (2010) Synthetic peptide-acrylate surfaces for long-term self-renewal and cardiomyocyte differentiation of human embryonic stem cells. Nat Biotechnol 28:606–610

Mitalipova M, Palmarini G (2006) Isolation and characterization of human embryonic stem cells. Methods Mol Biol 331:55–76

Peura T, Bosman A, Chami O, Jansen RP, Texlova K, Stojanov T (2008) Karyotypically normal and abnormal human embryonic stem cell lines derived from PGD-analyzed embryos. Cloning Stem Cells 10:203–216

Prokhorova TA, Harkness LM, Frandsen U, Ditzel N, Schoder HD, Burns JS, Kassem M (2009) Teratoma formation by human embryonic stem cells is site-dependent and enhanced by the presence of matrigel. Stem Cells Dev 18(1):47–54

Reubinoff BE, Pera MF, Fong CY, Trounson A, Bongso A (2000) Embryonic stem cell lines from human blastocysts: somatic differentiation in vitro. Nat Biotechnol 18:399–404

Rodin S, Antonsson L, Niaudet C, Simonson OE, Salmela E, Hansson EM, Domogatskaya A, Xiao Z, Damdimopoulou P, Sheikhi M, Inzunza J, Nilsson AS, Baker D, Kuiper R, Sun Y, Blennow E, Nordenskjold M, Grinnemo KH, Kere J, Betsholtz C, Hovatta O, Tryggvason K (2014) Clonal culturing of human embryonic stem cells on laminin-521/E-cadherin matrix in defined and xeno-free environment. Nat Commun 5:3195

Rodin S, Domogatskaya A, Ström S, Hansson EM, Chien KR, Inzunza J, Hovatta O, Tryggvason K (2010) Long-term self-renewal of human pluripotent stem cells on human recombinant laminin-511. Nat Biotechnol 28:611–615

Schwartz SD, Hubschman JP, Heilwell G, Franco-Cardenas V, Pan CK, Ostrick RM, Mickunas E, Gay R, Klimanskaya I, Lanza R (2012) Embryonic stem cell trials for macular degeneration: a preliminary report. Lancet 379(9817):713–720

Shetty R, Inamdar MS (2012) Derivation of human embryonic stem cell lines from poor quality embryos. Methods Mol Biol 873:151–161

Silva SS, Rowntree RK, Mekhoubad S, Lee JT (2008) X chomosome inactivation and epigenetic fluidity in human embryonic stem cells. Proc Natl Acad Sci U S A 105:4820–4825

Song M, Paul S, Lim H, Dayem AA, Cho SG (2012) Induced pluripotent stem cell research: a revolutionary approach to face the challenges in drug screening. Arch Pharm Res 35(2):245–260

Thomson JA, Itskovitz-Eldor J, Shapiro SS, Waknitz MA, Swiergiel JJ, Marshall VS, Jones JM (1998) Embryonic stem cell lines derived from human blastocysts. Science 282:1145–1147

Tropel P, Tournois J, Côme J, Varela C, Moutou C, Fragner P, Cailleret M, Laâbi Y, Peschanski M, Viville S (2010) High-efficiency derivation of human embryonic stem cell lines following pre-implantation genetic diagnosis. In Vitro Cell Dev Biol Anim 46:376–385

Unger C, Skottman H, Blomberg P, Dilber MS, Hovatta O (2008) Good manufacturing practice and clinicalgrade human embryonic stem cell lines. Hum Mol Genet 17:R48–R53

Villa-Diaz LG, Nandivada H, Ding J, Nogueira-de-Souza NC, Krebsbach PH, O'Shea KS, Lahann J, Smith GD (2010) Synthetic polymer coatings for long-term growth of humanembryonic stem cells. Nat Biotechnol 28:581–583

# Chapter 3
# Adipose-Derived Mesenchymal Stromal Cells

**Amanda Faria Assoni, Giuliana Castello Coatti, Juliana Plat Aguiar Gomes, Mayra Vitor Pelatti, and Mayana Zatz***

**Abstract** Adipose tissue is considered an important source of mesenchymal stromal cells since it is an abundant tissue with great distribution throughout the body and human adipose-derived stem cells (hASCs) have high proliferation rates *in vitro*. These adult stromal cells have the ability to differentiate into tissues including bone, cartilage and adipose *in vitro*, and also display some biological functions such as trophic, paracrine and immunomodulatory effects that may have the greatest therapeutic impact *in vivo* for diverse pathologies. The considerable therapeutic potential of hASCs turned them of substantial interest in many areas of research. hASCs transplantation is currently being tested in more than 70 clinical trials for many diseases, including bone fractures, graft *versus* host disease, multiple sclerosis, brain injury, acute respiratory distress syndrome, idiopathic pulmonary fibrosis and diabetes. In this chapter, we will describe the protocols for hASCs isolation and characterization, besides summarizing current reported studies about their potential use.

**Keywords** Adipose tissue • Mesenchymal stromal cells • Protocols • Isolation • Maintenance • Characterization • Karyotype

## 3.1 Introduction

Multipotent mesenchymal stromal cells (MSCs) were first identified as non-hematopoietic cells found in the bone marrow, comprising the multipotent precursors of the bone marrow stroma (Friedenstein et al. 1968). More recently, it was demonstrated that MSCs are derived from perivascular cells or pericytes (revised in Caplan and Sorrell 2015). Although mesenchymal stromal cells (MSCs) are partially defined by their ability to differentiate into tissues including bone, cartilage

*These authors contributed equally to this work.

A.F. Assoni • G.C. Coatti • J.P.A. Gomes • M.V. Pelatti • M. Zatz (✉)
Department of Genetics and Evolutionary Biology, Human Genome and Stem Cell Research Center, Biosciences Institute, University of São Paulo (USP),
Rua do Matão, 106, São Paulo, SP 05508-090, Brazil
e-mail: mayazatz@usp.br



H. Ulrich, P. Davidson Negraes, *Working with Stem Cells*,
DOI 10.1007/978-3-319-30582-0_3

and adipose *in vitro*, experiments in animal models revealed that it is their trophic, paracrine and immunomodulatory functions that might have the greatest therapeutic impact *in vivo* (Murphy et al. 2013; Caplan and Sorrell 2015).

According to the Mesenchymal and Tissue Stem Cell Committee of the International Society for Cellular Therapy, these fibroblast-like plastic-adherent cells, regardless of the tissue from which they are isolated, should be termed multipotent mesenchymal stromal cells (MSCs) (Horwitz et al. 2005). The committee also proposes 3 minimum criteria to define a cell population as MSCs: (1) cells must be plastic-adherent when maintained in standard culture conditions, (2) express CD105, CD73 and CD90, and lack expression of CD45, CD34, CD14 or CD11b, CD79 alpha or CD19 and HLA-DR surface molecules, and (3) must differentiate into osteoblasts, adipocytes and chondroblasts *in vitro* (Dominici et al. 2006).

MSCs from the adipose tissue, human adipose-derived stromal cells (hASCs), were described in 2001 by Zuk and coworkers (2001). In 2002, the same group demonstrated their differentiation capacity and clonogenicity, and reported hASCs as a new adult stromal cell population (Zuk et al. 2002). The first use of hASCs for a therapeutic purpose occurred in 2004 by combining these cells with bone grafts in an attempt to treat extensive craniofacial damage in a 7 year-old girl (Lendeckel et al. 2004).

Adipose tissue is considered an interesting source of MSCs, as it is an abundant tissue with great distribution throughout the body. Moreover, hASCs have high *in vitro* proliferation rates. These adult cells can be isolated from liposuction samples, a material usually discarded, or from biopsies, in which even a small amount of tissue can provide high yield of cells (Zhang et al. 2014). Since they are obtained from adult tissues, there are no ethical issues associated with their use, unlike other kinds of stem cells, such as embryonic stem cells or neural stem cells obtained from aborted fetuses.

MSCs and hASCs include a mixture of several heterogeneous populations characterized by the presence of specific cell surface markers which may indicate a particular function for that cell population (revised by Sorrell and Caplan 2010; Murphy et al. 2013). Therefore, MSCs may have different profiles according to the relative proportion of different sub-populations, or even due to the presence or absence of a particular population (Crisan et al. 2008; Zimmerlin et al. 2010, 2013; Li et al. 2011).

Similar to other MSCs, hASCs are immune privileged, as they lack expression of class II MHC and co-stimulatory molecules (CD80, CD86 or CD40) (Jacobs et al. 2013). This property allows the treatment of genetic disorders with cells from both healthy autologous or heterologous donors without immunosuppression. Interestingly, it was recently reported that the use of MSCs simultaneously with anti-inflammatories alters the activity of MSCs (Chen et al. 2014), suggesting that they should be applied without concomitant use of steroids.

At first it was believed that, since hASCs have the potential to differentiate into various cellular lineages *in vitro*, they could regenerate injured tissues *in vivo* (Vieira et al. 2008b, 2012). However, several studies, including xenogeneic transplants from our group (Valadares et al. 2014), reported functional recovery without significant differentiation of cells, supporting a new hypothesis according to which the beneficial effect of these cells occurs predominantly due to the production of trophic factors, cytokines, and antioxidants with the ability to promote diverse biologic responses, such as modulation of the immune system (Ichim et al. 2010; English et al. 2010; Gharaibeh et al. 2011; Singer and Caplan 2011; da Justa Pinheiro et al. 2012; Caplan and Sorrell 2015).

In short, the considerable therapeutic potential of hASCS turned them of great interest in many areas of research. hASCs transplantation are currently being tested in 74 clinical trials for many diseases and conditions such as ischemic revascularization (NCT01709279), diabetic foot ulcers (NCT02394385), bone/cartilage repair (NCT02307435) and brain injury (NCT01649700), among others (clinical trials.gov). hASCs have also been applied in trials for graft *versus* host disease (Fang et al. 2007). Here, we will describe the protocols for isolation, characterization and potential uses of hASCs.

## 3.2 Protocols

### 3.2.1 hASCs Separation from Adipose Tissue

This protocol was adapted from the original work of Zuk and coworkers (2001).

1. Obtain a sample of adipose tissue (lipoaspirate or biopsy) Sample must be collected in 50 ml tubes containing Dulbecco's modified Eagle's medium (DMEM) and 2 % antibiotic/antimycotic;
2. The sample can be processed immediately after collection or stored up to 24 h in the refrigerator at 4 °C in the tube with DMEM and antibiotic, until the processing (see ***Note 1***);
3. Wash sample with PBS (Phosphate Buffered Saline), with 2 % antibiotic/antimycotic. The volume of PBS should be twice the sample volume. Repeat this step two times (this step is important to remove the blood cells) (see ***Note 2***);
4. After the washing step, prepare the digestion solution: 0.075 % collagenase IA. Collagenase is diluted with PBS; the appropriate volume of this solution is approximately half the tissue sample volume;
5. Digest the sample for 30 min at 37 °C with digestion solution;
6. Enzyme activity is neutralized with DMEM containing 10 % FBS (Fetal Bovine Serum) and 1 % antibiotic/antimycotic, and centrifuged at 1200 RCF for 10 min to obtain a high-density pellet;
7. The pellet should be resuspended in DMEM containing 10 % FBS and 1 % antibiotic/antimycotic, plated and incubated overnight at 37 °C and 5 % $CO_2$. (Concentration: 5000 cells/cm$^2$);
8. Following incubation, the plates must be extensively washed with PBS with 2 % antibiotic/antimycotic to remove debris;
9. Incubate cells with growth medium (see ***Note 3***).

The same protocol was used by our group to isolate ASC from other species such as mouse, dog and cat. Adipose samples from healthy dogs and cats are obtained in animals submitted to castration surgery, under anesthesia. The adipose tissue is collected from the subcutaneous region without any harm to these animals.

***Note 1:*** *We tested a protocol for freezing samples immediately after collection, in 90 % FBS and 10 % DMSO while other groups reported freezing without cryopreservation or other frozen agents. According to Devitt and coworkers (*2015*), long-term storage of adipose tissue at −70 °C without cryopreservative agents can reduce viable cells yield, but these cells can be submitted to extended culturing, irrespective of cryopreservation duration or patient's age. Since these cells are cryopreserved for future use in plastic and reconstructive surgical procedures or cell therapy, they must be stored for a long period. According to Minonzio and coworkers (*2014*), adipose-derived stromal fraction retain 85 % of cell viability, normal proliferative capacity and differentiation potential when stored in liquid nitrogen using serum-free cryopreservation media.*

***Note 2:*** *Some protocols include an additional step of red blood cell lysing with a blood lysis buffer. When we compared the sample processing of a sample with or without the blood lysis step, we observed that cell viability may be reduced in the first option and thus we suggest removal of red blood cells (suspension cells) with culture passaging.*

***Note 3:*** *As firstly reported by Zuk and coworkers (*2001*), cells were maintained with a growth medium composed by DMEM, 10 % FBS and 1 % antibiotic/antimycotic. To our knowledge, both DMEM low glucose and DMEM high glucose are efficient in promoting cell growth (*Vieira et al. 2008a, b, 2012*). Recently, our group began to use also a medium richer in nutrients (DMEM F12 + glutamax, 20 % FBS, 1 % non-essential aminoacids and 1 % antibiotic/antimycotic).*

**Protocol variations:** Some variability can be found in published protocols such as collagenase type, and/or concentration. Some groups use cell strainers to help in removing debris. We observed that the use of these filters is not critical, particularly with small samples.

### *3.2.2 Maintenance*

#### 3.2.2.1 Growth/Expansion

1. Growth medium is composed of DMEM low glucose supplemented with 10 % FBS and 1 % antibiotic/antimycotic (see ***Note 3***);
2. Growth medium should be changed every 3–4 days until subculture is performed.

#### 3.2.2.2 Subculturing

1. Passage is done when cells reach 80–100 % of confluence;
2. Wash plates two times with PBS (3 ml for every 25 $cm^2$);
3. Incubate cells with trypsin at 37 °C for 5 min (1 ml for every 25 $cm^2$) (see ***Note 4***);

4. Add growth medium to inactivate trypsin (3 ml for every 25 $cm^2$).
5. Count cells;
6. For expansion, cells must be re-plated in concentration of 5000 cells/$cm^2$.

***Note 4:*** *Detachment of plate can be done incubating cells for 5 min with porcine derived trypsin-EDTA (in concentrations of 0.25 % or 0.05 %) or using recombinant and xenofree enzymatic products.*

#### 3.2.2.3 Freezing

1. Grow cells until they reach 80–100 % of confluence;
2. Wash plates two times with PBS (3 ml for every 25 $cm^2$);
3. Incubate cells with trypsin at 37 °C for 5 min (1 ml for every 25 $cm^2$);
4. Add growth medium to inactivate trypsin (3 ml for every 25 $cm^2$);
5. Count cells;
6. Centrifuge at 500 RCF for 5 min;
7. Discard supernatant;
8. Resuspend in cryopreservative medium containing 10 % DMSO and 90 % FBS (see ***Note 5***) and transfer cells to cryovial. Number of cells/vial may vary;
9. Quickly transfer cryovials to a freezing container, and then place it in −80 °C freezer overnight (see ***Note 6***);
10. Transfer cryovials to liquid nitrogen for long-term storage.

***Note 5:*** *Cryopreservative medium is made of 10 % dimethyl sulfoxide (DMSO) and with variable proportions of other components (90 % FBS or 80 % FBS + 10 % growth medium or 20 % FBS – 70 % growth medium).*

***Note 6:*** *Cells may be frozen by means of a programmable freezer with different controlled-rates or using a freezing container, which contains isopropanol alcohol, providing a 1 °C/min cooling rate in a −80 °C freezer. After overnight storage in a −80 °C freezer, cryotubes are transferred to liquid nitrogen at −196 °C and can be maintained for long-term storage.*

#### 3.2.2.4 Thawing

1. Cell thawing requires cryovials immersion in a waterbath at 37 °C for 2–3 min to gently defrost cryotubes (see ***Note 7***);
2. Mix defrosted cells with growth medium in a conical tube (5 ml of growth media for every 1 ml of defrosted cells);
3. Centrifuge at 500 RCF for 5 min;
4. Discard the supernatant;
5. Resuspend cell pellet in growth medium and plate cells in concentration of 5000 cells/$cm^2$.

***Note 7:*** *Cell thawing must be performed quickly and gently to avoid cell death caused by DMSO toxicity.*

## 3.2.3 *Characterization*

### 3.2.3.1 *In Vitro* Differentiation

One of the minimum criteria proposed for defining a multipotent stromal cell population is their ability to differentiate *in vitro* into chondroblasts, osteoblasts and adipocytes (Fig. 3.1). Differentiation medium are commercially available, or may be home-made prepared. All differentiation media, both commercial and home-made, have subtle variations, including different compounds or concentrations. Here, we summarize medium composition to induce *in vitro* differentiation as described by Zuk and coworkers (2002).

Besides quantitative methods specified below, RNA expression analysis through qRT-PCR may be also performed. The *in vivo* differentiation potential should be tested in a specific animal model that represents a disease, or in conditions that highlight the cell capacity of tissue regeneration. However, in practice, MSCs are not always tested for their *in vivo* differentiation potential.

Chondrogenic Differentiation

**Cell Preparation**

1. Grow cells until they reach 80–100 % of confluence;
2. Wash plates two times with PBS (3 ml for every 25 $cm^2$);
3. Incubate cells with trypsin at 37 °C for 5 min (1 ml for every 25 $cm^2$);
4. Add growth medium to inactivate trypsin (3 ml for every 25 $cm^2$);
5. Count cells;
6. Centrifuge at 500 RCF for 5 min;
7. Discard supernatant;
8. Resuspend cells to achieve droplets with high cell concentration (ranging from $1 \times 10^5$ to $1 \times 10^7$ cells per drop of 5 μl) (see ***Note 8***);
9. Drip every high cell concentration droplet in a 24-well plate, keeping the droplets spaced from each other;
10. Incubate at 37 °C for 2 h for cell adhesion;
11. Gently add growth medium in control wells and differentiation medium in experimental wells (see ***Note 9***);
12. Medium must be changed every 3–4 days until achieving 21 days of differentiation.

***Note 8:*** *Instead of high cell concentration droplets, one can also maintain a cell pellet (ranging from* $1 \times 10^5$ *to* $1 \times 10^7$ *cells) in a conic tube during differentiation. After 21 days, pellet must be collected and analyzed by histology.*

***Note 9:*** *Chondrogenic medium is composed of DMEM-LG supplemented with 1 % FSB, 6.25 μg/ml insulin, 10 ng/ml TGF-beta1, 50 nM ascorbate-2-phosphate, 1 % antibiotic/antimycotic.*

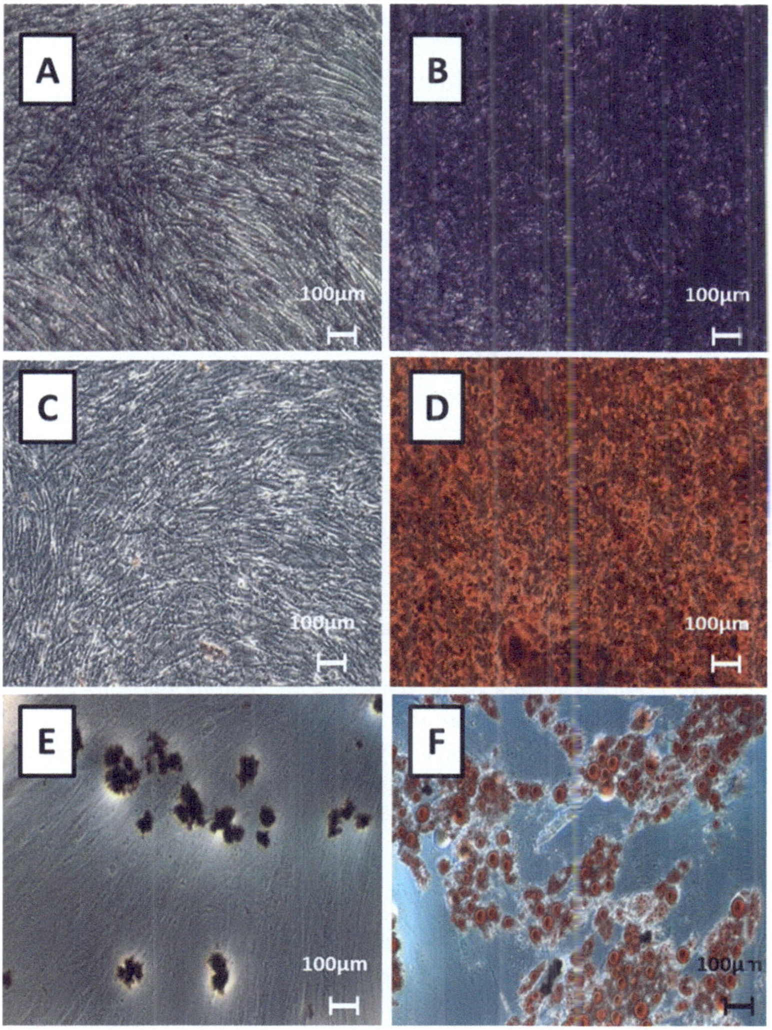


**Fig. 3.1** *In vitro* differentiation of hASC. Left and right columns represent evaluation performed with hASCs maintained for 21 days with growth media and with differentiation media, respectively. (**a**, **b**) chondrogenic differentiation visualized by toluidine blue staining; (**c**, **d**) osteogenic differentiation by Von Kossa staining; and (**e**, **f**) adipogenic differentiation by Oil Red staining

Chondrogenic differentiation is confirmed on day 21 by analysing presence of extracellular matrix mucopolysaccharides with toluidine blue staining.

**Staining with Toluidine Blue**

1. Prepare toluidine blue at 1 %;
2. Wash cells two times with PBS;

3. Fix cells with paraformaldehyde 4 % diluted in PBS for 30 min;
4. Discard solution;
5. Incubate with toluidine blue 1 % for 30 min;
6. Wash 3 times with PBS.

Osteogenic Differentiation

**Cell Preparation**

1. Grow cells until they reach 80 % of confluence in 24-well plate or larger;
2. Wash plates two times with PBS;
3. Add growth medium in control wells and differentiation medium in tested wells (see ***Note 10***);
4. Medium must be changed every 3–4 days until achieving 21 days of differentiation.

***Note 10:*** *Osteogenic differentiation medium is composed of DMEM-LG supplemented with 10 % FBS, 50 μM ascorbate-2-phosphate, 10 mM B-glycerophosphate and 0.1 μM dexamethasone.*

Osteogenesis is demonstrated by accumulation of mineralized calcium phosphate assessed by Von Kossa or Alizarin Red S staining. Quantitative analysis of osteogenic differentiation may be performed with alkaline phosphatase enzyme activity at 7 days of differentiation and at 14 or 21 days of differentiation with alizarin discoloration. Quantitative analysis should take into consideration the number of cells per well and the basal expression of alkaline phosphatase of non-differentiated cells in order to normalize results for comparison among samples.

**Alkaline Phosphatase Activity**

1. Prepare substrate buffer and substrate solutions (see ***Note 11***);
2. For each well, prepare a 1.5 ml tube;
3. Add a volume of 1 M NaOH equal to the volume of substrate that will be used;
4. Rinse cells twice with PBS;
5. Add an appropriate volume of substrate solution to each well size (0.5 ml per well of 12-well plate);
6. Incubate for 15 min;
7. Transfer the substrate solution to the NaOH-containing tubes;
8. Rinse cells twice with PBS and add differentiation medium to continue differentiation;
9. Set up a standard curve of p-Nitrophenol (see Table 3.1).

**Table 3.1** Components of the p-nitrophenol standard curve

| p-Nitrophenol concentration (nmol/ml) | Volume (ml) of diluted p-nitrophenol solution (100 nmol/ml) | Volume (ml) of 0.02 M NaOH |
|---|---|---|
| **9** | 0 1 | 1 |
| **18** | 0 2 | 0.9 |
| **36** | 0 4 | 0.7 |
| **54** | 0.6 | 0.5 |
| **72** | 0.8 | 0.3 |
| **90** | 1 | 0.1 |

## Note 11

- *Substrate buffer:* Dissolve 188 g of glycine and 0.1017 of $MgCl_2$ in 500 ml water. Adjust pH to 10.5 with 1 N NaOH.
- *Substrate solution:* Dissolve 1 tablet (5 mg) of phosphatase per 5 ml of substrate buffer.
- *p-Nitrophenol for standard curve:* Prepare 100 nmol/ml solution of p-nitrophenol by combining 100 µl of 10 µmol/ml p-nitrophenol standard solution with 9.90 ml 0.02 N NaOH. Prepare further dilutions as described in Table 3.1. As blank, use p-Nitrophenol and substrate solution. Read in spectrophotometer at 405 nm.

## Staining with Alizarin Red S

1. Prepare Alizarin Red S at 2 % and filtrate solution;
2. Wash cells two times with PBS;
3. Fix cells with ethanol 70 % diluted in PBS for 30 min;
4. Discard solution;
5. Wash once with Milli-Q water;
6. Incubate with alizarin red for 30 min;
7. Wash three times with PBS;
8. After microscopic analysis, quantitative evaluation may be performed (see ***Note 12***).

***Note 12:*** *To perform alizarin red discoloration, an alcoholic solution (20 % methanol, 10 % acetic acid, 70 % PBS) is used for 15 min incubation. The resultant solution can be quantified through spectrophotometer assay (at 450 nm).*

## Staining with Von Kossa

1. Carefully, wash cells with distilled water;
2. Add silver nitrate at 1 %;
3. Incubate under UV light for 40 min;
4. Carefully, wash cells with distilled water;
5. Remove silver nitrate excess using sodium thiosulfate 3 % or safranin O 1 % for 5 min;
6. Wash cells carefully, with distilled water;
7. Stain with Van Gieson solution (3.75 ml fuchsin 1 %, 46.25 ml saturated picric acid, 50 ml water) for 5 min;
8. Wash carefully with 100 % ethanol.

Adipogenic Differentiation

**Cell Preparation**

1. Grow cells until they reach 80 % of confluence in a 24-well plate or larger;
2. Wash plates two times with PBS;
3. Add growth medium in control wells and differentiation medium in tested wells (see ***Note 13***);
4. Medium must be changed every 3–4 days until achieving 21 days of differentiation.

***Note 13:*** *Adipogenic differentiation medium is composed of DMEM-LG supplemented with 10 % FBS, 1 μM dexamethasone, 500 μM 3-isobutyl-1-methylxanthine (IBMX), 200 μM indomethacin, 10 μM insulin and 1 % antibiotic/antimycotic.*

Adipogenic differentiation is confirmed on day 21 by intracellular accumulation of lipid-rich vacuoles stainable with Oil Red O.

**Staining with Oil Red O**

1. Prepare oil red stock solution diluting 0.15 g in 50 ml of isopropyl alcohol;
2. Prepare oil red work solution diluting 8.2 ml of stock solution in 6.8 ml of distilled water;
3. Wash cells two times with PBS;
4. Fix cells with para paraformaldehyde at 4 % diluted in PBS for 30 min;
5. Discard solution;
6. Incubate with oil red work solution for 30 min;
7. Wash three times with PBS;
8. After microscopic analysis, quantitative evaluation may be performed (see ***Note 14***).

***Note 14:*** *Quantitative analysis may be performed using a commercial colorimetric assay (AdipoRed, Lonza), where lipid droplets are stained and quantitatively measured in a fluorescent spectrophotometer, according to manufacturer's instructions.*

#### 3.2.3.2 Immunophenotyping

Cell surface proteins are used as markers for cell characterization. Mesenchymal stromal cells are positively marked for CD105, CD73 and CD90 (adhesion molecules) and lack the presence of CD45, CD34, CD14 or CD11b, CD79a or CD19 and HLA-DR. Immunophenotyping is performed using specific antibodies for these proteins, and presence or absence of antibody ligation is verified when cells are read in flow cytometer (Fig. 3.2).

1. Grow cells until they reach 80–100 % of confluence;
2. Wash plates two times with PBS (3 ml for every 25 $cm^2$);

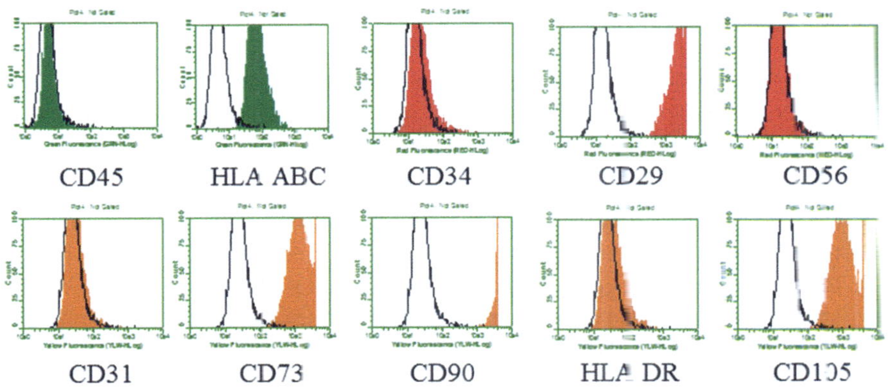


**Fig. 3.2** Exemplification of hASC immunophenotyping. In this assay, cells were evaluated in the 4th passage. Coloured peaks represent labeled population of hASCs with respective antibodies Comparison was performed with the same population of cells without any step of antibody incubation (white peaks)

3. Incubate cells with trypsin at 37 °C for 5 min;
4. Add growth medium to inactivate trypsin (6 ml for every 25 $cm^2$);
5. Count cells;
6. Resuspend with PBS in a concentration required in antibody manufacturer's instructions;
7. Incubate with antibody in dark room at 4 °C for 45 min, according to manufacturer's instructions (see ***Note 15***);
8. Wash with 500 μl PBS;
9. Centrifuge at 500 RCF for 5 min;
10. Discard supernatant;
11. Resuspend cells in 200 μl PBS (see ***Notes 15*** and ***16***);
12. Place cells in 96-wells plate with round bottom;
13. Perform reading in flow cytometer.

***Note 15:*** *If secondary antibody is required, incubate for another 15 min with secondary antibody after step 11. Then, repeat steps 8–11 and proceed to steps 12–13.*

***Note 16:*** *It is possible to fix cells after antibody labeling, allowing the analysis to be performed in the flow cytometer 2 days after labeling. To fix cells, instead of resuspending cells with PBS in step 11, resuspend with PFA at 4 % for 15 min, then repeat steps 9–10 and proceed to the next steps.*

#### 3.2.3.3 Karyotype

Karyotype analysis is crucial when cells are being considered for therapy. It has been reported that after long-term culture, cells may show chomosomal abnormalities (Roemeling-van Rhijn et al. 2013; Pan et al. 2014). Therefore, karyotype analysis should be performed in the same passage in which cells will be used for treatment.

1. Grow cells in a 75 $cm^2$ flask until they reach 70–80 % of confluence;
2. Incubate with colchicine for 4 h;
3. Before trypsinization, prepare a conical tube to save all the material that would normally be discarded during the procedure (such as the growth media and the PBS used to rinse cells);
4. Wash plates two times with PBS (9 ml each time);
5. Incubate cells with trypsin at 37 °C for 5 min (3 ml);
6. Add growth medium to inactivate trypsin (12 ml);
7. Centrifuge at 250 RCF for 10 min;
8. Resuspend cell pellet in KCl hypotonic solution and incubate them for 20 min (see ***Note 17***);
9. Centrifuge at 250 RCF for 10 min;
10. Resuspend with ice cold fixation solution and incubate for 1 min (see ***Note 18***);
11. Centrifuge at 250 RCF for 10 min;
12. Repeat steps 10 and 11 for three times;
13. Drop resuspended cells in slides;
14. Stain slides with giemsa or band for microscopic analysis (see ***Note 19***);

***Note 17*:** KCl hypotonic solution (KCl 0,075 M): 5,6 g KCl diluted in 1 L of distilled water.

***Note 18*:** Fixation solution: Mix 1 volume of glacial acetic acid with 3 volumes of methanol.

***Note 19*:** Giemsa/Banding: See manufacturer's instructions.

## 3.3 Potential Use

Historically, the therapeutic use of MSCs aimed to explore its regenerative potential into diverse tissues/organs such as liver (Theise et al. 2000) and heart (Zimmet and Hare 2005). A recent study reported that a proliferative population of myogenic progenitors was efficiently derived from ASCs. This population had characteristics similar to muscle satellite cells and was capable of terminal differentiation into myotubes. When transplanted into the *mdx* mice, the murine model for Duchenne muscular dystrophy, these progenitor cells successfully engrafted in skeletal muscle for up to 12 weeks, generating new muscle fibers and restoring dystrophin expression (Zhang et al. 2015).

Beside their regenerative potential, MSCs are capable of secreting paracrine factors with diverse biological functions such as immunomodulation, anti-scarring, chemoattraction, angiogenesis support and anti-apoptotic effect (reviewed in Meirelles et al. 2009). According to Caplan and Sorrell, the most notable immunomodulatory capacity of MSCs is that they can be used in allogenic and xenogenic transplants, in a variety of diseases (Caplan and Sorrell 2015).

MSCs can promote immunomodulation through several mechanisms. Proliferation of naïve T lymphocytes and memory lymphocytes can be suppressed by MSCs in the presence of alloantigens, mitogens, CD3 or CD28. Suppression was observed with autologous and allogeneic co-cultures and thus seems to be independent of MHC interaction (Di Nicola et al. 2002; Krampera et al. 2003; Le Blanc et al. 2003; Tse et al. 2003; Le Blanc et al. 2004; Meisel et al. 2004; Aggarwal and Pittenger 2005; Klyushnenkova et al. 2005). Additionally, the immunosuppressive property is maintained when MSCs and lymphocytes are co-cultured separated by a semi-permeable membrane, pointing out the involvement of soluble factors (Tse et al. 2003; Meisel et al. 2004; Glennie et al. 2005; Maccario et al. 2005; Mellor 2005; Plumas et al. 2005; Rasmusson et al. 2005; Selmani et al. 2009; Sioud et al. 2011).

In comparison to bone marrow and umbilical cord derived MSCs (BM-MSC and UC-MSC, respectively), hASCs display a more potent suppressor effect in dendritic cells differentiation and NK cell activation. Moreover, hASCs were shown to block the T cell activation process in an earlier phase than BM or UC-MSCs, yielding a greater proportion of T cells in the non-activated state. Concerning B cells and NK cells, hASCs and BM displayed an inhibitory effect while UC-MSCs did not influence their activation kinetics (Puissant et al. 2005; Ribeiro et al. 2013).

Immunosuppression is a desirable function to treat some inflammatory diseases, such as kidney diseases and graft *versus* host disease (GvHD). Animal models for severe acute renal failure have shown that administration of MSCs improves renal structure and function through paracrine effects that inhibit pro-inflammatory cytokines and stimulate anti-inflammatory cytokines (Tögel et al. 2005). A phase I clinical trial for treating idiopathic Focal Segmental Glumero Sclerosis (FSGS) with allogenic adipose derived MSCs (NCT02382874) was recently created and will start recruiting patients. Moreover, for patients diagnosed with GvHD, hASCs have been used in a number of Phase I and II trials with success (Herrmann and Sturm 2014).

In addition to immunomodulatory properties, the anti-fibrotic potential of hASCs was also demonstrated. Administration of hASCs on irradiated rats (model for radiation-induced pulmonary fibrosis), preserved the architecture of lungs without marked activation of fibroblasts or collagen deposition within the injured sites. The effect was mediated via host secretions of hepatocyte growth factor (HGF) and prostaglandin E2 (PGE2) (Dong et al. 2015). Based on this effect, a multicentric Phase I/II Clinical Trial with hASCs transplantation is currently recruiting participants in order to evaluate the safety and efficacy of autologous treatment of patients with diagnosis of idiopathic pulmonary fibrosis (NCT02135380).

Cell therapy is also being considered to treat some neurodegenerative diseases such as Parkinson's disease (as reviewed in Nanette et al. 2010) and Amyotrophic Lateral Sclerosis (as revised in Coatti et al. 2015). Although transdifferentiation of MSCs (mesodermal origin) into neuronal cells (ectodermal origin) is a still very debated issue, it was observed that in addition to the paracrine factors mentioned above, some MSCs subpopulations can also secrete neuroregulatory factors such as *brain derived neurotrophic factor* (BNDF) and *nerve growth factor* (NGF) (Crigler et al. 2006). In this scenario, a multi-center phase I trial (NCT02326935) is currently recruiting patients for an autologous transplantation of hASCs in patients diagnosed with multiple sclerosis, an incurable neurodegenerative disease.

To the present, more than 70 human clinical trials are listed in the clinical trial database (www.clinicaltrial.gov), for pathologies such as graft *versus* host disease (NCT01222039), severe brain injury (NCT01649700) and osteoarthitis (NCT01809769). Banking of hASCs is being performed by various companies such as the American Bank Celletex.

The classical steps to obtain a MSC lineage results in a heterogeneous population. Attempts to isolate specific subpopulations with known biological function from the stromal fraction are ongoing. Zimmerlin and coworkers (2013) reported different subpopulations of cells in human adipose stromal fractions, according to immunophenotyping verified by immunostaining and flow cytometry. Thee major population were described in this fraction: pericytes (CD45−, CD31−, CD146+), endothelial progenitor cells (CD45−, CD31+, CD34+), and supra adventitial adipose stromal cells (CD45−, CD31-, CD146-, CD34+) (Zimmerlin et al. 2013).

The therapeutic potential of hASCs and pericytes were investigated by our group in murine and canine models for muscular dystrophy, without immunosuppression. In the *SJL* mouse, the murine model for dysferlinopathy (or LGMD2B muscular dystrophy), hASCs transplantation revealed a functional benefit in transplanted animals when compared to untreated controls (Vieira et al. 2008b). The clinical effect as well as the safety of repeated hASCs injections from different donors was also tested in Golden Retriever Muscular Dystrophy (GRMD) dogs (Vieira et al. 2012). More recently, we compared the clinical effect of human pericytes obtained from different sources (adipose tissue, fallopian tube, muscle and endometrium) from a single female donor, in a double knockout mouse model (mdx/utr-) for Duchenne Muscular Dystrophy. It was observed that only pericytes obtained from adipose tissue, but not from other sources, were able to significantly increase the lifespan of treated mice (Valadares et al. 2014).

Cell therapy is also being considered for treating some pets' diseases. In a recent study, ten lame dogs with severe hip osteoarthitis (OA) and a control group of five dogs received intra-articular transplantation of autologous ASCs. Mean values of diseases scores were significantly improved within the first 3 months post-treatment in the OA group (Vilar et al. 2014). In humans, intra-articular injection of $1 \times 10^8$ hASCs into the osteoarthitic knee improved function and pain of the knee joint without causing adverse events, and reduced cartilage defects by regeneration of hyaline-like articular cartilage (Jo et al. 2014).

Besides its therapeutic use, hASCs can be used for disease modeling. Dossena and coworkers proposed the use of Spinal and Bulbar Muscular Atrophy (SBMA) patients' adipose derived MSCs as a new human *in vitro* model. They found that hASCs from SBMA patients showed less growth potential and differentiated only into adipocytes. Moreover, lower expression of key receptors was also found. The authors propose the use of hASCs from SMBA patients to test novel therapeutic approaches (Dossena et al. 2014).

**Table 3.2** Manufacturer information and catalog number of specific reagents/molecules cited in the protocols

| Reagent name | Manufacturer | Catalog number |
|---|---|---|
| 3-Isobutyl-1-methylxanthine | Sigma-Aldrich | I5879 |
| Adipogenesis differentiation kit | Gibco | A10070-01 |
| AdipoRed™ adipogenesis assay reagent | Lonza | PT-7009 |
| Alizarin Red S | Sigma-Aldrich | A5533 |
| Antibiotic-Antimycotic (100×) | Gibco | 15240062 |
| Antibody anti- HLA-DR | BD Biosciences | BD555812 |
| Antibody anti-CD105 | BD Biosciences | BD560839 |
| Anti-body anti-CD11b | BD Biosciences | BD340937 |
| Antibody anti-CD14 | BD Biosciences | BD560919 |
| Antibody anti-CD19 | BD Biosciences | BD557921 |
| Antibody anti-CD34 | BD Biosciences | BD340666 |
| Antibody anti-CD45 | BD Biosciences | BD347463 |
| Antibody anti-CD73 | BD Biosciences | BD550257 |
| Antibody anti-CD79a | BD Biosciences | BD555934 |
| Antibody anti-CD90 | BD Biosciences | BD555596 |
| Ascorbic acid 2-phosphate | Sigma-Aldrich | A8960 |
| Chondrogenesis differentiation kit | Gibco | A10071-01 |
| Colchicine (Colcemid™ solution) | Gibco | 15210-040 |
| Collagenase IA | Sigma-Aldrich | C2674 |
| Dexamethasone | Sigma-Aldrich | D8893 |
| DMEM high glucose | Gibco | 12100-046 |
| DMEM low glucose | Gibco | 31600-034 |
| DMEM/F-12, GlutaMAX™ supplement | Gibco | 10565018 |
| DMSO | LGC Biotecnologia | BR2650-01 |
| Fetal bovine serum | Gibco | 10437-028 |
| Fuchsin | Sigma-Aldrich | 3244-88-0 |
| Giemsa's azur eosin methylene blue solution | Merck | 109204 |
| Indomethacin | Sigma-Aldrich | I7378 |
| Insulin | Sigma-Aldrich | I1882 |
| Lysis solution for blood | Sigma-Aldrich | L3289 |
| MEM non-essential amino acids solution (100×) | Gibco | 11140050 |
| Oil Red O | Sigma-Aldrich | O0625 |
| Osteogenesis differentiation kit | Gibco | A10072-01 |
| PBS, pH 7.4 | Gibco | 10010049 |
| Phosphatase substrate | Sigma-Aldrich | 104-105 |
| Picric acid solution | Sigma-Aldrich | P6744 |

(continued)

**Table 3.2** (continued)

| Reagent name | Manufacturer | Catalog number |
|---|---|---|
| p-Nitrophenol standard | Sigma-Aldrich | 104-1 |
| Safranin O | Sigma-Aldrich | S8884 |
| Silver nitrate | Sigma-Aldrich | 209139 |
| Sodium thiosulfate | Sigma-Aldrich | S7026 |
| TGF-beta 1 protein | R&D Systems | 240-B |
| Toluidine blue O | Sigma-Aldrich | 198161 |
| TrypLE™ express enzyme | Gibco | 12605-010 |
| Trypsin-EDTA (0.05 %) | Gibco | 25300054 |
| β-Glycerophosphate disodium salt hydrate | Sigma-Aldrich | G6251 |

In short, MSCs and hASCs have multiple biological functions as well as regenerative, anti-fibrotic and immunomodulatory properties. Such plasticity, in association with the facility to obtain and handle them, makes hASCs an important tool for future use in medicine. To achieve this goal, scientific experiments should be carefully designed and implemented.

**Reagent specification** (See Table 3.2)

## References

Aggarwal S, Pittenger MF (2005) Human mesenchymal stem cells modulate allogeneic immune cell responses. Blood 105:1815–1822

Caplan AI, Sorrell JM (2015) The MSC curtain that stops the immune system. Immunol Lett Euro Fed Immunol Societ 168:136–9. Available from: http://linkinghub.elsevier.com/retrieve/pii/S0165247815001030.

Chen X, Gan Y, Li W, Su J, Zhang Y, Huang Y, Roberts AI, Han Y, Li J, Wang Y, Shi Y (2014) The interaction between mesenchymal stem cells and steroids during inflammation. Cell Death Dis 5:e1009

Coatti GC, Beccari MS, Olávio TR, Mitne-Neto M, Okamoto OK, Zatz M (2015) Stem cells for amyotrophic lateral sclerosis modeling and therapy: myth or fact? Cytom Part A 87:197–211

Crigler L, Robey RC, Asawachaicharn A, Gaupp D, Phinney DG (2006) Human mesenchymal stem cell subpopulations express a variety of neuro-regulatory molecules and promote neuronal cell survival and neuritogenesis. Exp Neurol 198:54–64

Crisan M, Yap S, Casteilla L, Chen C-W, Corselli M, Park TS, Andriolo G, Sun B, Zheng B, Zhang L, Norotte C, Teng P-N, Traas J, Schugar R, Deasy BM, Badylak S, Buhing H-J, Giacobino J-P, Lazzari L, Huard J, Péault B (2008) A perivascular origin for mesenchymal stem cells in multiple human organs. Cell Stem Cell 3:301–313

Da Justa Pinheiro CH, de Queiroz JCF, Guimarães-Ferreira L, Vitzel KF, Nachbar RT, de Sousa LGO, de Souza-Jr AL, Nunes MT, Curi R (2012) Local injections of adipose-derived mesenchymal stem cells modulate inflammation and increase angiogenesis ameliorating the dystrophic phenotype in dystrophin-deficient skeletal muscle. Stem Cell Rev 8:363–374

Devitt SM, Carter CM, Dierov R, Weiss S, Gersch RP, Percec I (2015) Successful isolation of viable adipose-derived stem cells from human adipose tissue subject to long-term cryopreservation: positive implications for adult stem cell-based therapeutics in patients of advanced age Stem Cells Int 2015:1–11

Di Nicola M, Carlo-Stella C, Magni M, Milanesi M, Longoni PD, Matteucci P, Grisanti S, Gianni AM (2002) Human bone marrow stromal cells suppress T-lymphocyte proliferation induced by cellular or nonspecific mitogenic stimuli. Blood 99:3838–3843

Dominici M, Le Blanc K, Mueller I, Slaper-Cortenbach I, Marini F, Krause D, Deans R, Keating A, Prockop D, Horwitz E (2006) Minimal criteria for defining multipotent mesenchymal stromal cells. The International Society for Cellular Therapy position statement. Cytotherapy 8:315–317

Dong L-H, Jiang Y-Y, Liu Y-J, Cui S, Xia C-C, Qu C, Jiang X, Qu Y-Q, Chang P-Y, Liu F (2015) The anti-fibrotic effects of mesenchymal stem cells on irradiated lungs via stimulating endogenous secretion of HGF and PGE2. Sci Rep 5:8713

Dossena M, Bedini G, Rusmini P, Giorgetti E, Canazza A, Tosetti V, Salsano E, Sagnelli A, Mariotti C, Gellera C, Navone SE, Marfia G, Alessandri G, Corsi F, Parati EA, Pareyson D, Poletti A (2014) Human adipose-derived mesenchymal stem cells as a new model of spinal and bulbar muscular atrophy. PLoS One 9:e112746

English K, French A, Wood KJ (2010) Mesenchymal stromal cells: facilitators of successful transplantation? Cell Stem Cell 7:431–442

Fang B, Song Y, Zhao RC, Han Q, Lin Q (2007) Using human adipose tissue-derived mesenchymal stem cells as salvage therapy for hepatic graft-versus-host disease resembling acute hepatitis. Transplant Proc 39:1710–1713

Friedenstein AJ, Petrakova KV, Kurolesova AI, Frolova GP (1968) Heterotopic of bone marrow Analysis of precursor cells for osteogenic and hematopoietic tissues. Transplantation 6:230–247

Gharaibeh B, Lavasani M, Cummins JH, Huard J (2011) Terminal differentiation is not a major determinant for the success of stem cell therapy—cross-talk between muscle-derived stem cells and host cells. Stem Cell Res Ther 2:31

Glennie S, Soeiro I, Dyson PJ, Lam EW-F, Dazzi F (2005) Bone marrow mesenchymal stem cells induce division arrest anergy of activated T cells. Blood 105:2821–2827

Herrmann R, Sturm M (2014) Adult human mesenchymal stromal cells and the treatment of graft versus host disease. Stem Cells Cloning Adv Appl 7:45–52

Horwitz EM, Le Blanc K, Dominici M, Mueller I, Slaper-Cortenbach I, Marini FC, Deans RJ, Krause DS, Keating A (2005) Clarification of the nomenclature for MSC: The International Society for Cellular Therapy position statement. Cytotherapy 7:393–395

Ichim TE, Alexandrescu DT, Solano F, Lara F, Campion RDN, Paris E, Woods EJ, Murphy MP, Dasanu CA, Patel AN, Marleau AM, Leal A, Riordan NH (2010) Mesenchymal stem cells as anti-inflammatories: implications for treatment of Duchenne muscular dystrophy. Cell Immunol 260:75–82

Jacobs SA, Roobrouck VD, Verfaillie CM, Van Gool SW (2013) Immunological characteristics of human mesenchymal stem cells and multipotent adult progenitor cells. Immunol Cell Biol 91:32–39

Jo CH, Lee YG, Shin WH, Kim H, Chai JW, Jeong EC, Kim JE, Shim H, Shin JS, Shin IS, Ra JC, Oh S, Yoon KS (2014) Intra-articular injection of mesenchymal stem cells for the treatment of osteoarthitis of the knee: a proof-of-concept clinical trial. Stem Cells 32:1254–1266

Klyushnenkova E, Mosca JD, Zernetkina V, Majumdar MK, Beggs KJ, Simonetti DW, Deans RJ, McIntosh KR (2005) T cell responses to allogeneic human mesenchymal stem cells: immunogenicity, tolerance, and suppression. J Biomed Sci 12:47–57

Krampera M, Glennie S, Dyson J, Scott D, Laylor R, Simpson E, Dazzi F (2003) Bone marrow mesenchymal stem cells inhibit the response of naive and memory antigen-specific T cells to their cognate peptide. Blood 101:3722–3729

Le Blanc K, Tammik L, Sundberg B, Haynesworth SE, Ringdén O (2003) Mesenchymal stem cells inhibit and stimulate mixed lymphocyte cultures and mitogenic responses independently of the major histocompatibility complex. Scand J Immunol 57:11–20

Le Blanc K, Rasmusson I, Götherström C, Seidel C, Sundberg B, Sundin M, Rosendahl K, Tammik C, Ringdén O (2004) Mesenchymal stem cells inhibit the expression of CD25 (interleukin-2

receptor) and CD38 on phytohaemagglutinin-activated lymphocytes. Scand J Immunol 60:307–315
Lendeckel S, Jödicke A, Chistophis P, Heidinger K, Wolff J, Fraser JK, Hedrick MH, Berthold L, Howaldt H-P (2004) Autologous stem cells (adipose) and fibrin glue used to treat widespread traumatic calvarial defects: case report. J Cranio-Maxillofacial Surg 32:370–373
Li H, Zimmerlin L, Marra KG, Donnenberg VS, Donnenberg AD, Rubin JP (2011) Adipogenic potential of adipose stem cell subpopulations. Plast Reconstr Surg 128:663–672
Maccario R, Podestà M, Moretta A, Cometa A, Comoli P, Montagna D, Daudt L, Ibatici A, Piaggio G, Pozzi S, Frassoni F, Locatelli F (2005) Interaction of human mesenchymal stem cells with cells involved in alloantigen-specific immune response favors the differentiation of CD4 + T-cell subsets expressing a regulatory/suppressive phenotype. Haematologica 90:516–525
Meirelles LDS, Fontes AM, Covas DT, Caplan AI (2009) Mechanisms involved in the therapeutic properties of mesenchymal stem cells. Cytokine Growth Factor Rev 20:419–427
Meisel R, Zibert A, Laryea M, Göbel U, Däubener W, Dilloo D (2004) Human bone marrow stromal cells inhibit allogeneic T-cell responses by indoleamine 2,3-dioxygenase-mediated tryptophan degradation. Blood 103:4619–4621
Mellor A (2005) Indoleamine 2,3 dioxygenase and regulation of T cell immunity. Biochem Biophys Res Commun 338:20–24
Minonzio G, Corazza M, Mariotta L, Gola M, Zanzi M, Gandolfi E, De Fazio D, Soldati G (2014) Frozen adipose-derived mesenchymal stem cells maintain high capability to grow and differentiate. Cryobiology 69:211–216
Murphy MB, Moncivais K, Caplan AI (2013) Mesenchymal stem cells: environmentally responsive therapeutics for regenerative medicine. Exp Mol Med 45:e54
Nanette J, Geralyn A, Louisa W, Scott O, Gerhard B, Nolta J, Joyce N, Annett G, Wirthlin L, Olson S, Bauer G, Nolta A (2010) Mesenchymal stem cells for the treatment of neurodegenerative disease. Regen Med 5:933–946
Pan Q, Fouraschen SM, Ruiter PED, Dinjens WN, Kwekkeboom J, Tilanus HW, Laan LJVD (2014) Detection of spontaneous tumorigenic transformation during culture expansion of human mesenchymal stromal cells. Exp Biol Med 239:105–115
Plumas J, Chaperot L, Richard M-J, Molens J-P, Bensa J-C, Favrot M-C (2005) Mesenchymal stem cells induce apoptosis of activated T cells. Leukemia 19:1597–1604
Puissant B, Barreau C, Bourin P, Clavel C, Corre J, Bousquet C, Taureau C, Cousin B, Abbal M, Laharrague P, Penicaud L, Casteilla L, Blancher A (2005) Immunomodulatory effect of human adipose tissue-derived adult stem cells: comparison with bone marrow mesenchymal stem cells. Br J Haematol 129:118–129
Rasmusson I, Ringdén O, Sundberg B, Le Blanc K (2005) Mesenchymal stem cells inhibit lymphocyte proliferation by mitogens and alloantigens by different mechanisms. Exp Cell Res 305:33–41
Ribeiro A, Laranjeira P, Mendes S, Velada I, Leite C, Andrade P, Santos F, Henriques A, Grãos M, Cardoso CMP, Martinho A, Ml P, da Silva C, Cabral J, Trindade H, Paiva A (2013) Mesenchymal stem cells from umbilical cord matrix, adipose tissue and bone marrow exhibit different capability to suppress peripheral blood B, natural killer and T cells. Stem Cell Res Ther 4:125
Roemeling-van Rhijn M, de Klein A, Douben H, Pan Q, van der Laan LJW, Ijzermans JNM, Betjes MGH, Baan CC, Weimar W, Hoogduijn MJ (2013) Culture expansion induces non-tumorigenic aneuploidy in adipose tissue-derived mesenchymal stromal cells. Cytotherapy 15:1352–1361
Selmani Z, Naji A, Gaiffe E, Obert L, Tiberghien P, Rouas-Freiss N, Carosella ED, Deschaseaux F (2009) HLA-G is a crucial immunosuppressive molecule secreted by adult human mesenchymal stem cells. Transplantation 87:S62–S66
Singer NG, Caplan AI (2011) Mesenchymal stem cells: mechanisms of inflammation. Annu Rev Pathol Mech Dis 6:457–478
Sioud M, Mobergslien A, Boudabous A, Fløisand Y (2011) Mesenchymal stem cell-mediated T cell suppression occurs through secreted galectins. Int J Oncol 38:385–390

Sorrell JM, Caplan AI (2010) Topical delivery of mesenchymal stem cells and their function in wounds. Stem Cell Res Ther 1:30

Theise ND, Nimmakayalu M, Gardner R, Illei PB, Morgan G, Teperman L, Henegariu O, Krause DS (2000) Liver from bone marrow in humans. Hepatology 32:11–16

Tögel F, Hu Z, Weiss K, Isaac J, Lange C, Westenfelder C (2005) Administered mesenchymal stem cells protect against ischemic acute renal failure through differentiation-independent mechanisms. Am J Physiol Renal Physiol 289:F31–F42

Tse WT, Pendleton JD, Beyer WM, Egalka MC, Guinan EC (2003) Suppression of allogeneic T-cell proliferation by human marrow stromal cells: implications in transplantation. Transplantation 75:389–397

Valadares MC, Gomes JP, Castello G, Assoni A, Pellati M, Bueno C, Corselli M, Silva H, Bartolini P, Vainzof M, Margarido PF, Baracat E, Péault B, Zatz M (2014) Human adipose tissue derived pericytes increase life span in Utrn (tm1Ked) Dmd (mdx)/J mice. Stem Cell Rev 10(6):830–840

Vieira NM, Brandalise V, Zucconi E, Jazedje T, Secco M, Nunes VA, Strauss BE, Vainzof M, Zatz M (2008a) Human multipotent adipose-derived stem cells restore dystrophin expression of Duchenne skeletal-muscle cells in vitro. Biol Cell 100:231–241

Vieira NM, Bueno CR, Brandalise V, Moraes LV, Zucconi E, Secco M, Suzuki MF, Camargo MM, Bartolini P, Brum PC, Vainzof M, Zatz M (2008b) SJL dystrophic mice express a significant amount of human muscle proteins following systemic delivery of human adipose-derived stromal cells without immunosuppression. Stem Cells 26:2391–2398

Vieira NM, Valadares M, Zucconi E, Secco M, Bueno CR, Brandalise V, Assoni A, Gomes J, Landini V, Andrade T, Caetano HVA, Vainzof M, Zatz M, Bueno Junior CR, Brandalise V, Assoni A, Gomes J, Landini V, Andrade T, Caetano HVA, Vainzof M, Zatz M (2012) Human adipose-derived mesenchymal stromal cells injected systemically into GRMD dogs without immunosuppression are able to reach the host muscle and express human dystrophin. Cell Transplant 21:1407–1417

Vilar JM, Batista M, Morales M, Santana A, Cuervo B, Rubio M, Cugat R, Sopena J, Carrillo JM (2014) Assessment of the effect of intraarticular injection of autologous adipose-derived mesenchymal stem cells in osteoarthitic dogs using a double blinded force platform analysis. BMC Vet Res 10:143

Zhang N, Dietrich MA, Lopez MJ (2014) Therapeutic doses of multipotent stromal cells from minimal adipose tissue. Stem Cell Rev 10:600–611

Zhang Y, Zhu Y, Li Y, Cao J, Zhang H, Chen M, Wang L, Zhang C (2015) Long-term engraftment of myogenic progenitors from adipose-derived stem cells and muscle regeneration in dystrophic mice. Hum Mol Genet 24(21):6029–6040

Zimmerlin L, Donnenberg VS, Pfeifer ME, Meyer EM, Péault B, Rubin JP, Donnenberg AD (2010) Stromal vascular progenitors in adult human adipose tissue. Cytom Part A 72:22–30

Zimmerlin L, Donnenberg VS, Rubin JP, Donnenberg AD (2013) Mesenchymal markers on human adipose stem/progenitor cells. Cytom Part A 83A:134–140

Zimmet JM, Hare JM (2005) Emerging role for bone marrow derived mesenchymal stem cells in myocardial regenerative therapy. Basic Res Cardiol 100:471–481

Zuk PA, Zhu M, Mizuno H, Huang J, Futrell JW, Katz AJ, Benhaim P, Lorenz HP, Hedrick MH (2001) Multilineage cells from human adipose tissue: implications for cell-based therapies. Tissue Eng 7:211–228

Zuk PA, Zhu M, Ashjian P, De Ugarte DA, Huang JI, Mizuno H, Alfonso ZC, Fraser JK, Benhaim P, Hedrick MH (2002) Human adipose tissue is a source of multipotent stem cells. Mol Biol Cell 13:4279–4295

# Chapter 4
# Neural Differentiation of Rodent Neural Progenitor Cells and Isolation and Enrichment of Human Neural Progenitor/Stem Cells

**Antonio H. Martins, Jose L. Roig-Lopez, and Maxine Nicole Gonzalez**

**Abstract** Neural progenitor cells (NPC) are multipotent and give rise to neurons, astrocytes and oligodendrocytes. NPC possess the ability of *in vivo* and *in vitro* proliferation and can therefore be expanded for research proposes. Cell proliferation generates a three dimensional aggregate called neurospheres; posteriorly, these neurospheres can be differentiated into multiple cell types and become an important tool for understanding the mechanisms regarding cell differentiation and neuronal regeneration. In this chapter, we describe the extraction of rat neural progenitor cells from embryonic telencephalon (14 days) and generation of neurospheres. Next, we describe protocols to isolate and proliferate human fetal stem cells derived from elective abortion, recognizing the ethical dilemmas. The advantage of using the neurospheres-rat model is that the differentiation can be achieved by removing the growth factors and allowing cell differentiation in poly-D-lysine/laminin coated plates. The cells can then be dissociated and grafted in animals or be differentiated and grafted, depending on the research hypothesis. The endogenous cells can be followed by staining with lipophilic dyes or analyzed by adding 5-Bromo-2′-deoxyuridine (BrdU) using the posterior double immunofluorescence technique, which identifies the post-mitotic neural progenitor cells and the fate acquired. The methods here described will help the researcher to perform cell extraction, differentiation without cell enrichment, track the proliferating neural progenitor cells and perform characterization after the cell graft, as well as isolate and maintain human fetal stem cells.

**Keywords** Neurosphere • Differentiation • Rat • Human

---

A.H. Martins (✉)
Pharmacology and Toxicology Department, School of Medicine
University of Puerto Rico Medical Science Campus, San Juan, PR, USA

Department of Biochemistry, Universidad Central del Caribe, Bayamón, PR, USA
e-mail: antonio.martins@upr.edu

J.L. Roig-Lopez
School of Science and Technology, Universidad Del Este, AGMUS, Carolina, PR, USA

M.N. Gonzalez
Department of Biochemistry, Universidad Central del Caribe, Bayamón, PR, USA



H. Ulrich, P. Davidson Negraes, *Working with Stem Cells*,
DOI 10.1007/978-3-319-30582-0_4

## 4.1 Introduction

The notion that the central nervous system (CNS) was unable to produce new neurons was once held certain by the scientific community. Discoveries over the past 20 years of neuronal progenitor cells (NPC) and neural stem cells (NSC) in adult animals have caused euphoria, bringing the possibility to repopulate the brain with new neurons capable of reestablishing connections, either physiologically lost with age or due to a variety of pathologic conditions such as traumatic brain injury, stroke, Alzheimer and amyotrophic lateral sclerosis (Einstein and Ben-Hur 2008). NPCs are defined as multipotent and self-renewing, give rise to neurons, oligodendrocytes, and astrocytes without generating other cells from embryonic layers other than the ectotherm, thereby showing more restriction. In other words, NPC are less committed than NSC and exist in low amounts in the CNS.

Neural progenitor cells are basically localized in thee niches: the olfactory bulb, the subventricular zone (SVZ), and the subgranular zone of dentate gyrus (SGZ). These niches are distinct regarding the ability to respond to different stimuli or in the ability to produce a specific type of neuron. For example, several groups have demonstrated that the adult human NPCs at SVZ are the primary site of proliferation and neurogenesis in patients with neurodegenerative diseases (Nait-Oumesmar et al. 2007). At the same time, NPC derived from dentate gyrus is more suitable to proliferation regarding memory formation (Eckenhoff and Rakic 1988). To study differentiation processes, NPC derived from the embryonic brain from day 14 was used in several works (for review see Trujillo et al. 2009). However, even with a great capability for proliferation, the total number of NPC remains scarce. To increase the total number of cells for cell graft or to study the process of differentiation, the NPC can be cultured *in vitro* in tridimensional agglomerates, called neurospheres, in the presence of epidermal growth factor (EGF) and fibroblast growth factor (FGF). The easy induction of differentiation by removing the growth factors and the ability to drive the cell differentiation into neurons (using retinoic acid), oligodendrocytes (with insulin-like growth factor 1, IGF-1), and astrocytes (using bone morphogenetic protein), make NPC important in the fields of both differentiation and neurorepair.

Here, we will focus on the extraction, proliferation, maintenance and analyses of cell differentiation by cell markers, using immunofluorescence and flow cytometry for neurospheres from rats and mice. In this chapter, we also describe human NPC purification and expansion.

## 4.2 Protocols

### 4.2.1 Preparation of Non-adherent Cell Culture Flasks

1. Use non-adherent culture flasks from brands of your preference or,
2. Prepare an anti-adherent solution of 12 % of 2-hydroxyethyl methacrylate MW 20,000 (6 g of 2-hydroxyethyl methacrylate (Sigma-Aldrich, Cat. No. 128635)

in 50 ml of 95 % ethanol). Because it is difficult to dissolve, keep it covered in shacker at 37 °C overnight;

3. Cover the flask bottom with the solution, remove the excess (which can be re-used) and leave the flask open inside the cell culture hood to completely dry;
4. After it is dry, wash three times with phosphate-buffered saline (PBS). It is then ready to use.

***Note:*** *Use sterilized beaker, spatula, and so on to prepare the solution even if 2-hydroxyethyl methacrylate is not sterile. This will decrease the chances of contamination by reducing the number of bacterial or fungi spores. Remember that ultraviolet light does not penetrate plastic or glass. The flasks can be stored closed for several months at room temperature.*

### *4.2.2 Animals*

1. Time-pregnant rats (for instance: Wistar, Sprague Dawley with 14–16 weeks of age) with 14.5 days of pregnancy purchased from the company of your choice or, paired one male and one female for approximately 12 h; after this period of time perform a vaginal smear using distillated water. *Do not use PBS or other buffer because of drying and crystal formation. Use a wet swab and a clean slide; this procedure can be performed in rats but not in mice, because in the latter you will remove the vaginal plug.*
2. Introduce a wet swab in the vagina and then make a smear in a clean slide The presence of spermatozoids is a potential sign that the rat will become pregnant. Sometimes, despite the presence of spermatozoids, the rat may not be pregnant. *Non-observance also does not rule out the possibility of pregnancy.* We roughly calculate an 80 % chance of pregnancy.

### *4.2.3 Surgery (Instruments and Materials)*

1. Sterilize the instruments in the autoclave using a steel box or an instant sealing sterilization pouch.

*For removing the head*

2. One operating scissor 5.5″ (Roboz Company, Cat. No. RS-6812)—For cutting the rat's head;

*For holding the skin of the rat while the skull is removed*

3. Curved forceps 7″ (Roboz Company, Cat. No. RS-5271)

*For meninges peeling*

4. Two micro-dissecting forceps 4″ slightly curved 0.8 (Roboz Company; Catalog number RS-5135)—For placing the cortices in the petri dishes containing medium, four micro-dissecting tweezers (Roboz Company, Cat. No. RS-4972);

*For separating the cortices*

5. Two Micro-dissecting forceps 4″ angled fine sharp (Roboz Company, Cat. No. RS 5095).

### *4.2.4 Culture Materials*

- Serological pipettes 5, 10, and 25 ml.
- Thee 60 mm petri dishes.
- Sterile syringe of 5 ml.
- Invert phase microscopy.
- Different micropipette tips (0.2 and 1 ml).
- Cell strainer 40 μm BD Falcon.
- Laminin.
- Centrifuge conical tubes of 15 and 50 ml.
- Stomacher 80 Seward® Lab System.
- Model 80 Bags for Stomacher.
- Trypsin/EDTA.
- Fibroblast growth factor.
- Epidermal growth factor.
- Poly-D-lysine.
- Heparin.
- Dulbecco's Modified Eagle's Medium (DMEM) and F12 medium.
- Fetal bovine serum (FBS).
- B27 supplements 50× Gibco Life technologies (Cat. No. 12587-010).
- Trypan blue solution (for measurement of cell viability).

## 4.3 Methods

### *4.3.1 Neurospheres from Rat Embryonic Telencephalon*

My collaborators and I have used for several years the protocol described below. We developed it to obtain NPCs and to generate neurospheres from mice and rats. Although the cells can be placed in any non-adherent cell culture flasks or dishes after extraction, we recommend flasks, which decreases the chances of contamination. After distributing the cells, carefully look inside the flasks to find any contaminants such as fungi or bacteria. It is not necessary to anesthetize the fetuses for the surgery since at this stage of development they do not register pain and the head will be taken out following the NIH guidelines for euthanasia.

1. Use time-pregnant rats (14.5 days or mice of 12.5 days);
2. There are several protocols for animal preparation for the surgery; (a) euthanize using an overdose of Isoflurane (5 % until the rat stop breathing) or (b) $CO_2$ chamber;

***Note:*** *Use a specific set of instruments for opening the skin and another set to remove the uterus containing the fetuses. This practice is important to avoid contamination. Remember that the skin is disinfected but not sterilized.*

3. Perform an antiseptic cleaning of the skin using ethanol 70 % or iodine solution (10 % povidone-iodine);
4. Open the skin using scissors and forceps;
5. Remove the uterus using a different forceps and put it in 60 mm petri dishes;
6. In another plate pour 5 ml of DMEM/F12 70 %/30 % medium;
7. Open the uterus and let the fetuses felt in the liquid;
8. Grasp the fetuses and decapitate those (RS 6812);
9. Pour 5 ml of medium in a second petri dish;
10. Place the head in the petri dish and discard the body in a plastic bag;
11. Using two micro-dissecting tweezers (RS 5882), sustain the animals and make an incision between the eyes;
12. Then, go laterally with the tweezers as follows: one blade of the tweezer will be inside the skin and the other, outside. Using the tweezer lift and remove the portion that will be the future skull;
13. Remove the meninges completely, then the cerebellum and everything from underneath each telencephalon;
14. Pour 2 ml of medium inside the Model 80 bags for Stomacher 80, leaving space to seal;
15. Place the bag into the Stomacher 80 and close the door. The cell dissociation is performed during 2.5 min at high speed or, gently triturate the tissue using trypsin/EDTA through different series of descending micropipette tips to make single cell suspension. Add FBS to inactivate the trypsin;
16. Filter the bag content using a 40 $\mu$m cell strainer coupled to a sterile syringe;
17. The volume should be dispensed into a 15 ml conical tube;
18. Centrifuge for 5 min at $200 \times g$, add 10 ml of fresh medium and centrifuge again;
19. Repeat Step 18 twice.

### 4.3.2 Neurosphere Formation

At this point, there are mixtures of cells smaller than 40 µm. The following protocol is used to expand the neural progenitor cells in the neurosphere shape.

**Prepare the Plates for Plating Neurospheres**

1. Count the cells using Trypan-blue and add 180,000–200,000 cells/ml of viable cells into the anti-adherent flasks or into coated flasks using the following medium: DMEM/F12 70 %/30 %, 100 IU/ml of penicillin, 100 μg/ml of streptomycin, 2 mM L-glutamine 20 ng/ml fibroblast growth factor, 20 ng/ml epidermal growth factor, 5 μg/ml of heparin (Sigma-Aldrich, Cat. No. 1304005), 2 % B27 at 37 °C in 5 % $CO_2$;
2. Allow the cells to grow for 10 days changing half of the medium every 3–4 days.

#### 4.3.2.1 Neurosphere Phenotype Analysis

After 10 days, the neurospheres can be analyzed. Due to variation in size, we recommend to use neurospheres with similar diameter to avoid structural differences. For example, large neurospheres possess more cell death in the center.

**Examples of Analysis**

A. Reverse Transcription-polymerase chain reaction (RT-PCR) to analyze gene expression;
B. Western blot to evaluate protein expression;
C. Flow cytometry analysis to quantify the number of neural progenitor cells and a specific protein of interest;
D. Immunofluorescence to analyze protein expression and localization.

Neurosphere Immunofluorescence

**Preparation**

1. Collect neurosphere samples of similar size;
2. Wash twice to remove the excess medium;
3. Fix using 4 % paraformaldehyde (PFA 4 %) at room temperature (25 °C) for 1 h;
4. Afterwards, wash thee times with PBS and then transfer the neurospheres to a solution of PBS/sucrose 10 % (w/v) for 1 h, followed by PBS/sucrose 20 % (w/v) for 1 h and finally PBS/sucrose 30 % (w/v) for 16–18 h at 4 °C;
5. Put the neurospheres in HistoPrep (Fisher Scientific, Cat. No. SH75-125D) and freeze in dry ice for 5 min, followed by freezing at −80 °C overnight;

***Note:*** *Place the neurospheres at the bottom and very carefully add the HistoPrep, in order to keep the neurospheres at the same plane.*

6. Make slices of 10 μm using a Cryostat and put them onto signalized Superfrost slides (Fisher Scientific, Cat. No. 48311-703).

**Proceed for immunofluorescence detection**

There are different antibody dilutions and different blockers that could be used, depending on the protein of interest. Nonetheless, we will give a general protocol for permeabilization and blocking to analyze nestin, glial fibraly acidic protein (GFAP) and β3-tubulin expression patterns.

7. Prepare a permeabilization and blocking solution using 0.1 % Triton X-100 (Sigma Aldrich, Cat. No. X100) and 3 % fetal bovine serum in PBS; incubate the neurospheres at 25 °C for 30 min;
8. Add the primary antibody (1:250) using PBS and 3 % of fetal bovine serum and incubate for 2 h at 25 °C;
9. Wash thee times with PBS;
10. Add the secondary antibody (1:250) and incubate for 2 h at 25 °C. *Remember to use antibodies raised in different species in case of double staining*;
11. Wash three times with PBS;
12. Place a mounting medium for immunofluorescence and add a coverslip.

#### 4.3.2.2 Neurosphere Differentiation

Prepare dishes or 6-well plates with poly-D-lysine (Sigma-Aldrich, Cat. No. P7280) and laminin (Sigma Aldrich, Cat. No. L4544), as following:

i. Prepare a solution of 0.05 mg/ml of poly-D-lysine, MW 30–70 K, and then add 0.5 ml/well 16–18 h (6-well plate). Remove the poly-D-lysine and wash the wells with sterile distilled $H_2O$ twice. Let it dry for 1 h.
ii. Prepare a laminin solution and use it at final concentration of 20 μg/ml in water. Add 200 μl into the previously coated poly-D-lysine dishes making circles using a pipette (recommended concentration by Sigma-Aldrich is 1–2 μg/$cm^2$). The circles will decrease the amount of volume used in each dish. Do not allow the laminin to drain to the edges. Incubate the dishes or 6-well plates at 37 °C for 1 h. Wash twice using cell culture medium. We recommend not to store the plates coated with laminin.

1. Allow the neurospheres to decant and transfer to a 15 ml conical centrifuge tube; wash the neurospheres three times with 10 ml of medium (70 %/30 % DMEM/F12) to remove any trace of growth factors;
2. Place the neurospheres in the dishes or 6-well plates prepared with poly-D-lysine and laminin;
3. Let them to differentiate for 7 days. A radial pattern of cell differentiation can be visualized; next, analyze the protein of interest.

### 4.3.3 *Characterization of the Cells Obtained Using the Described Protocol*

To verify proper cell differentiation, we recommend analysis of the cells using immunofluorescence and/or flow cytometry. After differentiation, the cells should express β3-tubulin, GFAP, S100β, and decreased expression of nestin (Martins et al. 2008).

#### 4.3.3.1 For Immunofluorescence

1. Fix the differentiated neurospheres using 4 % paraformaldehyde at 4 °C for 30 min. Wash the cells twice using PBS for 5 min;
2. Permeabilize and block the cells using a solution of 0.1 % Triton and 10 % goat serum in PBS at 25 °C for 30 min;
3. Add the primary antibody (1:750 anti-GFAP Cy3, Sigma-Aldrich, Cat. No. G3893 or 1:1000 anti-β3-tubulin (Abcam, Cat. No. 18207) for 2 h at 25 °C using PBS and 1 % of goat serum), or at 4 °C overnight;
4. Wash thee times with PBS (5 min each);
5. Add the secondary antibody (1:1000) at 25 °C for 1 h;
6. Wash thee times with PBS (5 min each);
7. Add 4′:6-Diamidino-2-Phenylindole; dihydrochloride- (DAPI) (Life Technologies, Cat no. D1306) solution to counterstain the nuclei;
8. Put a mounting medium for immunofluorescence and add a coverslip. Observe under a inverted immunofluorescence microscope.

***Note:*** *Remember to use primary antibodies raised in different species and secondary with different colors in case of double staining.*

Several different markers are listed in Table 4.1. Some cellular markers are shared by human stem/progenitor cells and neuron and glial cells, listed in Table 4.2.

#### 4.3.3.2 Flow Cytometry

1. Dissociate the neurospheres with Accutase (Sigma-Aldrich, Cat. No. A6964) or trypsin into single cells using different decreasing micropipette tip size. When using trypsin, inactivate the trypsin with 10 % fetal bovine serum (final concentration) (BioAnalytical Instruments, Cat. No. 16000-044);
2. Centrifuge at 200 × *g* for 5 min;
3. Fix the cells in ice-cold 1 % PFA in PBS for 20 min, then wash with PBS;
4. Incubate the cells using 2 % of fetal bovine serum in PBS and 0.1 % Triton for 30 min;

**Table 4.1** Neural progenitor/stem cells markers during development and *in vitro* differentiation

| Target | Markers | Reference |
|---|---|---|
| Neural stem cells | CXCR4 | Itoh et al. (2009) |
| | FABP7/B-FABP | Yun et al. (2012) |
| | ID2 | Park et al (2013) |
| | Musashi-1 | Kirik et al. (2013) |
| | Nestin | Park et al (2010) |
| | Notch-1 | Nagato et al. (2005) |
| | SOX-1 | Venere et al. (2012) |
| | SOX-2 | Ellis et al (2004) |
| | SSEA-1 | Son et al. (2009) |
| Neural progenitor cells | Activin A | Rodriguez-Martinez et al. (2012) |
| | FABP7/B-FABP | Arai et al. (2005) |
| | GFAP | Liu et al. (2010) |
| | NCAM-1/CD56 | Marmur et al. (1998) |
| | Nestin | Suzuki et al. (2010) |
| | ROR2 | Endo et al. (2012) |
| | SOX-2 | Graham et al. (2003) |
| | Vimentin | Kim et al. (2009) |
| Glial progenitor cells | Nestin | Messam et al. (2000) |
| | S100B | Raponi et al. (2007) |
| Young neurons | β3-Tubulin | Braun et al. (2002) |
| | Calretinin | Brandt et al. (2003) |
| | MAP2 | Arai et al. (2005) |
| | NCAM-1/CD56 | Terkelsen et al. (1992) |
| Mature neurons | Calbindin | Young et al. (2000) |
| | ID2 | Neuman et al. (1993) |
| | NeuN | Seki (2002) |
| | Synaptophysin | Gingras et al. (2007) |
| Oligodendrocyte-like cells | Olig1 | Ohnishi et al. (2003) |
| | O4 | Lu et al. (2000) |
| Astrocyte-like cells | EAAT1/GLAST-1 | Miralles et al. (2001) |
| | GFAP | Brahmachari et al. (2006) |

5. Incubate with primary antibodies specific for neurons (β3-tubulin, Abcam, Cat. No. 18207), astrocytes (1:500 anti-GFAP Cy3 Sigma-Aldrich, Cat. No. G3893) and mature neurons microtubule associated protein 2 (1:500 MAP-2, Cell Signaling Technology, Cat. No. 4542) for 2 h;
6. After 2 h, wash twice with PBS, and then incubate the cells with 1:500 Alexa Fluor 488- or 555-conjugated secondary antibodies (Life Technologies and Molecular Probes, Cat. No. A11001 and A21428, respectively);

**Table 4.2** Human neural progenitor/stem cells markers during development and *in vitro* differentiation

| Target | Markers (human) | Reference |
|---|---|---|
| Neural stem cells/neural progenitor cell | CD133⁺ (Prominin) | Uchida et al. (2000) |
| | CD34⁻ | |
| | CD45⁻ | |
| | Nestin | Tohyama et al. (1992) |
| | GFAP | |
| | Integrin subunit α6 | Hall et al. (2006) |
| | Integrin subunit β1 | |
| | Pax-6 | Hansen et al. (2010) |
| | SOX-2 | |
| | TRBR2 (EOMES) | |
| | Musashi-1 | Dhara et al. (2008) |
| Glial progenitor cells | Nestin | Hansen et al. (2010) |
| | CD133 radial glia | Pfenninger et al. (2007) |
| Young neurons | β-3Tubulin | Liu et al. (2010) |
| | TBR2 (EOMES) | Hansen et al. (2010) |
| | Calbindin | Eriksson et al. (1998) |
| Mature neurons | MAP2 | Anacker et al. (2011) |
| | Synaptophysin | Gingras et al. (2007) |

7. Analyze in flow cytometer (Fc500, Beckman Coulter, Fullerton, CA) using an argon laser line for fluorescence excitation (FL1 525 nm and FL2 575 nm, band pass filter). Between 10,000 and 50,000 events per sample should be acquired with fluorescence measured in logarithmic scales. Set up the background fluorescence using unlabeled cells and those labeled with secondary antibody alone. Use it to set gating parameters between antigen-positive and negative cell populations;
8. Use forward and side light-scatter gates for exclusion of cell aggregates and small debris;
9. Use Cyflogic software (http://www.cyflogic.com) and plot the results in a histogram format.

## 4.4 Use of Neural Progenitor Cells After Stroke

The majority of cerebral strokes are caused by the occlusion of a blood vessel in the brain, which decreases oxygen supply to the affected area. An early onset neuronal death begins within a minute after the beginning of an ischemic stroke; it occurs in the area most affected by the ischemia. This area, called the infarct core, has either a low or no blood flow. The resulting oxygen and glucose deprivation is followed by a decrease in ATP, thus triggering an excitotoxicity cascade. The energy failure stops the sodium and potassium pumps and promotes collapse of the neuronal transmembrane ionic gradient, $Ca^{2+}$ influx and release of glutamate, which causes excitotoxicity (Barone et al. 1997; Dirnagl et al. 1999). The ischemic brain injury also increases the synthesis of inflammatory cytokines that spread the inflammation into the surrounding area (Huang et al. 2006; Iadecola and Alexander 2001). The brain tissue that surrounds the core is called the penumbra and can be saved if neuroprotective drugs or thombolytic drugs are administrated timely. The penumbra neurons can either recover and survive, or be recruited to the ischemic core and die. The period during which the penumbra neurons can be salvaged is called "the therapeutic window of opportunity". Therefore, the neuroprotective interventions that target delayed processes such as apoptosis and inflammation in the penumbra have a potential for clinical treatment in acute stroke (Xu et al. 2004).

Interestingly, recent evidence suggests that global and focal ischemia enhance the proliferation and differentiation into neurons in several zones including dentate gyrus, the anterior subventricular zone, and the posterior periventricular zone, adjacent to the hippocampus. Moreover, Arvidsson et al. (2002) showed that the initial generation of neuroblasts following a stroke, far exceeds the final number of surviving new neurons, showing that number could be increased. The plasticity could allow brain repair without the need of cell graft (Andres et al. 2011) and much evidence exists describing the involvement of NPCs in stroke treatment (Andres et al. 2011; for review see Barkho and Zhao 2011). Grafted NPCs are "known to have the ability of migrating long distances to damage sites after brain injury and differentiate into new neurons" (for a review see Barkho and Zhao 2011).

## 4.4.1 *Protocol to Observe Cell Migration/Proliferation After Stroke*

### 4.4.1.1 Middle Cerebral Artery Occlusion (MCAO)

Surgical procedures should be performed using sterile/aseptic techniques in accordance with institutional guidelines and the initial assistance of the veterinarian. Sprague-Dawley rats weighing 250–300 g will fast for no more than 4 h to avoid hyperglycemia after administration of nitric oxide and isoflurane. The animals will be subjected to permanent or transient MCAO as previously described (Xu et al. 2004; Martins et al. 2015).

**Middle Cerebral Artery Tools**

- Rodent Brain Matrices (Stoelting, Cat. No. 51388).
- 2 U Forceps (Stoelting, Cat. No. 52110-31).
- Silk Sutures (4-0 & 6-0) (Ashaway).

- 2 U Microvascular Clamp (Fine Science Tools, Cat. No. 18055-040).
- Cautery/Bipolar coagulation Forceps.
- Micro-Serrafine Clamp applicator (Fine Science Tools, Cat. No. 18056-14).
- Dual Gooseneck Illuminator (Stoelting, Cat. No. 59259).
- Anesthesia system (Stoelting).
- Micro-dissecting Forceps (Stoelting, Cat. No. 52102-54 and 52102-35).
- Anesthesia scavenger (Stoelting).
- Spring Scissors (Stoelting, Cat. No. 52130-00).
- Micro-dissecting scissors (Stoelting, Cat. No. 52132-42).
- Needle Holder (Stoenting. Cat. No. 52181-84).
- Heating Pad.
- Shaver.
- Surgery Microscope.
- Cotton tipped applicators.

***To observe endogenous neural progenitor cell proliferation after stroke***

The NPCs could be analyzed by using a pulse or a cumulative protocol of 5-bromo-2′-deoxyuridine (BrdU), a thymidine analogue (Sigma-Aldrich, Cat. No. B5002) diluted in 0.9 % saline, depending on the researcher interest.

1. As an example, inject 50 mg/kg of BrdU intraperitoneal every 4 h for 12 h in the same day of the animal sacrifice (Zhang et al. 2001) to obtain a temporal profile;
2. To observe all proliferating NPCs, inject intraperitoneal 50 mg/kg of BrdU, saline immediately after the surgery. Inject 50 mg/kg every 24 h until euthanasia;
3. Remove and freeze the brain using HistoPrep;
4. Cut the brain into 40 μm slices and put them into Superfrost slides;
5. Fix the slices using 4 % of paraformaldehyde for 30 min, then wash using PBS;
6. Prepare a permeabilization and blocking solution using 0.1 % Triton, 5 % FBS or BSA in PBS for 10 min; add the primary antibodies in 1 % FBS in PBS at room temperature for 16–18 h; wash twice with PBS;
7. Add the secondary antibodies: for instance, anti-mouse Alexa-Fluor 488 and anti-rabbit Alexa-Fluor 555 for 2 h.

***Note:*** *The primary antibodies should be raised in different species in order to make a double staining. We recommend using Anti-BrdU plus a different combination of antibodies to detect newly generated neural progenitor cells that differentiate into astrocyte, (BrdU$^{+}$GFAP$^{+}$ cells), mature neurons (BrdU$^{+}$NeuN$^{+}$ cells) or oligodendrocyte (BrdU$^{+}$ Anti-oligodendrocyte specific protein (Abcam, Cat. No. ab53041) positive cells), migrating newly neural progenitor cells or migrating newly immature neurons (BrdU$^{+}$ Doublecotin DCX$^{+}$, Abcam, Cat. No. ab18723).*

#### 4.4.1.2 Injection of Neural Progenitor Cells

Prepare a stock solution of lipophilic $DiOC_{18}(3)$ in dimethyl sulfoxide (DMSO) at 1 mg/ml.

***Note:*** *Dimethylformamide (DMF) is the solvent indicated by the company, however our laboratory dilutes in DMSO without complications.*

1. Prepare the NPCs as described above;
2. Stain the cells using a lipophilic dye 3,3′dioctadecyloxacarbocyanine perchlorate $DiOC_{18}(3)$ (ThermoFischer Scientific, Cat. No. D-275) as following;
3. Prepare $2.5 \times 10^6$ of dissociated NPCs; pass the cells through a 40 μm cell strainer to avoid cell clamps and put them in complete NPC medium;
4. Add 5 μl of a stock solution of 2.0 mg/ml to the cells and incubate at 37 °C for 1 h;
5. Wash the cells twice using fresh medium;
6. Check under fluorescence microscope the quality of the staining (intensity and if all cells are stained);
7. Graft of 120,000 cells/2 μl will be injected by means of the cisterna magna (intracranial) using a glass micropipette or a 1 cc insulin syringe, as previously described by Mitome et al. (2001);
8. After a period of time, which will vary according to the research interest, remove the brain and perform a slice preparation as described above;
9. Chcek the fluorescence under an immunofluorescence microscope;
10. The researcher has also the possibility to verify the type of cell that was grafted and posteriorly differentiated;
11. Use antibodies, such as anti-NeuN (Abcam, Cat. No. ab104225) to detect the differentiation of NPCs into neurons, anti-GFAP for astrocytes, and anti-Nestin (Abcam, Cat. No. ab6142) for undifferentiated cells. Remember to use different colors for staining cells; for instance use DiO and DiD (Molecular Probes) and distinct antibodies.

## 4.5 Human Fetal Tissue as Source of Neural Stem Cells/Neural Progenitor Cells

Fetal samples for enrichment and isolation of NSCs are usually discarded as biohazard samples from elective abortions performed in the USA and other countries To properly obtain these samples, researchers must consult with an institutional biosafety and compliance officer to develop an Institutional Review Board (IRB) Protocol at least 6 months prior to experimental procedures. Other options to acquire tissue samples depend on research centers and/or non-profit organizations. For CNS samples, it will usually be crude brain tissue from the end of the first to the second trimester. Despite multiple moral and ethical issues, established protocols make this source of human neural stem cells very attractive.

Fetal human adult or somatic neural stem cells come from mitotically active regions of the CNS during development, after 5 weeks of gestation, to adult human regions of the CNS such SVZ and outer SVZ (OSVZ) along the ventricles. These NSCs share the basic characteristic of self-renovation and the capacity, upon differentiation, to give rise to neurons, glia and oligodendrocytes (Vescovi et al. 1999). At 10.5 weeks of gestation from a diencephalon sample, 1 % of 1000 cells form neurospheres using the regular serum free media with EGF/FGF-2 and supplemental hormones/salt mix. These NSCs, shaped as neurospheres, are multipotent upon differentiation, which involves the removal of growth factors (EGF/FGF-2). The cells will attach to the substrate via the poly-ornithine coating. Such cultures can be maintained up to 2 years with a subculturing or passage rate every 2 weeks. Finally, and more importantly, differentiated neurons and glial cells survive the transplantation procedure into the striatum of immunosuppressed rats.

The nature of the human neural stem cells was discovered by enrichment, using cell sorting of human fetal samples with antibodies previously used to isolated human Mesenchymal/Hematopoietic Stem cells. The NSC in fetal samples was found to be CD133+ (Prominin), CD34−, and CD45−. Single cells with these lineage markers form a neurosphere under standard proliferative conditions (Uchida et al. 2000; Hall et al. 2006). It was later established that CD133 stains the radial glial/ependymal cell type at the early postnatal stage, and ependymal cells in the adult brain, but not neurogenic astrocytes in the adult human SVZ (Pfenninger et al. 2007). In another work by Hall et al. (2006), human fetal tissue (8–10 weeks of gestation) enriched by the use of integrin subunits $\alpha 6$ and $\beta 1$, was capable to form NSC cultures. The presence of others markers of NSCs, such as Sox2, Sox3, Nestin, Bmi1 and Musashi1 (Sun et al. 2009) were found to corroborated the presence of such cells.

The developing fetal human forebrain has a massively expanded outer region throught to contribute to cortical size and complexity of the prospective human cortex. However, SVZ cytoarchitecture was presumed to be similar to rodent species and progenitor cell types; its contribution to neurogenesis is still not well understood. A meticulous analysis showed that large numbers of radial glia-like cells and intermediate progenitor cells populate the human fetal SVZ (Hansen et al. 2010). A second population of radial glia-like cells, named OSVZ, was found to have long basal process but, surprisingly, are non-epithelial as they lack contact with the ventricular surface. By real-time imaging, clonal analysis with mitotic markers and using viral markers, the Krigelstein group (Hansen et al. 2010) showed that these cells can undergo proliferative divisions and self-renewing asymmetric divisions to generate NPCs that can further proliferate. This OSVZ with their population of outer Radial Glia (oRG) may be isolated from human fetal samples and expanded in cultures due to the fact that the oRG are more numerous and prominent at the second trimester of gestational week of the fetus.

### 4.5.1 *Isolation, Enrichment and Culture of Human Fetal Neurospheres*

Fetal discarded tissue from elective abortions can be obtained from Advanced Bioscience Resources, a nonprofit organization that collects and sends human fetal tissue for clinical and biomedical purposes, and follows all state and federal regulatory procedures. Under the laws of the US Commonwealth of Puerto Rico, fetal tissue is classified as a cadaveric donation. In our experiments we used complete brain samples from 16 to 18 weeks of gestation. Every order required proper clearance and approbation of IRB and Bio-Safety Institutional compliance.

**Materials**

- Accutase 1× (Sigma-Aldrich, Cat. No. 6964).
- B-27 supplement. **No vitamin A**. (Cat. No. 12587-010). *Note: contains insulin, antioxidants cocktail among others hormones.*
- Micropipette P1000, Blue Tips, sterile (Fisher Scientific, Cat. No. 02-707-405).
- DMEM/F12 1:1 (Sigma-Aldrich, Cat. No. D8437).
- Sterile 10 ml 13-678-11E, Tubes 50 ml Falcon (Fisher Scientific, Cat. No 352098).
- GlutaMax (Invitrogen-Life Technologies, Cat. No. 35050061).
- 6-Well Plate Falcon (Fisher Scientific, Cat. No. 353046).
- Leukemia Inhibitory Factor, final concentration 5 ng/ml. (Peprotech, Cat. No. 300-05).
- Recombinant Human EGF (Catalog number AF 100-15), final concentration 10 ng/ml.
- Neurospheres Proliferation Media (NPM): D-MEM/F12 1:1, EGF, FGF-basic, LIF, Glutamax, antibiotic-antimycotic, and B-27 supplement.
- Recombinant Human FGF-basic (154 a.a.) (Catalog No 100-18B) final 10 ng/ml.
- Antibiotic-antimycotic 100×. (Gibco, Cat. No. 15240-062).

<u>**To isolated and enrich Neurospheres from tissue**</u>

1. Decant shipping media, put tissue in sterile cold basal media with 2× antibiotic-antimycotic and a mix of wide spectrum antibiotics in a 50 ml conical tube. Repeat the process once with fresh media with antibiotic-antimycotic mix;
2. After 5 min, transfer the tissue using a 25 ml pipette to another 50 ml tube; this will start the mechanical trituration of the tissue. Add around 25 ml of Accutase 1×;
3. Triturate the tissue using a 10 ml pipette with ten times up and down movements in the same 50 ml with Accutase. Leave the 50 ml tube in the incubator (37 °C) for 10 min;
4. Repeat the trituration process using a 10 ml pipette ten more times;
5. Set a P1000 to 800 μl and with sterile tips, pipette up and down the tissue. If the sample clog the pipette, repeat Step 4. Try to use P1000 to pipette up and down

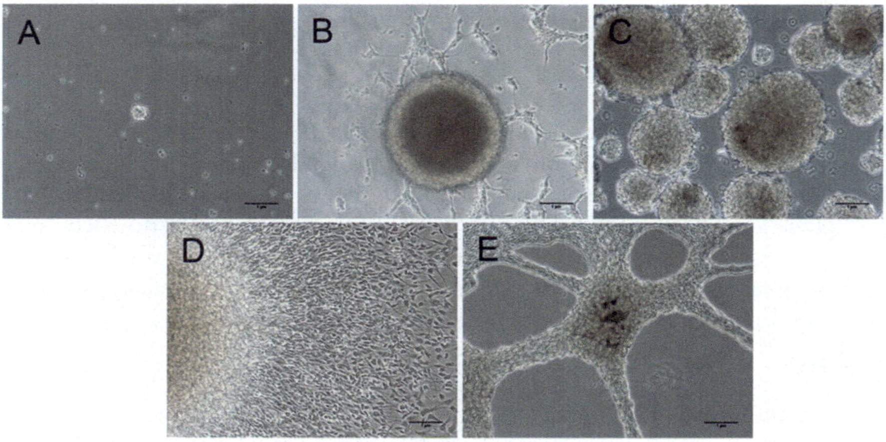


**Fig. 4.1** Human neurospheres at different stages in culture. Neurospheres derived from rodent and humans look very similar. The differences occur at distinct time points in the culture, the main one is the doubling time and how long the cells can be maintained *in vitro*. For example: human fetal samples can last more than 2½ years with passages every 2–3 weeks. (**a**) After the dissociation process of the tissue sample, the cells need 1–2 days to reprogram and start to proliferate in both rodent and human species. (**b**) After 14 days *in vitro*, proliferating spheres form a big bright golden sphere. (**c**) Human neurospheres at 21 days *in vitro* are ready to be split. At this point, a combination of small, medium and large spheres can be observed. From day 0 (or the initial isolation for human samples), this step takes 16–21 days. Embryonic samples from rodent take only 2 weeks for initial passage P0 to P1. (**e**) Upon differentiation, neurospheres attach to the precoated surface with Matrigel and start to migrate out from the center in radial patter, and differentiate from the sphere; a similar migration pattern also occurs in rodent. (**e**) After 3 days *in vitro*, the neurospheres attach and, supported by growth and neurotrophic factors, form a network of proliferating spheres with extensions of cells that connect some spheres with others. Scale bar = 1 μm

approximately 20 times, then leave the sample at 37 °C for 5 min, pipette again 30 more times; filter the tissue and cells using a 40 μm filter.

6. Dilute the volume in 15 ml of basal media, sediment at 1200 rpm (277 × *g*), decant and plate ~2,000–10,000 cells/cm$^2$ in 2 ml of NPM with FGF/EGF/LIF/B-27 in a 6-well plate. *We strongly recommend the 6-well plate, because paracrine factors help cells to reprogram and proliferate more efficiently if they are in close proximity*;
7. Feed the cells at day 3 and at day 5 or 6, adding FGF, EGF, B-27 and antibiotic-antimycotic. At this point, you should see small proliferation aggregates (Fig. 4.1a). Change half media (50 %) for the first time at day 8 or 9 (Fig. 4.1b);
8. After neurospheres become larger, regular feeding will be once a week; change half media, 2–3 days later, exclusively add FGF/EGF/LIF/B-27 and antibiotic-antimycotic. This feeding schedule applies to the rest of the neurosphere culture;
9. Initial generation and enrichment of the "mature" neurospheres at P0 will take at least 14 to 21 days *in vitro*; after this time, dissociated cells for the next passage.

Subsequent generations of passages are done every 2–3 weeks depending on the growing conditions (Fig. 4.1c).

**Chemical and mechanical dissociation of Neurospheres**

10. After the neurospheres float in neural proliferation medium for 14–21 days (Fig. 4.1c), transfer them to a sterile 50 ml tube (up to four 6-well plate per 50 ml tube) and centrifuge at 277 × *g* for 7 min;

    ***Note:*** *Collect supernatant in a sterile tube as this will be used for the next plating after dissociating neurospheres*
11. Add 1 ml Accutase 1× at room temperature to the cells;
12. Set a P1000 automatic pipetter to 800 μl and with sterile tips, gently pipette up and down the neurospheres for approximately 30 times. Incubate the sample at 37 °C for 10 min ($CO_2$ incubator), and pipette 30 more times. At this moment, the solution of Accutase and cells will look milk-like;
13. To wash the cells, add 15 ml basal media (no growth factors), sediment at 277 × *g*, decant and split at 1:2 ratio by using another 50 ml tube. Add 50 ml of fresh basal media to the cells, mix, transfer 25 ml with half of the cells in another 50 ml tube. Add 25 ml of used media to each 50 ml tube; you will use half of the cells plus half fresh media and conditioned media. Plate into a 6-well plate. Add FGF/EGF/LIF/B-27 (2,000–10,000 cells/cm$^2$ in 2 ml NPM);

***Note:*** *Neurosphere are mostly aggregated. We usually do not count cells after passaging and just make a 1:2 split ratio in terms of cm$^2$ for every 6-well plate or flask that is full of neurospheres, diluting cells and media in two, so that they will grow better in high densities than in higher split ratios.*

14. Once a week, change half media and add FGF/EGF/LIF/B-27 and antibiotic-antimycotic and, 2–3 days later add FGF/EGF/LIF/B-27 and antibiotic-antimycotic (without changing the media);
15. Repeat the dissociation every 2–3 weeks, depending on the number of neurospheres (Fig. 4.1c). As can be seen in Fig. 4.1d, e, the cells migrate and differentiate perpendicular to the center of the neurosphere. The differentiated neurons in the periphery can be analyzed by patch clamp studies to identify variations in the ionic currents under different experimental conditions.

## 4.6 Potential Use of Human NSCs/NPCs

One of the most compelling arguments for isolation and characterization of human NSCs is its importance in biomedical knowledge and the enormous potential for translational studies in neurodegenerative diseases, including possible cell replacement therapy with an autologous cell source. Thee sources are available to obtain tissue samples for biomedical research and translational studies: (1) fetal samples, (2) post-mortem biopsies and (3) surgical biopsies from elective operations of olfactory epithelia, curative neuro-surgical interventions such as in epilepsy.

Postmortem samples have long been the prime source of human NSCs; some research scientists often argue that the resource is underutilized. Advances in stem cell culture procedures allow postmortem samples to grow stem/progenitor cells in a greater number and for longer periods of time than before. The use of postmortem brain for complex neuropsychiatric diseases is a good example of underutilized biomedical resource (McCullumsmith et al. 2014). Cells from the human olfactory mucosa are another tissue that can be harvested from living human subjects. They regenerate throughout life from neural stem cells. A simple biopsy performed by an otorhinolaryngologist can produce 10–25 $cm^2$ of epithelial mucosa. This olfactory epithelium can be used for cultures that can be expanded into the large quantities required for molecular and functional analyses, compared to patient-derived *in vitro* model systems of complex neural disease (Matigian et al. 2010).

**Acknowledgments** This work was supported by NIH (U54NS083924) for AHM and MNG. JLRL received support from the NIGMS (8P20GM103475-12) and MBRS-RISE at UNE (2R25GM066250-05A1) from NIH.

## References

Anacker C, Zunszain PA, Cattaneo A, Carvalho LA, Garabedian MJ, Thuret S, Pariante CM (2011) Antidepressants increase human hippocampal neurogenesis by activating the glucocorticoid receptor. Mol Psychiatry 16:738–750

Andres RH, Horie N, Slikker W, Keren-Gill H, Zhan K, Sun G, Manley NC, Pereira MP, Sheikh LA, McMillan EL (2011) Human neural stem cells enhance structural plasticity and axonal transport in the ischaemic brain. Brain 134:1777–1789

Arai Y, Funatsu N, Numayama-Tsuruta K, Nomura T, Nakamura S, Osumi N (2005) Role of Fabp7, a downstream gene of Pax6, in the maintenance of neuroepithelial cells during early embryonic development of the rat cortex. J Neurosci 25:9752–9761

Arvidsson A, Collin T, Kirik D, Kokaia Z, Lindvall O (2002) Neuronal replacement from endogenous precursors in the adult brain after stroke. Nat Med 8:963–970

Barkho BZ, Zhao X (2011) Adult neural stem cells: response to stroke injury and potential for therapeutic applications. Curr Stem Cell Res Ther 6:327–338

Barone FC, Feuerstein GZ, Spera RP (1997) Calcium channel blockers in cerebral ischaemia. Expert Opin Investig Drugs 6:501–519

Brahmachari S, Fung YK, Pahan K (2006) Induction of glial fibrillary acidic protein expression in astrocytes by nitric oxide. J Neurosci 26:4930–4939

Brandt MD, Jessberger S, Steiner B, Kronenberg G, Reuter K, Bick-Sander A, Von Der Behens W, Kempermann G (2003) Transient calretinin expression defines early postmitotic step of neuronal differentiation in adult hippocampal neurogenesis of mice. Mol Cell Neurosci 24:603–613

Braun H, Schäfer K, Höllt V (2002) Beta III tubulin-expressing neurons reveal enhanced neurogenesis in hippocampal and cortical structures after a contusion trauma in rats. J Neurotrauma 19:975–983

Dhara SK, Hasneen K, Machacek DW, Boyd NL, Rao RR, Stice SL (2008) Human neural progenitor cells derived from embryonic stem cells in feeder-free cultures. Differentiation 76:454–464

Dirnagl U, Iadecola C, Moskowitz MA (1999) Pathobiology of ischaemic stroke: an integrated view. Trend Neurosci 22:391–397

Eckenhoff MF, Rakic P (1988) Nature and fate of proliferative cells in the hippocampal dentate gyrus during the life span of the rhesus monkey. J Neurosci 8:2729–2747

Einstein O, Ben-Hur T (2008) The changing face of neural stem cell therapy in neurologic diseases. Arch Neurol 65:452–456
Ellis P, Fagan BM, Magness ST, Hutton S, Taranova O, Hayashi S, McMahon A, Rao M, Pevny L (2004) SOX2, a persistent marker for multipotential neural stem cells derived from embryonic stem cells, the embryo or the adult. Dev Neurosci 26:148–165
Endo M, Doi R, Nishita M, Minami Y (2012) Ror family receptor tyrosine kinases regulate the maintenance of neural progenitor cells in the developing neocortex. J Cell Sci 125:2017–2029
Eriksson PS, Perfilieva E, Björk-Eriksson T, Alborn AM, Nordborg C, Peterson DA, Gage FH (1998) Neurogenesis in the adult human hippocampus. Nat Med 4:1313–1317
Gingras M, Champigny MF, Berthod F (2007) Differentiation of human adult skin-derived neuronal precursors into mature neurons. J Cell Physiol 210:498–506
Graham V, Khudyakov J, Ellis P, Pevny L (2003) SOX2 functions to maintain neural progenitor identity. Neuron 39:749–765
Hall PE, Lathia JD, Miller NG, Caldwell MA, Ffrench-Constant C (2006) Integrins are markers of human neural stem cells. Stem Cells 24:2078–2084
Hansen DV, Lui JH, Parker PRL, Kriegstein AR (2010) Neurogenic radial glia in the outer subventricular zone of human neocortex. Nature 464:554–561
Huang J, Upadhyay UM, Tamargo RJ (2006) Inflammation in stroke and focal cerebral ischemia. Surg Neurol 66:232–245
Iadecola C, Alexander M (2001) Cerebral ischemia and inflammation. Curr Opin Neurol 14:89–94
Itoh T, Satou T, Ishida H, Nishida S, Tsubaki M, Hashimoto S, Ito H (2009) The relationship between SDF-1alpha/CXCR4 and neural stem cells appearing in damaged area after traumatic brain injury in rats. Neurol Res 31:90–102
Kim M, Habiba A, Doherty JM, Mills JC, Mercer RW, Huettner JE (2009) Regulation of mouse embryonic stem cell neural differentiation by retinoic acid. Dev Biol 328:456–471
Kirik OV, Vlasov TD, Korzhevskii DE (2013) Neural stem cell markers nestin and musashi-1 in rat telencephalon cells after transient focal ischemia. Neurosci Behav Physiol 43:587–591
Liu Y, Namba T, Liu J, Suzuki R, Shioda S, Seki T (2010) Glial fibrillary acidic protein-expressing neural progenitors give rise to immature neurons via early intermediate progenitors expressing both glial fibrillary acidic protein and neuronal markers in the adult hippocampus. Neurosci.166:241–251
Lu QR, Yuk D, Alberta JA, Zhu Z, Pawlitzky I, Chan J, McMahon AP, Stiles CD, Rowitch DH (2000) Sonic hedgehog—regulated oligodendrocyte lineage genes encoding bHLH proteins in the mammalian central nervous system. Neuron 25:317–329
Marmur R, Mabie PC, Gokhan S, Song Q, Kessler JA, Mehler MF (1998) Isolation and developmental characterization of cerebral cortical multipotent progenitors. Dev Biol 204:577–591
Martins AH, Alves JM, Trujillo CA, Schwindt TT, Barnabe GF, Motta FLT, Guimaraes AO, Casarini DE, Mello LE, Pesquero JB, Ulrich H (2008) Kinin-B2 receptor expression and activity during differentiation of embryonic rat neurospheres. Cytometry 73:361–368
Martins AH, Hu J, Xu Z, Mu C, Alvarez P, Ford BD, El Sayed K, Eterovic VA, Ferchmin PA, Hao J (2015) Neuroprotective activity of (1S,2E,4R,6R,-7E,11E)-2,7,11-cembratriene-4,6-diol (4R) in vitro and in vivo in rodent models of brain ischemia. Neuroscience 291:250–259
Matigian N, Abrahamsen G, Sutharsan R, Cook AL, Vitale AM, Nouwens A, Bellette B, An J, Anderson M, Beckhouse AG, Bennebroek M, Cecil R, Chalk AM, Cochane J, Fan Y, Feron F, McCurdy R, McGrawth JJ, Murrell W, Perry C, Raju J, Ravishankar S, Silburn PA, Sutherland GT, Mahler S, Mellick GD, Wood SA, Sue CM, Wells CA, Mackay-Sim A (2010) Disease-specific, neurosphere-derived cells as models for brain disorders. Dis Model Mech 3:785–798
McCullumsmith RE, Hammond JH, Shan D, Meador-Woodruff JH (2014) Postmortem brain: an underutilized substrate for studying severe mental illness. Neuropsychopharmacology 39:65–87
Messam CA, Hou J, Major EO (2000) Coexpression of nestin in neural and glial cells in the developing human CNS defined by a human-specific anti-nestin antibody. Exp Neurol 161:585–596
Miralles VJ, Martínez-López I, Zaragozá R, Borrás E, García C, Pallardó FV, Viña JR (2001) Na+ dependent glutamate transporters (EAAT1, EAAT2, and EAAT3) in primary astrocyte cultures: effect of oxidative stress. Brain Res 922:21–29

Mitome M, Low HP, van den Pol A, Nunnari JJ, Wolf MK, Billings-Gagliardi S, Schwartz WJ (2001) Towards the reconstruction of central nervous system white matter using neural precursor cells. Brain 124:2147–2161

Nagato M, Heike T, Kato T, Yamanaka Y, Yoshimoto M, Shimazaki T, Okano H, Nakahata T (2005) Prospective characterization of neural stem cells by flow cytometry analysis using a combination of surface markers. J Neurosci Res 80:456–466

Nait-Oumesmar B, Picard-Riera N, Kerninon C, Decker L, Seilhean D, Hoglinger GU, Hirsch EC, Reynolds R, Baron-Van Evercooren A (2007) Activation of the subventricular zone in multiple sclerosis: evidence for early glial progenitors. Proc Natl Acad Sci U S A 104:4694–4699

Neuman T, Keen A, Zuber MX, Kristjansson GI, Gruss P, Nornes HO (1993) Neuronal expression of regulatory helix-loop-helix factor Id2 gene in mouse. Dev Biol 160:186–195

Ohnishi A, Sawa H, Tsuda M, Sawamura Y, Itoh T, Iwasaki Y, Nagashima K (2003) Expression of the oligodendroglial lineage-associated markers Olig1 and Olig2 in different types of human gliomas. J Neuropathol Exp Neurol 62:1052–1059

Park D, Xiang AP, Mao FF, Zhang L, Di CG, Liu XM, Shao Y, Ma BF, Lee JH, Ha KS (2010) Nestin is required for the proper self-renewal of neural stem cells. Stem Cells 28:2162–2171

Park HJ, Bronson RT, Frankel WN, Hong M, Israel MA, Yun K (2013) Elevated Id2 expression results in precocious neural stem cell depletion and abnormal brain development. Stem Cells 31:1010–1021

Pfenninger CV, Roschupkina T, Hertwig F, Kottwitz D, Englund E, Bengzon J, Jacobsen SE, Nuber UA (2007) CD133 is not present on neurogenic astrocytes in the adult subventricular zone, but on embryonic neural stem cells, ependymal cells, and glioblastoma cells. Cancer Res 67:5727–5736

Raponi E, Agenes F, Delphin C, Assard N, Baudier J, Legraverend C, Deloulme JC (2007) S100B expression defines a state in which GFAP-expressing cells lose their neural stem cell potential and acquire a more mature developmental stage. Glia 55:165–177

Rodriguez-Martinez G, Molina-Hernandez A, Velasco I (2012) Activin A promotes neuronal differentiation of cerebrocortical neural progenitor cells. PloS One 7:e43797

Seki T (2002) Expression patterns of immature neuronal markers PSA-NCAM, CRMP-4 and NeuroD in the hippocampus of young adult and aged rodents. J Neurosci Res 70:327–334

Son MJ, Woolard K, Nam DH, Lee J, Fine HA (2009) SSEA-1 is an enrichment marker for tumor-initiating cells in human glioblastoma. Cell Stem Cell 4:440–452

Sun Y, Kong W, Falk A, Hu J, Zhou L, Pollard S, Smith A (2009) CD133 (Prominin) negative human neural stem cells are clonogenic and tripotent. PloS One 4:e5498

Suzuki S, Namiki J, Shibata S, Mastuzaki Y, Okano H (2010) The neural stem/progenitor cell marker nestin is expressed in proliferative endothelial cells, but not in mature vasculature. J Histochem Cytochem 58:721–730

Terkelsen OB, Stagaard Janas M, Bock E, Møllgård K (1992) NCAM as a differentiation marker of postmigratory immature neurons in the developing human nervous system. Int J Dev Neurosci 10:505–516

Tohyama T, Lee VM, Rorke LB, Marvin M, McKay RD, Trojanowski JQ (1992) Nestin expression in embryonic human neuroepithelium and in human neuroepithelial tumor cells. Lab Invest 66:303–313

Trujillo CA, Schwindt TT, Martins AH, Alves JM, Mello LE, Ulrich H (2009) Novel perspectives of neural stem cell differentiation: from neurotransmitters to therapeutics. Cytometry A 75:38–53

Uchida N, Buck DW, He D, Reitsma MJ, Masek M, Phan TV, Weissman IL (2000) Direct isolation of human central nervous system stem cells. Proc Natl Acad Sci 97:14720–14726

Venere M, Han YG, Bell R, Song JS, Alvarez-Buylla A, Blelloch R (2012) Sox1 marks an activated neural stem/progenitor cell in the hippocampus. Development 139:3938–3949

Vescovi AL, Parati EA, Gritti A, Poulin P, Ferrario M, Wanke E, Galli R (1999) Isolation and cloning of multipotential stem cells from the embryonic human CNS and establishment of transplantable human neural stem cell lines by epigenetic stimulation. Exp Neurol 156:71–83

Xu Z, Jiang J, Ford G, Ford BD (2004) Neuregulin-1 is neuroprotective and attenuates inflammatory responses induced by ischemic stroke. Biochem Biophys Res Commun 322:440–446

Young MJ, Ray J, Whiteley SJO, Klassen H, Gage FH (2000) Neuronal differentiation and morphological integration of hippocampal progenitor cells transplanted to the retina of immature and mature dystrophic rats. Mol Cell Neurosci 16:197–205
Yun SW, Leong C, Zhai D, Tan YL, Lim L, Bi X, Lee JJ, Kim HJ, Kang NY, Ng SH, Stanton LW, Chang YT (2012) Neural stem cell specific fluorescent chemical probe binding to FABP7. Proc Natl Acad Sci 109:10214–10217
Zhang RL, Zhang ZG, Zhang L, Chopp M (2001) Proliferation and differentiation of progenitor cells in the cortex and the subventricular zone in the adult rat after focal cerebral ischemia. Neuroscience 105:33–41

# Chapter 5
# Mice Post-natal Subventricular Zone Neurospheres: Derivation, Culture, Differentiation and Applications

**Laura Sardá-Arroyo, Clarissa Schitine, Sara Alves Xapelli, and Henning Ulrich**

**Abstract** Neurogenesis is a highly regulated process, by which the brain is able to produce new neural cells mainly under normal conditions or after injury. Neurogenesis is restricted to specific areas in the brain, known as niches, where neural stem cells (NSCs) are generated. Two main regions have been described as neurogenic in mammalian brain, the subventricular zone (SVZ) in the anterolateral ventricles, and the subgranular zone (SGZ) of the dentate gyrus in the hippocampus. Under appropriate microenviromental circumstances, NSCs divide and give rise to neural progenitor cells (NPCs) that will generate mature neurons or glia. These precursors can be obtained from rodent animal models to produce neurospheres as an *in vitro* model for neurogenesis. Neurospheres are aggregates of NPC that grow in suspension and are capable of differentiating into every lineage of neural cells, including neurons and glia. This capacity shows the high potential of neurospheres as an *in vitro* model to study neurogenesis and alternatives methodologies to investigate neurodegenerative disorders. Here, we present an easy and highly reproducible protocol for working with SVZ neurospheres from post-natal mice.

**Keywords** Neurospheres • Adult neurogenesis • SVZ • Newborn • Post-natal

---

L. Sardá-Arroyo • H. Ulrich (✉)
Department of Biochemistry, Institute of Chemistry, University of São Paulo, Av. Prof. Lineu Prestes 748, São Paulo, SP 05508-000, Brazil
e-mail: laurasarda@usp.br; henning@iq.usp.br

C. Schitine
Institute of Biophysics Carlos Chagas Filho, Federal University of Rio de Janeiro, Rio de Janeiro, RJ 21941-902, Brazil
e-mail: cschitine@gmail.com

S.A. Xapelli
Instituto de Farmacologia e Neurociências and Instituto de Medicina Molecular, Faculdade de Medicina da Universidade de Lisboa, Av. Prof. Egas Moniz, Lisbon 1649-028, Portugal
e-mail: sxapelli@gmail.com



H. Ulrich, P. Davidson Negraes, *Working with Stem Cells*,
DOI 10.1007/978-3-319-30582-0_5

## 5.1 Introduction

### *5.1.1 History of Post-natal Neurospheres*

The generation of new neurons in the mammalian central nervous system (CNS) was throught, by the majority of neuroscientists until the 1990s, to occur only during development. The great specialization of the nervous system was not associated with the possibility of brain plasticity. This concept followed the dogma of the immutable brain, first postulated by the prominent histologist Santiago Ramon y Cajal. He was awarded the Nobel Prize for Medicine and Psychology in 1906 together with Camillo Golgi, given for their brilliant discoveries and detainments about the CNS. Cajal established the neuronal doctrine postulating that "once the development was ended, the founts of growth and regeneration of the axons and dendrites dried up irrevocably. In the adult centers, the nerve paths are something fixed, ended, and immutable. Everything may die, nothing may be regenerated. It is for the science of the future to change, if possible, this harsh decree." Similarly, Giulio Bizzozero, the most prestigious Italian histologist and mentor of Camillo Golgi, named the tissues of the human body as "labile, stable and perennial". Among these tissues, nerve cells were throught unable to proliferate and generate new cells in the post-natal brain.

The incapacity of the brain to generate new cells was perpetuated mainly by four concepts at that time (Colucci-D'Amato et al. 2006). First, neurological pathologies such as Parkinson's disease, Alzheimer's disease, brain traumas or cerebrovascular diseases present little functional recovery, and in the majority of cases, they are neurodegenerative, in agreement with the absence of neurogenesis. Second, the neural functions are so specialized, that generation of new neurons could alter the existing circuitry. The third reason was based on learning and memories processes. The idea at that time was that memories could be retrieved from stable neural circuits; therefore, the formation of new neural cells would be incompatible with new memories. Today it is well known that learning and memory are events associated with neurogenesis in the dentate gyrus of the hippocampus (Gould et al. 1999). The fourth reason was the lack of appropriate techniques and tools, such as autoradiography, high-resolution microscopy and molecular markers for different cell types and cell cycle phases.

This scenario began to change in the sixties, with Joseph Altman's work, showing the presence of dividing cells in the hippocampus. He performed lesions in adult rats combined with $^{3}$H-thymidine injections in brain lesions and found many labeled nuclei of numerous glia cells in brain regions associated with the traumatized areas. In addition, some neurons and neuroblasts showed labeling, suggesting the possibility of proliferation of neurons (Altman 1962). However, the lack of specific neuronal markers did not allow the demonstration of new neurons in the adult dentate gyrus (Colucci-D'Amato et al. 2006).

In order to confirm the genesis of new neuronal cells in the adult brain, Michel Kaplan in the seventies reproduced Altman's experiments coupling the autoradiography technique to electron microscopy (Kaplan and Hinds 1977). Indeed, the work demonstrated that neurons were labeled with $^{3}$H-thymidine, showing the genesis of new neuronal cells in the adult rat olfactory bulb and dentate gyrus. Nevertheless, the scientific community remained skeptical about the new paradigm, sustaining that neuroblasts, by definition, were incapable of cell division (Kaplan 2001). In contrast, new studies kept breaking the immutable brain dogma. In a series of elegant experiments, starting in the early 1980s, Nottebohm and his colleagues showed that, indeed, thousands of new neurons were added every day in the brain of songbirds, which marked a turning point showing that there is neurogenesis in the adult brain (reviewed by Nottebohm 2002). The pioneering work of Nottebohm and collaborators showed the presence of seasonal neurogenesis in adult songbirds and the integration of the newborn neurons in the pre-existing neural circuits that control song. These effects depend on social changes, experience and learning, associating for the first time neurogenesis with physiological functional processes (Nottebohm 2002). Still, the evidence of neurogenesis was viewed as an exotic specialization related to their seasonal cycles of singing birds.

During the early 1990s, several achievements finally established the concept of neurogenesis. A very important development was the introduction of the synthetic thymidine analogue BrdU (5-bromo-3′-deoxyuridine), which can be incorporated into the newly synthesized DNA, current concept of neurogenic niches and dividing progenitors. This technique allows stereological estimation of the number of proliferating cells as well as it allows co-immunostaining against cell-specific marker proteins (Gross 2000). At the same time, different laboratories demonstrated that neural stem cells (NSCs) could be isolated from adult mammalian brain and expanded *in vitro*, using specific cocktails of growth factors and morphogens.

The identification of neurospheres, achieved by Reynolds and Weiss (1992), opened a new avenue in neural stem cell research and their manipulation *in vitro*. A neural stem cell is, by definition, able to self-renew and to give rise to all three neural populations of the nervous system: neurons, astrocytes and oligodendrocytes. In fact, NSCs can divide in three different ways: symmetrically, originating two new neural stem cells (expansion, symmetric self-renewal); asymmetrically, originating one neural stem cell and one differentiated cell (maintenance, asymmetrical self-renewal); or also symmetrically, originating two differentiated cells (extinction, symmetrical commitment) (Dirks 2008). The discovery of neural stem cells (NSCs) and their ability to grow *in vitro* has provided a tool to study and identify factors and molecular events, such as signaling pathways that underlay neural differentiation and the generation of specified neuronal phenotypes.

Although studies of adult neurogenesis encountered strong opposition in the scientific community in the past, nowadays this field is object of extensive research, searching for novel strategies of treatment of a growing number of neurological diseases.

### 5.1.2 Adult Neurogenesis in the SVZ and SGZ

Adult neurogenesis is a dynamic, finely tuned process, subject to modulation by various physiological, pathological, pharmacological and environmental conditions. Moreover, adult NSCs/neural progenitor cells (NPCs) are also regulated by many genes and signaling pathways (Kriegstein and Alvarez-Buylla 2009).

Until recently, active adult neurogenesis was spatially restricted to two specific "neurogenic" brain regions, the SGZ in the dentate gyrus of the hippocampus, where new dentate granular cells are generated; and the SVZ of the anterolateral ventricles, where generated neuroblasts migrate through the rostral migratory stream (RMS) to the olfactory bulb to become mature interneurons (Gage 2000; Ming and Song 2011). However, during the last decade, many works have provided evidences suggesting that adult neurogenesis is not only restricted to these two regions but also happens in niches located along the extent of the cerebral ventricular system (Bennett et al. 2010).

NSCs in the brain resides in special places called niches where cell-cell interactions and microenvironmental signals interact in the regulation of self-renewal and/or neural differentiation. The largest population of NSCs is present within an epithelium called the ventricular-SVZ (V-SVZ) in the walls of the lateral ventricle (Lim and Alvarez-Buylla 2014). The primary precursors in the SVZ correspond to type B cells, a subpopulation of slowly dividing astroglial cells, adjacent to a layer of ependymal cells (E cells) that separates cerebrospinal fluid (CSF) from the brain epithelia (Doetsch et al. 1999). In proper environmental conditions, type B cells can divide and produce type C cells, a type of transit-amplifying cell that divides rapidly to produce neuroblasts or type A cells (Lim and Alvarez-Buylla 2014) (Fig. 5.1). Initially, it was throught that cells in this lineage were separated from the ventricle by a layer of ependymal cells, however, several studies have been suggesting that most, if not all, B cells actually contact the ventricle through small, specialized apical processes (Mirzadeh et al. 2008; Lim and Alvarez-Buylla 2014).

NPCs express many radial and other glial markers, including BLPB (brain lipid-binding protein), GLAST (glutamate aspartate transporter) and GFAP (Glial fibrillary acidic protein) (Lim and Alvarez-Buylla 2014). Dividing B cells give rise to C cells, which express Dlx2 (Distal-Less Homeobox 2) instead of GFAP. Neuroblasts or A cells, which will originate mature neurons, initially express PSA-NCAM (polysialylated neural cell adhesion molecule) and other neuronal markers such as Dcx (doublecortin) (Alvarez-Buylla and Garcia-Verdugo 2002). These neuroblasts migrate in chains through the RMS and finally get integrated into neural circuits in the olfactory bulb as interneurons, where they express specific mature markers such as NeuN (neuronal nuclei) (Alvarez-Buylla and Garcia-Verdugo 2002; Doetsch 2003; Mignone et al. 2004; Brazel et al. 2005; Merkle et al. 2007; Kriegstein and Alvarez-Buylla 2009). Neurogenesis in this niche is essential for olfactory discrimination/memory (Imayoshi et al. 2008).

As previously mentioned, the SGZ is located in the dentate gyrus of the hippocampus. This niche is comprised by cells similar to the ones present in the SVZ, but they are structured differently because of their different location. The SGZ does not

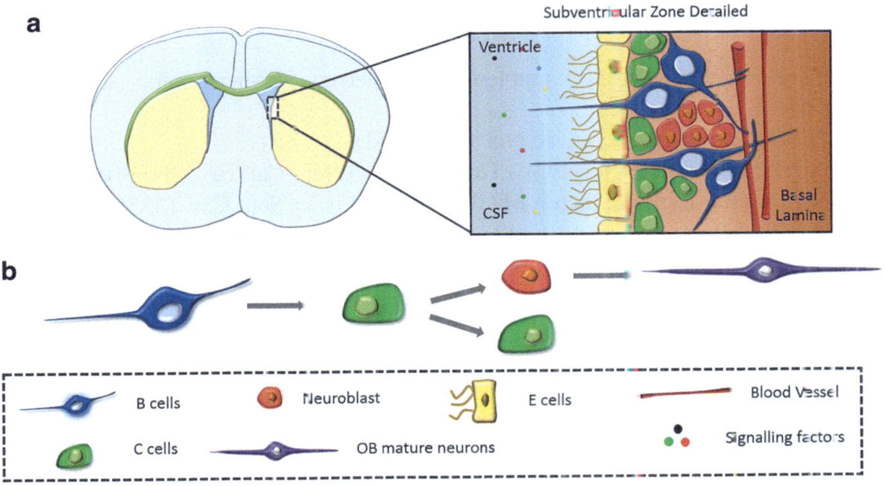


**Fig. 5.1** Scheme of the SVZ neurogenic niche. The figure illustrates: (**a**) Coronal mouse brain section showing the lateral ventricles and the SVZ; (**b**) Cells involved in SVZ neurogenesis. *CSF* cerebrospinal fluid, *OB* olfactory bulb

have any E cells, because the niche is not in contact with any brain ventricle. However, the hippocampal sulcus could serve as an access to CSF factors, which are essential for maintenance and dynamics of the niche. SGZ stem cells are called radial astrocytes or type 1 cells (Seri et al. 2001; Kriegstein and Alvarez-Buylla 2009; Bonaguidi et al. 2012; Fuentealba et al. 2012). When specific signaling pathways are triggered, they can divide and give rise to intermediate progenitors called type-2a and -2b cells, which express specific neural markers, such as Dcx or Ascl-1 (Achaete-scute homolog 1). Intermediate progenitors migrate to the dentate granule layer and get integrated into hipoccampal circuits as mature neurons (Basak and Taylor 2009; Lim and Alvarez-Buylla 2014). Therefore, this event has essential functions in memory formation and consolidation.

## 5.2 Protocol

### 5.2.1 Material and Methods

#### 5.2.1.1 Animals

The present protocol can be applied to C57BL/6J newborn mice between postnatal day 1 (P1) and day 3 (P3), regardless of sex.

### 5.2.1.2 Reagents

- DMEM/F-12/GlutaMAX supplement (Thermo Fisher Sci., Cat. No. 10565-018).
- B27 (Thermo Fisher Sci., Cat. No. 17504-044).
- Epidermal Growth Factor, EGF (Thermo Fisher Sci., Cat. No.53003-018).
- Basic fibroblast growth factor, bFGF (Thermo Fisher Sci., Cat. 13256-029).
- Penicillin/Streptomycin (10,000 U/ml) (Thermo Fisher Sci., Cat. No. 15140-122).
- Hank's Balanced Salt Solution, HBSS (Thermo Fisher Sci., Cat. No. 14175-053).
- Trypsin/EDTA 0.05 % (Thermo Fisher Sci., Cat. No. 25300-054).
- Fetal Bovine Serum, FBS (Thermo Fisher Sci., Cat. No. 10270-106).
- Poly-D-Lysine hydrobromide (Sigma Aldrich. Cat. No. P7886).
- Dimethyl sulfoxide, DMSO (Sigma-Aldrich, Cat.No. D2650).
- Sterile autoclaved MilliQ water.
- Sterile phosphate buffered saline PBS (1×), pH 7.4 (Thermo Fisher Sci., Cat. No. 10010-023).
- Paraformaldehyde (Sigma-Aldrich, Cat. No. P6148).
- Triton X-100 (Thermo Fisher Sci., Cat. No. 85111).
- Bovine serum albumin, BSA (Sigma Aldrich, Cat. No. A9647).

*Antibodies*

- DCX (Cell Signalling, Cat. No. 4604).
- Nestin (Millipore, Cat. No. AB6142).
- Nestin Millipore, Cat. No. AB7659).
- SOX2 (SantaCruz, Cat. No. SC17320).
- Beta-III Tubulin (Cell Signalling, Cat. No. 4466).
- GFAP (Cell Signalling, Cat. No. 3670).
- Olig2 (Millipore, Cat. No. AB9610).
- O4 (Millipore, Cat. No. MAB345).
- MAP2 (Sigma-Aldrich, Cat. No. M4430).

### 5.2.1.3 Materials and Equipment

#### Brain Dissection

- McIlwain Tissue chopper (TED PELLA INC., Cat. No. U0800).
- Stereomicroscope (Zeiss, Cat. No. SteREO Discovery.V8).

#### Microdissecting Instruments

- Tweezer (Fine Science Tools, Cat. No. 11253-20).
- Scissors (Fine Science Tools, Cat. No. 14068-12).
- Forceps (Fine Sciente Tools, Cat. No. 11253-20).
- Forceps (Fine Science Tools, Cat. No. 11271-30).

- Iris Spatula (Fine Science Tools, Cat. No 10093-13).

**Others**

- Water bath set at 37 °C.
- Laminar flow hood.
- $CO_2$ Incubator (set at 37 °C, 5%$CO_2$ and 95 % air).
- Fume hood.
- Table top centrifuge.
- Hemocytometer.
- CellCooler SV2 (Biocision).
- 10, 200 and 1000 µl automatic pipetters.
- 5, 10 and 50 ml pipettes.
- 1.5 ml tubes, 15 ml and 50 ml polypropylene conical tubes.
- 60 mm Petri dishes (Corning, Cat. No. 430166).
- 24-Well tissue culture plates.
- 12 mm Glass Coverslips (Marienfeld, Cat. No. 631-0666).
- Parafilm (Parafilm, Cat. No. PM999).
- 0.2 µm filter (Corning Life Sciences, Cat. No. CLS431218-50EA).
- Cryotubes.

**Media**

- *Dissection media (DM):* HBSS supplemented with 100 U/ml penicillin and 100 µg/ml streptomycin.
- *Neurosphere generation media (NGM):* DMEMF12/Glutamax supplemented with 100 U/ml penicillin and 100 µg/ml streptomycin, 10 ng/ml EGF, 5 ng/ml bFGF and 1 % B27.
- *Neurosphere plating and differentiation media (NPDM)*: DMEM F12/Glutamax supplemented with 100 U/ml penicillin/streptomycin and 1 % B27.
- *Cryopreservation media (CRM)*: DMSO supplemented with 5 % FBS.

## *5.2.2 Procedure*

### 5.2.2.1 NPC Extraction

Good manual practices for animal handling are needed to develop this protocol. International guidelines for Euthanasia of Rodent Neonates (newborns) must be followed strictly to avoid animal suffering.

Preparing the Material and Equipment

Prepare all the required materials before starting the protocol. Extraction and subsequent steps of this protocol must be developed under a sterile atmosphere in a laminar flow hood.

1. Dispose all surgery instruments, tissue chopper and pipettes under ultraviolet (UV) light under a laminar flow hood for at least 15 min to sterilize (Fig. **5.2**a);
2. Warm DM in a 37 °C bath;

Preparing Animals

3. Carefully separate postnatal mice between P1-P3 from the mother;
4. Keep littermates together in a clean container with paper or in any other material that will maintain animal bodies' temperature around 22 ± 2 °C;
5. Removing the offspring of mothers is a stressful event, especially as it inhibits nursing, so keeping littermates together will help to avoid this stress;

Decapitate newborn mice under a laminar flow hood

6. Spray 70 % ethanol over newborn mouse neck;

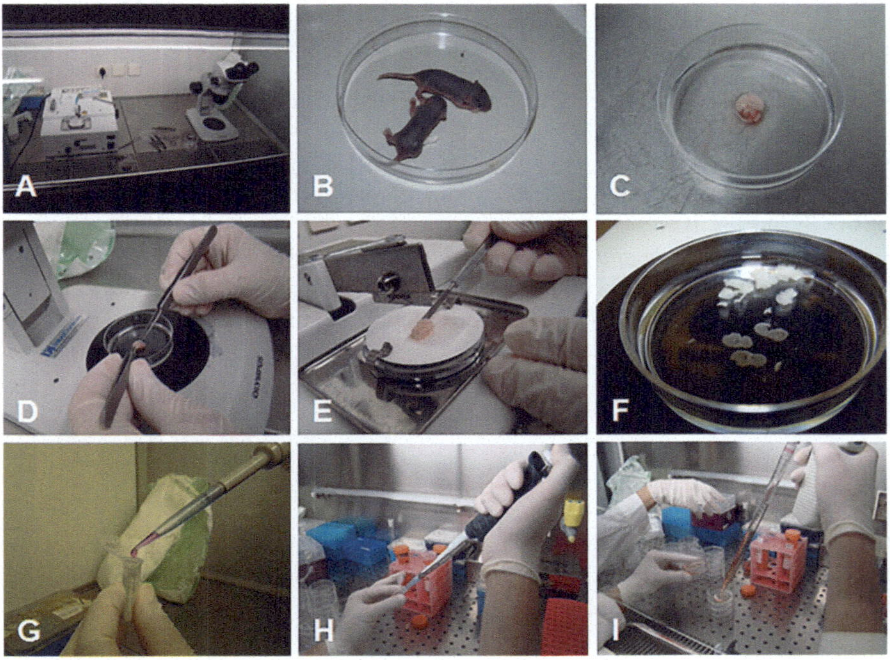


**Fig. 5.2** Details of SVZ dissection procedure. (**a**) Dissection materials inside a culture hood; (**b**) Post-natal day 3 C57BL/6J mice; (**c**) Dissected mouse brain in *DM*; (**d**) Procedure of removing olfactory bulbs and meningeal tissue from brain; (**e**) Brain disposal on the tissue chopper dispositive for slicing; (**f**) Brain slices comprising the SVZ region; (**g**) Digestion of SVZ tissue with 0.5 % tripsin/EDTA; (**h**) Manual digestion of SVZ tissue; (**i**) SVZ cells plating in 60 mm untreated Petri dishes

7. Cut the head with sterilized scissors;
8. Transfer the head to a 60 mm Petri dish;
9. Remove the overhead skin with scissors in a posterior-anterior direction;
10. Set aside the skin to make the brain visible;

### Removing Brain from Skull

11. Introduce a spatula under the cerebellum and cut the cranial nerves;
12. Transfer the brain to a new 100 mm petri dish with fresh DM (Fig. 5.2b);
13. Repeat steps C and D with the other animals;

*Optional: Scissors and spatula should be cleaned with 70 % ethanol between newborns to avoid contamination.*

### Brain Preparation and Slicing

14. Carefully remove the olfactory bulbs and meninges with sharp tweezers (Fig. 5.2c);

    ***Note:** Residues of meningeal tissues will interfere with brain slicing.*
15. Adjust the tissue chopper device to obtain coronal brain sections of 450 μm thickness (Fig. 5.2d);
16. Align brains in the tissue chopper. Turn the instrument on for slicing;
17. Transfer whole sliced brains to a new 60 or 100 mm plate with fresh 5 or 10 ml DM respectively;

### NPC Extraction

18. Put the plate under the stereoscope and very carefully separate slices with tweezers, for not damaging the tissue;
19. Select those slices that comprise the lateral ventricles (Figs 5.1 and 5.2e, f) and discard the others;
20. Cut the tissue belonging to the SVZ using forceps and with a P200 automatic pipetter aspirate the piece of tissue and put it into a 1.5 ml tube with 1 ml of DM;
21. Repeat the same approach as in step 20 with the other ventricle slices;

***Note:** If you are using knockout mice, after genotyping you can put wild-type samples together and do the same with the knockout ones.*

### NPC Processing

22. Add to each 1.5 ml tubes, 2–3 drops of 0.05 % Trypsin/EDTA. Carefully mix solutions and incubate them for 10–15 min at 37 °C (Fig. 5.2g);

23. Following incubation cells will be at the bottom of the tube. Aspirate the supernatant without disturbing the cells;
24. Wash the cells twice with 1000 μl of DM, discarding the media between washes;
25. Digest the cells mechanically by pipetting softly up and down with a P1000 automatic pipetter (Fig. 5.2h).

#### 5.2.2.2 Neurosphere Generation and Culturing

1. Resuspend the cells in 1 ml of prewarmed NGM and plate them in a confluence of 3000 cells/cm$^2$ in petri dishes non-treated for adhesion.

   ***Note:*** *We recommend to plate cells in 60 mm diameter Petri dishes (surface area = 20 cm$^2$).*
2. Incubate the dishes for 6 days at 37 °C, 5 % $CO_2$ (optimal conditions for neurosphere generation).

#### 5.2.2.3 Neurosphere Passaging

Neurospheres can be passaged for at least ten times. More cycles of passaging will increase genetic instability.

1. After 6–7 days of culture in suspension, collect the media with primary neurospheres and transfer the volume to a 15 ml conical tube;
2. Centrifuge at 200 × *g*, 5 min at room temperature.
3. Carefully discard the media and add 1 ml of 0.5 % trypsin/EDTA. Incubate at 37 °C, 3–4 min;
4. Dissociate the spheres with a P1000 pipette;
5. Block trypsin activity with 3 ml of *NGM*;
6. Pellet the cells at 200 × *g* for 5 min, room temperature;
7. Count the number of viable cells with 0.4 % Trypan Blue in a hemocytometer chamber;
8. Plate cells at a 3,000 cells/cm$^2$ in uncoated petri dishes. Incubate at 37 °C, 5 % $CO_2$.

#### 5.2.2.4 Neurosphere Differentiation

Direct Differentiation

1. Neurospheres are obtained after 6 days in culture.
2. Neurospheres must be spherical and of similar size;
3. Put the dish under the laminar flow hood and swirl the plate carefully with circular movements to concentrate the cells in the center of the plate;

   ***Note:*** *A black surface will help to see the neurospheres during collection.*
4. Plate approximately 200 neurospheres in a 6-well plate, previously treated with poly-D-lysine (see *Supplementary Information*). For 24-well plates use glass coverslips treated with poly-D-lysine and plate 50 neurospheres per well;

Gently extend the volume of cells in the wells with culture medium and incubate them for 15 min at 37 °C.

***Note:*** *Incubation times over 15 min will result in drying of the solution.*

5. Carefully add 2 ml or 500 µl of NPDM to each well of the 6- or 24-well plate, respectively;

   ***Note:*** *in this step, cells easily detach if the media is added very fast.*
6. Incubate the cells at 37 °C, 5 % $CO_2$;
7. After 2 days, a pseudomonolayer of cells will be obtained comprising undifferentiated and neural cells (neurons, astrocytes and oliogodendrocytes at different stages of differentiation);
8. Remove culture media and add new NPDM supplemented with the compounds, whose effects on differentiation and phenotype determination shall be tested.
9. Incubate the cells at 37 °C, 5 % $CO_2$. (for a further time period dependent on the analysis, e.g.: 2 days for detecting marker proteins for NPCs or 7 days for studying the progress of neural differentiation). The cells can be maintained in culture for at least 3 weeks.

Specific Differentiation

Specific protocols must be applied in order to obtain a population enriched in a desired type of cells, such as for example motor neurons or dopamine neurons. Optimization of these protocols is required to adapt them from the one we present here (Morizane et al. 2006; Wu et al. 2012; Garcia-Bennett et al. 2014).

#### 5.2.2.5 Cryopreservation of Neurospheres

Neurospheres can be freeze after 6 days in culture;

1. Swirl the dish to collect the cells with circular movements;
2. Transfer the cells to a new 15 ml tube and centrifuge at 800 × *g*, 5 min;
3. Aspirate the supernatant and add 1 ml of CRM;
4. Slowly mix the pellet (do not vortex);
5. Transfer the media with cells into cryotubes;
6. Put the tubes into a freezing disposal at −80 °C, 48–72 h. Later, transfer them into liquid nitrogen.

### *5.2.3 Characterization of the Cells Obtained Using the Described Protocol*

#### 5.2.3.1 Immunocytochemistry

Immunocytochemistry should be performed in cells plated in glass coverslips.

Cells can be fixed on day 8–9 for immunostaining for marker proteins of immature cells, or on day 15 to detect neuronal or glial mature marker proteins (Fig. 5.3).

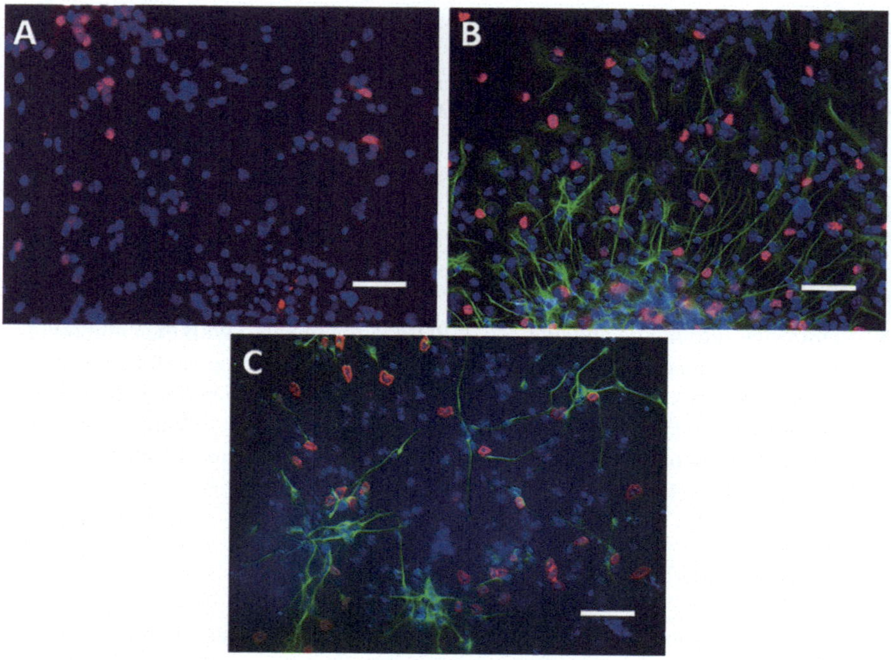


**Fig. 5.3** Immunofluorescence images of SVZ cells expressing neural markers. Representative images of (**a**) Neuronal nuclei (NeuN)-positive cells (*red*), a marker of mature neurons, and Hoechst staining (*blue* nuclei); (**b**) GFAP-positive astrocytes (*green*), oligodendrocyte transcription factor 2 (olig2)-positive cells (*red*) and Hoechst staining (*blue* nuclei); (**c**) Beta-III tubulin-positive immature neurons (*green*), bromodeoxyuridine (BdrU)-positive proliferative cells (*red*) and Hoechst staining (*blue* nuclei). Scale bar: 50 μm

## Fixation

1. Remove culture media and wash once with PBS (1×);
2. Discard PBS and add 4 % paraformaldehyde to cover the entire surface of the culture well;
3. Incubate at room temperature for 15–30 min;
4. Discard paraformaldehyde in a proper container and wash two times with PBS (1×);
5. Fixed cells can be store at 4 °C in PBS (1×). We recommend wrapping the culture vessels with parafilm before stocking;

### Permeabilization and Blocking

6. Remove the media and wash three times with PBS (1×) for 5 min each;
7. Discard PBS and incubate 1 h 30 min in blocking solution (0.5 % Triton X-100, 3 % BSA) (see *Supplementary Information*);

### Incubation

8. Without washing, incubate the antibody in incubation solution at 4 °C overnight (see Table 5.1);

**Table 5.1** Antibodies for neural characterization

| | | Antibody | Company | Cat. No. | Dilution |
|---|---|---|---|---|---|
| Immature markers | NPCs markers | DCX | Cell signalling | 4604 | 1:200 |
| | | Nestin | Millipore | AB6142 | 1:200 |
| | | | Millipore | AB7659 | 1:200 |
| | | SOX2 | Santa cruz | SC17320 | 1:100 |
| | | Beta-III tubulin | Cell signaling | 4466 | 1:200 |
| | | GFAP | Cell signalling | 3670 | 1:200 |
| | | Olig2 | Millipore | AB9610 | 1:200 |
| Mature markers | Glial | GFAP | Cell signalling | 3670 | 1:200 |
| | | O4 | Millipore | MAB345 | 1:100 |
| | Neural | MAP2 | Sigma-Aldrich | M4403 | 1:200 |

*Note 9: Dilution of each antibody must be adjusted to the protocol.*

9. On the next day, wash three times with PBS (1×) for 5 min each;
10. Incubate secondary antibody in PBS (1×) for 1 h at room temperature (see ***Note 9***);
11. Wash three times with PBS (1×) for 5 min each;
12. Incubate with Hoechst 33342 exclude (6 mg/ml in PBS) for nuclei staining;
13. Cover slides with fluorescence mounting medium.

## 5.3 Neurosphere Applications

NSCs offer a unique and powerful tool for basic research and regenerative medicine. The isolation of NSCs from their natural niches and their expansion in culture have been challenging issues, because the requirements to maintain these cells in their physiological state are yet poorly understood. As discussed in the previous section, isolation and expansion of the putative NSCs from the adult murine brain was first described by Reynolds and Weiss in 1992 using a chemically defined serum-free culture system known as the Neurospheres assay (NSA) (Reynolds and Weiss 1992; Azari et al. 2010). In this assay, the majority of differentiated cell types die within a few days of culture but a small population of growth factor responsive precursor cells undergo active proliferation in the presence of epidermal growth factor (EGF) and/or basic fibroblastic growth factor (bFGF) (Azari et al. 2010). These cells form colonies of undifferentiated cells, the so-called neurospheres, which in turn can expand the pool of neural stem cells (NSCs) and differentiate, generating the three major cell types of the CNS (neurons, astrocytes and oligodendrocytes). Since then, many studies reported that NSCs could be isolated from various regions of rodent (mouse and rat) and human brain at several developmental stages, as well as from germinative areas of the adult brain.

The neurosphere assay (NSA) provides a powerful tool to supply a consistent, renewable source of undifferentiated CNS precursors, which can be used for *in vitro* studies of neurogenesis and neural development and for therapeutic purposes. Most studies on the genetic and molecular control of regional specification of neural precursors have been performed *in vivo*. Complimentary *in vitro* approaches are useful to determine the degree of intrinsic specification present in neural precursors at various developmental time points, as well as to study the full potential of the cells when removed from extrinsic cues provided by their normal environment (Jensen and Parmar 2006). Neurospheres are a good system for such studies, because they are maintained under defined serum-free conditions, where the environmental cues are limited to surrounding cells. Additionally, it is easy to manipulate the extrinsic cues, to which the cells are exposed during their development by the changing culture environment during either the expansion or the differentiation phase. This can be done by simply adding precise and variable amounts of factors of interest to the media or by culturing neurospheres together with other types of cells. This method is also very useful to test different drugs and compounds modulating neurogenesis events such as proliferation, migration and differentiation. Another advantage is that the intrinsic properties of the neurosphere-expanded cells can be easily and efficiently modulated by viral transduction using retroviral vectors.

Although NSA is the most common method to isolate and expand NSCs, similar to all technologies, it also presents its limitations (Azari et al. 2011a, b; Reynolds and Rietze 2005). Given that both stem and progenitor cells are capable of generating neurospheres, and that the number of progenitor cells far outweigh the number of stem cells in the SVZ, the NSA cannot accurately measure NSC frequency in a mixed population of NPCs (stem and progenitors), once there is not a one-to-one relationship between the number of neurospheres and the number of stem cells (Azari et al. 2011a). Neurosphere assay can be used to determine the existence of NSCs when combined with extensive serial subculturing experiments. However, this is not useful as a measure of stem cell frequency. As further complication of the analysis neurospheres aggregate in high density cultures (Ahmed 2009). This phenomenon has been observed most directly by differential labelling of cells, using actin promoter driven EYFP and dsRed, followed by neurosphere color analysis (Coles-Takabe et al. 2008, Ahmed 2009). The lack of clonality of bulk cultures prevents the use of the method for modelling purposes as well as for multipotency assays, unless using low-density cultures or immobilized neurosphere assays to determine self-renewal and multipotency.

However, the neurosphere assay is a very powerful methodology as it allows the identification of NSCs. Its conceptual simplicity and robustness turns the neurosphere technique into a starting point for studying molecular mechanisms of neurodevelopment, as well as or activity screening of numerous compounds for developmental and adult brain diseases.

## 5.4 Supplementary Information

<u>*EGF Stock Solution*</u>

- For reconstitution of lyophilized EGF dilute the product in sterile autoclaved Milli-Q water to reach a final concentration of 1 mg/ml;

- Under a laminar flow hood, filter the solution with a 0.22 μm filter;
- Aliquot and store in polypropylene vials at −5 to −20 °C;
- Store for 2 months.

*Alternative*: In order to increase the stability of the aliquots, the product can be diluted in a solution of 0.1 % BSA in sterile autoclaved Milli-Q water (0.1 mg BSA in 100 ml Milli-Q water). EGF should not be diluted to concentrations lower than 10 μg/ml. These solutions can be stored at −20 °C for more than 2 months.

*bFGF stock solution*

bFGF must be reconstituted with a solution of 10 mM Tris, pH 7.6

- *To prepare 50 ml of 10 mM Tris, pH 7.6*: weigh 60.57 mg of Tris ($(HOCH_2)_3CNH_2$), and dilute it in 40 ml of sterile Milli-Q water. Adjust the pH to 7.6 and them complete to 50 ml with sterile Milli-Q water;
- *To prepare 10 ml of 0.1 % BSA in 10 mM Tris pH 7.6*: weigh 10 mg of BSA and dilute it in 10 ml of 10 mM Tris, pH 7.6;
- Reconstitute the product with 1000 μl of 0.1 % BSA in 10 mM Tris pH 7.6 for 10 μg of FGF (final concentration of 10 μg/ml);
- Filter the solution with 0.22 μm filter under a laminar flow hood;
- Aliquot the solution into polypropylene tubes and keep them at −20 °C for a maximum of 6 months.

*Poly-D-Lysine hydrobromide stock solution*

- *0.1 M Borate Buffer, pH 8.2:* To prepare 500 ml of 0.1 M borate buffer solution, dissolve 3.09 g of boric acid ($H_3BO_3$) in sterile autoclaved MilliQ water. Adjust the pH to 8.2 with NaOH. Sterilize by filtering through a 0.22 μm filter and store it for several months at 4 °C;
- Reconstitute the vial of poly-L-Lysine with 0.1 M borate buffer, pH 8.2, to reach a final concentration of 1 mg/ml. Aliquot and store at −20 °C.

*Poly-D-Lysine coating plates procedure*

- Under a laminar flow hood, open the poly-D-lysine stock solution. Dilute it in 166.6 mM borate buffer (pH 8.2) to obtain a working concentration of 0.1 mg/ml;
- Add the optimal amount of the working solution to the culture (14 ml for T75 or 7 ml for T25 cell culture flasks, 3.5 ml for 60 mm dishes, 2 ml for 35 mm dishes, or 500 μl for 24-well plates);
- Incubate them for at least 2 h at 37 °C. Do not sterilize with UV light;
- Discard the poly-D-lysine solution;
- Rinse the plates thee times with sterile autoclaved MilliQ water;
- Aspirate the water;
- Treated plates can be used in this moment or they can be air-dried and then stored at 4 °C.

***Note:*** *Wrap coated plates with parafilm before removing them from the laminar flow hood. Store them at 4 °C.*

*4 % Paraformaldehyde*

Paraformaldehyde is a toxic substance; therefore, it should be prepared under a fume hood.

- Provide 100 ml of PBS (1×) in into a measuring cylinder;
- Weigh 4 g of paraformaldehyde under the fume hood and add to the PBS solution;
- Dissolve paraformaldehyde using a magnetic stirrer and cover the measuring cylinder with parafilm;
- Under the fume hood adjust the solution to a moderate temperature under stirring. Avoid overheating (max: 70 °C);
- When the solution appears to be clear, allow to cool down to room temperature. Adjust to pH 7.4;
- Then aliquot and store the solution at 4 °C or −20 °C.

*Blocking solution (0.5 % TritonX-100, 3 %BSA)*
*10 % BSA stock solution*

- To prepare 50 ml of a 10 % BSA stock solution, measure 5 g of BSA and dilute it in 50 ml of PBS (1×). Store at 4 °C, 1 week.

***Blocking Solution Preparation***

- To prepare 10 ml of 3 % BSA solution, dilute 3 ml of 10 % BSA in 7 ml of PBS (1×);
- Dissolve 500 μl of Triton X-100 (10 %) in 9.5 ml of the 3 % BSA solution. Store at 4 °C (1 week).

*Incubation solution (0.1 % Triton X-100, 0.3 % BSA)*

**0.3 % BSA Solution Preparation**

- To prepare 10 ml of 0.3 % BSA solution, dilute 300 μl of 10 % BSA in 9.7 ml of PBS (1×). Store at 4 °C, 1 week.

**Incubation Solution Preparation**

To prepare 10 ml of incubation solution, dilute 100 μl of Triton X-100 (10 %) in 9.9 ml of 0.3 % BSA. Store at 4 °C, 1 week.

**Acknowledgements** HU acknowledges grant support by Fundação de Amparo à Pesquisa do Estado de São Paulo (FAPESP Project No. 2012/50880-4) and Conselho Nacional de Desenvolvimento Científico e Tecnológico (CNPq Project No. 486294/2012-9 and 467465/2014-2), Brazil. LSA's doctoral thesis research is supported by a fellowship from FAPESP (Project No. 2013/25338-4). CS is recipient of a CAPES-FAPERJ Postdoc fellowship. SX postdoctoral research is supported by a fellowship from FCT-Portugal (SFRH/BPD/ 76642/2011). We would like to acknowledge Ana Moreira and Rui Rodrigues for their kind help with photos and images for Fig. 5.2.

## References


Ahmed S (2009) The culture of neural stem cells. J Cell Biochem 106:1–6

Altman J (1962) Are new neurons formed in the brains of adult mammals? Science 135:1127–1128

Alvarez-Buylla A, Garcia-Verdugo JM (2002) Neurogenesis in adult subventricular zone. J Neurosci 22:629–634

Azari H, Rahman M, Sharififar S, Reynolds BA (2010) Isolation and expansion of the adult mouse neural stem cells using the neurosphere assay. J Vis Exp 20:2393

Azari H, Louis SA, Sharififar S, Vedam-Mai V, Reynolds BA (2011a) Neural-colony forming cell assay: an assay to discriminate bona fide neural stem cells from neural progenitor cells. J Vis Exp 6:2639

Azari H, Sharififar S, Rahman M, Ansari S, Reynolds BA (2011b) Establishing embryonic mouse neural stem cell culture using the neurosphere assay. J Vis Exp 11:2457

Basak O, Taylor V (2009) Stem cells of the adult mammalian brain and their niche. Cell Mol Life Sci 66:1057–1072

Bennett LB, Cai J, Enikolopov G, Iacovitti L (2010) Heterotopically transplanted CVO neural stem cells generate neurons and migrate with SVZ cells in the adult mouse brain. Neurosci Lett 475:1–6

Bonaguidi MA, Song J, Ming GI, Song H (2012) A unifying hypothesis on mammalian neural stem cell properties in the adult hippocampus. Curr Opin Neurobiol 22:754–761

Brazel CY, Nuñez JL, Yang Z, Levison SW (2005) Glutamate enhances survival and proliferation of neural progenitors derived from the subventricular zone. Neuroscience 131:55–65

Coles-Takabe BL, Brain I, Purpura KA, Karpowicz P, Zandstra PW, Morshead CM, Van der Kooy D (2008) Don't look: growing clonal versus nonclonal neural stem cell colonies. Stem Cells 26:2938–2944

Colucci-D'Amato L, Bonavita V, di Porzio U (2006) The end of the central dogma of neurobiology: stem cells and neurogenesis in adult CNS. Neurol Sci 27:266–270

Dirks PB (2008) Brain tumour stem cells: the undercurrents of human brain cancer and their relationship to neural stem cells. Philos Trans R Soc Lond B Biol Sci 363:139–152

Doetsch F (2003) A niche for adult neural stem cells. Curr Opin Genet Dev 13:543–550

Doetsch F, Caillé I, Lim DA, García-Verdugo JM, Alvarez-Buylla A (1999) Subventricular zone astrocytes are neural stem cells in the adult mammalian brain. Cell 97:703–716

Fuentealba LC, Obernier K, Alvarez-Buylla A (2012) Adult neural stem cells bridge their niche. Cell Stem Cell 10:698–708

Gage FH (2000) Mammalian neural stem cells. Science 287:1433–1438

Garcia-Bennett AE, König N, Abrahamsson N, Kozhevnikova M, Zhou C, Trolle C, Pankratova S, Berezin V, Kozlova EN (2014) In vitro generation of motor neuron precursors from mouse embryonic stem cells using mesoporous nanoparticles. Nanomedicine (Lond) 9:2457–2466

Gould E, Beylin A, Tanapat P, Reeves A, Shors TJ (1999) Learning enhances adult neurogenesis in the hippocampal formation. Nat Neurosci 2:260–265

Gross CG (2000) Neurogenesis in the adult brain: death of a dogma. Nat Rev Neurosci 1:67–73

Imayoshi I, Sakamoto M, Ohtsuka T, Takao K, Miyakawa T, Yamaguchi M, Mori K, Ikeda T, Itohara S, Kageyama R (2008) Roles of continuous neurogenesis in the structural and functional integrity of the adult forebrain. Nat Neurosci 11:1153–1161

Jensen JB, Parmar M (2006) Strengths and limitations of the neurosphere culture system. Mol Neurobiol 34:153–161

Kaplan MS (2001) Environment complexity stimulates visual cortex neurogenesis: death of a dogma and a research career. Trends Neurosci 24:617–620

Kaplan MS, Hinds JW (1977) Neurogenesis in the adult rat: electron microscopic analysis of light radioautographs. Science 197:1092–1094

Kriegstein A, Alvarez-Buylla A (2009) The glial nature of embryonic and adult neural stem cells. Annu Rev Neurosci 32:149–184

Lim DA, Alvarez-Buylla A (2014) Adult neural stem cells stake their ground. Trends Neurosci 37:563–571

Merkle FT, Mirzadeh Z, Alvarez-Buylla A (2007) Mosaic organization of neural stem cells in the adult brain. Science 317:381–384

Mignone JL, Kukekov V, Chiang AS, Steindler D, Enikolopov G (2004) Neural stem and progenitor cells in nestin-GFP transgenic mice. J Comp Neurol 469:311–324

Ming GL, Song H (2011) Adult neurogenesis in the mammalian brain: significant answers and significant questions. Neuron 70:687–702

Mirzadeh Z, Merkle FT, Soriano-Navarro M, Garcia-Verdugo JM, Alvarez-Buylla A (2008) Neural stem cells confer unique pinwheel architecture to the ventricular surface in neurogenic regions of the adult brain. Cell Stem Cell 3:265–278

Morizane A, Takahashi J, Shinoyama M, Ideguchi M, Takagi Y, Fukuda H, Koyanagi M, Sasai Y, Hashimoto N (2006) Generation of graftable dopaminergic neuron progenitors from mouse ES cells by a combination of coculture and neurosphere methods. J Neurosci Res 83:1015–1027

Nottebohm F (2002) Why are some neurons replaced in adult brain? J Neurosci 22:624–628

Reynolds BA, Rietze RL (2005) Neural stem cells and neurospheres—re-evaluating the relationship. Nat Methods 2:333–336

Reynolds BA, Weiss S (1992) Generation of neurons and astrocytes from isolated cells of the adult mammalian central nervous system. Science 255:1707–1710

Seri B, García-Verdugo JM, McEwen BS, Alvarez-Buylla A (2001) Astrocytes give rise to new neurons in the adult mammalian hippocampus. J Neurosc 21:7153–7160

Wu CY, Whye D, Mason RW, Wang W (2012) Efficient differentiation of mouse embryonic stem cells into motor neurons. J Vis Exp 64:e3813

# Chapter 6
# Very Small Embryonic Like Stem Cells (VSELs) and Their Hematopoietic Specification

**Malwina Suszynska, Mariusz Z. Ratajczak, and Janina Ratajczak**

**Abstract** Evidence has accumulated that adult tissues, including bone marrow, contain developmentally early stem cells that remain in a dormant state as well as stem cells that are more proliferative, supplying tissue-specific progenitor cells, which, therefore, display an active role in the turnover of adult tissues. To support this idea, we have identified in adult tissues a population of dormant pluripotent very small embryonic/epiblast like stem cells (VSELs) that, as we hypothesize, are deposited at the begining of gastrulation in developing tissues and play an important role as backup population of tissue committed stem cells. We envision that during steady state conditions VSELs may be involved in tissue rejuvenation and in processes of regeneration/repair after organ injuries. These small cells, similarly to epiblast-derived migrating primordial germ cells, change the epigenetic signature of some of the parentally imprinted genes and remain quiescent in adult tissues, not growing into teratoma. In this chapter, we will present FACS based strategies on how to correctly detect, enumerate and sort these cells in murine bone marrow. We will also provide a protocol for hematopoietic specification of VSELs in a co-culture system using adherent OP-9 supportive cell line.

**Keywords** Adult stem cells • VSELs • Primordial germ cells • Hematopoietic stem cells • Hematopoietic specification • Tissue engineering

---

M. Suszynska • J. Ratajczak
Stem Cell Institute at James Graham Brown Cancer Center, University of Louisville, 500 S. Floyd Street, Rm. 107, Louisville, KY 40202, USA

M.Z. Ratajczak (✉)
Stem Cell Institute at James Graham Brown Cancer Center, University of Louisville, 500 S. Floyd Street, Rm. 107, Louisville, KY 40202, USA

Department of Regenerative Medicine, Warsaw Medical University, Warsaw, ul. Zwirki and Wigury 61, Poland
e-mail: mzrata01@louisville.edu


H. Ulrich, P. Davidson Negraes, *Working with Stem Cells*,
DOI 10.1007/978-3-319-30582-0_6

## 6.1 Introduction

The rapidly developing field of regenerative medicine is searching for safe and therapeutically efficient sources of pluripotent stem cells (PSCs) that give rise to cells from all thee germ layers and therefore could be employed to regenerate damaged organs/tissues. It has been postulated by several investigators that PSCs could be isolated from the postnatal tissues, Although the hierarchy within the adult stem cell compartment is still incompletely understood and various types of stem cells residing in postnatal tissues that possess more than one germline specification potential have been described (Anjos-Afonso and Bonnet 2007; Asahara et al. 1997; Beltrami et al. 2007; D'Ippolito et al. 2004; Jiang et al. 2002; Kogler et al. 2004; Li and Clevers 2010; Ling et al. 2006; Petersen et al. 1999; Prockop 1997; Serafini et al. 2007; Wakao et al. 2011; Yu et al. 2006).

We have identified and isolated from adult tissues a population of pluripotent $Lin^{-}CD45^{-}Sca\text{-}1^{+}$ very small embryonic-like stem cells (VSELs). These cells were purified from adult murine bone marrow (BM), murine fetal livers (FLs) and several adult murine organs including brain, liver, kidney, lungs skeletal muscles and retina (Kucia et al. 2006, 2007b, 2008; Zuba-Surma et al. 2008). Similar population of small $Lin^{-}CD45^{-}CD133^{+}$ cells has been identified in human umbilical cord blood (UCB) and bone marrow (BM) (Kucia et al. 2007a; Sovalat et al. 2011; Zuba-Surma et al. 2010). VSELs express several morphological markers (e.g. relatively large nuclei containing euchomatin) and molecular markers (e.g. expression of SSEA-1, Oct4, Nanog, Rex1) that are characteristic for embryonic stem cells (ESCs), induced pluripotent stem cells (iPSCs), epiblast stem cells (EpiSCs) and primordial germ cells (PGCs) (Shin et al. 2010a, b). Moreover, careful molecular analysis of gene libraries established from single sorted VSELs revealed that these cells are, in fact, somewhat heterogeneous (Shin et al. 2012). We found at least thee different types of libraries generated from single, sorted VSELs, and some of these libraries exhibited a strong EpiSCs- or PGCs-like gene expression pattern. In support of this, we observed that murine BM-derived VSELs express several genes that are characteristic of EpiSCs (*Gbx2, Fgf5,* and *Nodal*) and, more importantly, of germline specification and migrating PGCs (*Stella, Prdm14, Fragilis, Blimp1, Nanos3,* and *Dnd1*) (Shin et al. 2010a, b). The expression of some of these crucial genes has subsequently been confirmed by demonstrating the presence of transcriptionally active promoters in these genes. The fact that gene expression array analysis revealed that some purified VSELs express several genes that are characteristic for PGCs is quite intriguing, because PGCs are the first stem cells specified from epiblast in the developing embryo in both rodents and humans.

We have recently provided some evidence that supports a potential developmental link between PGCs, VSELs and hematopoietic stem/progenitor cells (HSPCs) (Mierzejewska et al. 2015; Ratajczak et al. 2011a; Shin et al. 2010b; Suszynska et al. 2014). We postulated, in agreement with some investigators, that HSPCs could become specified during development from a population of migrating PGCs (Mierzejewska et al. 2015; Ohtaka et al. 1999; Rich 1995; Yoshimoto et al. 2009). In support of this intriguing possibility, HSPCs and PGCs share similar properties such as for example both are highly migratory cells, express α-chemokine receptor CXCR4 and migrate in response to stromal derived factor-1 (SDF-1) gradient. Moreover, specification of the first primitive HSPCs in yolk sac blood islands as well as subsequently the origin of definitive HSPCs in the aorta-gonado-mesonephos (AGM) region are chonologically and anatomically correlated with the developmental migration of PGCs in extra- and intra-embryonic tissues. In addition, several papers have described the sharing of chomosomal aberrations between germline tumors and leukemias or lymphomas, and expression of splat-like zinc finger transcription factor (Sall4), which suggests their common clonal origin (Chaganti et al. 1989; Nichols et al. 1985; Woodruff et al. 1995).

Based on this, we postulate that VSELs are an intermediate population of developmental early stem cells between PGCs and HSPCs. To support this, VSELs similarly to PGCs-derived cells, express several functional receptors for sex hormones (SexH) (Mierzejewska et al. 2015) and more importantly, murine VSELs respond *in vivo* by proliferation after prolonged administration of SexH (Mierzejewska et al. 2015). We reported that Although both murine and human VSELs right after isolation do not show any hematopoietic potential, they might become specified into hematopoietic cells in co-cultures with OP-9 adherent cell line (Ratajczak et al. 2011a, b). Of note, OP-9 stroma cells are also well known to promote hematopoietic specification of ESCs and iPSCs in co-culture systems.

In this chapter, we will present FACS based strategies on how to correctly detect, enumerate and sort these cells in murine bone marrow. We will also provide a protocol for hematopoietic specification of VSELs in co-culture with adherent OP-9 supportive cell line, using a similar protocol that is employed for ESCs and iPSCs.

## 6.2 FACS-Based Isolation of VSELs from Murine Bone Marrow

### 6.2.1 *Materials*

#### 6.2.1.1 Animals

Femora and tibia from ~6–10 C57Bl/6 mice (4–8 weeks-old). *Please, note that Balb-C mice do not express Sca-1, which is a positive marker for identification of murine VSELs.*

#### 6.2.1.2 Reagents

- RPMI-1640 Medium, 1× (HyClone, Cat. No. SH30027).
- Penicillin Streptomycin Solution, 100× (P/S) (Lonza, Cat. No. 17-602E).
- Fetal bovine serum (Seradigm, Cat. No. 1300-500; Heat Inactivated: 56 °C, 30 min).
- Lysing buffer, BD Pharmlyse (BD Biosciences, Cat. No. 555899).

**Table 6.1** List of antibodies against murine epitopes used in identification of murine VSELs

| Antibody | Fluorochome | Clone | Vendor | Cat. No. |
|---|---|---|---|---|
| Lineage markers:<br>Anti-CD45R/B220<br>Anti-Gr1 (Ly-6G and Ly-6C)<br>Anti-TCRβ<br>Anti-γδ T Cell<br>Anti-CD11b<br>Anti-TER-119 | PE<br>PE<br>PE<br>PE<br>PE<br>PE | RA3-6B2<br>RB6-8C5<br>H57-597<br>GL3<br>M1/70<br>TER-119 | BD Pharmingen<br>BD Pharmingen<br>BD Pharmingen<br>BD Pharmingen<br>BD Pharmingen<br>BD Pharmingen | 553089<br>553128<br>553172<br>553178<br>557397<br>553673 |
| Anti-CD45 | APC-Cy7 | 30-F11 | BD Pharmingen | 557659 |
| Anti-Ly-6A/E (Sca-1)<br>Streptavidin | Biotin<br>PE-Cy5 | E13-161.7 | BD Pharmingen<br>BD Pharmingen | 553334<br>554062 |
| Isotype controls:<br>Hamster IgG2, κ<br>Rat IgG2b, κ<br>Hamster IgG2, λ1<br>Rat IgG2a, κ<br>Rat IgG2a, κ<br>Rat IgG2b, κ | PE<br>PE<br>PE<br>PE<br>Biotin<br>APC-Cy7 | B81-3<br>A95-1<br>Ha4/8<br>R35-95<br>R35-95<br>A95-1 | BD Pharmingen<br>BD Pharmingen<br>BD Pharmingen<br>BD Pharmingen<br>BD Pharmingen<br>BD Pharmingen | 550085<br>553989<br>553965<br>553930<br>553928<br>552773 |

- Monoclonal antibodies against murine epitopes (Table 6.1).
- Flow Cytometry Size Calibration Kit microspheres (Invitrogen, Cat. No. F13838).

#### 6.2.1.3 Equipment

- Sterile tweezers and scissors.
- Cell culture Dishes, PS, 100×20 mm (CELLSTAR, Cat. No. 664160).
- 25G×5/8 in needles (BD, Cat. No. 305122).
- 3 ml Syringe (BD, Cat. No. 309657).
- Cell Strainer 40 μm Nylon (Falcon, Cat. No. 352340).
- 50 ml Polypropylene Conical Tubes (Falcon, Cat. No. 352098).
- 5 ml Polystyrene Round-Bottom Tubes (Falcon, Cat. No. 352054).
- Centrifuge.
- Cell sorter.

### *6.2.2 Methods*

#### 6.2.2.1 Preparation of BM-Nucleated Cell Suspension

1. Euthanize mice using the institution's animal care committee-approved protocol and spray the animal skin with 70 % ethanol. Subsequent experimental steps will be conducted in a laminar air flow to maintain tissue sterility;

2. Carefully isolate tibias and femora from mice, removing the remaining muscles and not breaking the bone ends. Place the bones into a 100 mm cell culture dish containing 10 ml of cold RPMI 1640 Medium supplemented with Penicillin/Streptomycin;
3. Cut off the ends of the bones and use a 3 ml syrgine filled with RPMI 1640 Medium and 25 G needle to flush the contents of marrow from both ends of the bone shafts into another culture dish containing 10 ml of cold RPMI 1640 Medium supplemented with Penicillin/Streptomycin. To get a single-cell suspension disperse flushed from bones BM tissue by gently aspirating it into syringe and needle few times;
4. Collect the single-cell suspension into a sterile 50 ml tube, by passing it through a 40-μm cell strainer. Fill the 50 ml tube with RPMI 1640 Medium supplemented with Penicillin/Streptomycin and centrifuge the cell suspension at 500 × *g* for 10 min at 4 °C;
5. Remove the supernatant and lyse the red blood cells by resuspending the cell pellet in 3 ml of prepared, prewarmed lysing buffer (1×) and incubate the cell suspension at room temperature for 8–10 min;
6. After lysis, stop the reaction by diluting lysing buffer with RPMI 1640 medium with 2 % FBS (with P/S) up to 50 ml and centrifuge immediately at 500 × *g*, 4 °C for 10 min;
7. Decant the supernatant and wash the BM-nucleated cells by resuspending the pellet in 25 ml of RPMI 1640 medium with 2 % FBS (with P/S). Centrifuge again at 500 × *g* for 10 min at 4 °C. Finally, resuspend the cells in 0.5 ml of RPMI 1640 medium with 2 % FBS (with P/S) for further staining using monoclonal antibodies.

#### 6.2.2.2 Staining of BM-Nucleated Cells with Monoclonal Antibodies for VSELs Sorting

1. Transfer cells into 5 ml Round-Bottom Tubes; also prepare control samples of BM-nucleated cells for FACS settings, including: (a) negative samples (no stain), (b) similarly labeled non-specific primary antibody samples, ideally of the same isotype as the specific antibody and, (c) positive samples for each different labeled monoclonal antibody (single-color);
2. Combine the recommended quantities of each primary antibody against Sca-1, CD45, and Lin markers (see Table 6.1), add to cells and pulse vortex gently. Stain bone marrow mononuclear cells for 30 min on ice, protect from light;
3. Wash the cells by filling 5 ml Round-Bottom Tubes with RPMI 1640 medium with 2 % FBS (with P/S) and pellet the cells by centrifugation at 500 × *g* for 10 min at 4 °C. Repeat for a total of two washes, discarding supernatant between washes;
4. Resuspend cells to be sorted in RPMI 1640 medium with 2%FBS (with P/S) and divide the single-cell suspension into 4–5 5 ml Round-Bottom Tubes by passing

it through a 40-μm cell strainer. Add RPMI 1640 medium with 2 % FBS (with P/S) to 4 ml of final volume;

5. Stained cells should be stored on ice until sorting, and must be filtered again prior to sorting using a 40 μm cell strainer.

#### 6.2.2.3 Setting the Instrument for Sorting

1. Run cells and adjust FSC and SSC voltages in linear scale so that bone marrow mononuclear cells are visible on scale. Make sure to choose FSC/SSC settings, that will keep smallest cells on scale roughly centered;
2. Run Flow Cytometry Size Calibration beads (with standard diameters of 1, 2, 4, 6, 10, and 15 μm) to adjust FSC threshold to see all objects as small as 2 μm. Please, note that the side scatter voltage may need to be lowered to view the beads on scale. If so, note the difference in the voltage setting;
3. Create a gate between the showed events for 2 and 10 μm size beads on the FSC/SSC dot-plot;
4. Run the unstained sample to adjust voltage for each color so that the negative population is in the first scale decade;
5. Run the single-color stained samples of BM-NCs to compensate for spectral overlap;
6. Run the isotype control stained samples of BM-NCs to estimate the level of background staining with specific isotypes of antibodies;
7. To identify murine BM-VSELs, create the gating strategy to visualize $Lin^-$/$CD45^-$/$Sca\text{-}1^+$ population based on negative, positive and isotype stained samples (as shown in Fig. 6.1);
8. Run the stained sample and proceed to sort using single cell mode and speed lower than 20,000 events/s to obtain a highly purified population of VSELs.

#### 6.2.2.4 Preparation of Classical Gating Strategy for Sorting Murine BM-VSELs

1. Create a gate (R1) on the FSC vs. SSC dot plot as described above, including small events ranging from 2 to 10 μm (as initially set up with Flow Cytometry Size Calibration beads). Adjust 2–10 μm gate exclusively on lymphocytes to include only agranular objects (SSC low). Setting up an enlarged gate on FSC vs. SSC results in enrichment of sorted fractions with artifacts reflected in further analyses;
2. In order to increase the purity of sorted population of VSELs gate out doublets and clumps by setting gates on FSC-H vs. FSC-W (R2) and SSC-H vs. SSC-W (R3) dot-plots;
3. The single-cell fraction from gate R3 is further visualized on Sca-1 vs. Lin expression dot plot. As VSELs and HSCs are rare fractions among sorted cells, we recommend using dot-plots, not contour plots. A gate including $Sca\text{-}1^+$/$Lin^-$

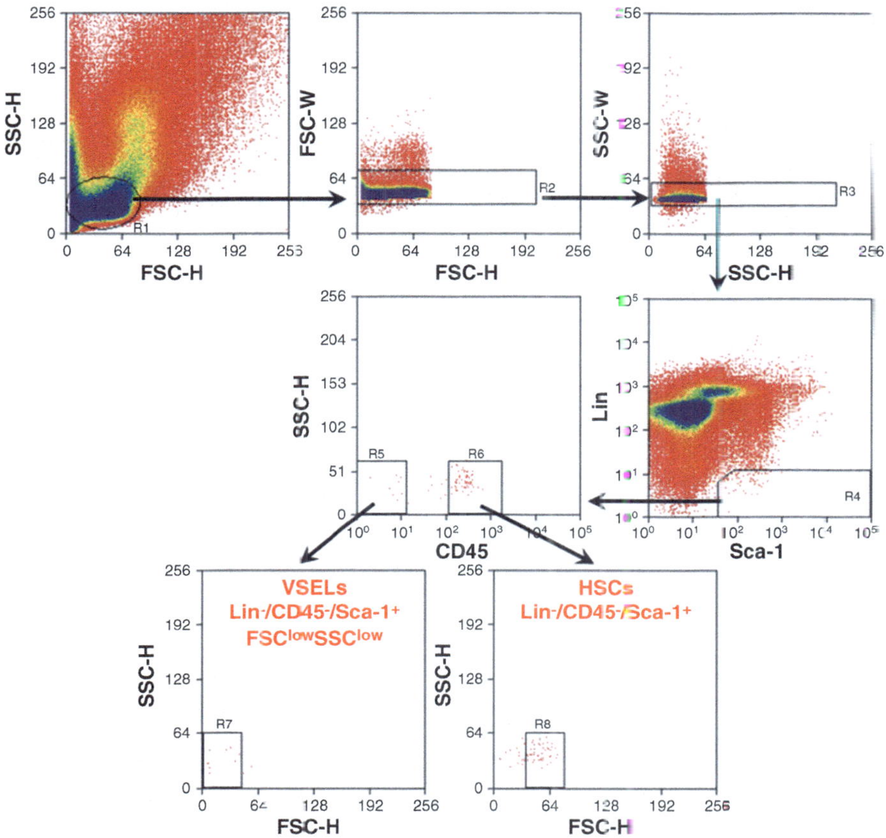


**Fig. 6.1** Classical sorting strategy for murine BM-derived VSELs by fluorescence-activated cell sorting. Gate small, agranular events ranging from 2 to 10 μm (as initially set up with Flow Cytometry Size Calibration beads) on the FSC vs. SSC dot plot - region R1. To increase the purity, gate out doublets and clumps by setting gates on FSC-H vs. FSC-W (R2) and SSC-H vs. SSC-W (R3) dot-plots. Visualize the single-cell fraction from gate R3 on Sca-1 vs. Lin expression dot plot and create R4 gate including Sca-1⁺/Lin⁻ cells. Cells from region R4 are next analyze for CD45 expression on CD45 vs. SSC dot plot, separating non-hematopoietic CD45⁻ fraction (R5) and hematopoietic CD45⁺ fraction (R5). Events from R5 and R6 gates are visualized by back-gating again on FSC vs. SSC dot plots and use of additional size gates, including smaller events for VSELs (R7) and larger events for HSCs (R8) to increase VSELs and HSCs purity. Sorted VSELs may be evaluated for purity by RQ-PCR for Oct-4 expression

(R4) cells should be set up strictly for Lin negative objects. Enlarging the gate into Lin$^{dim}$ objects to increase VSEL yield, may greatly enrich the sorted population in erythoblasts (Fig. 6.2) and thus, significantly dilute the VSELs and "contaminate" sorted cells for some erythoblasts;

4. Sca-1⁺/Lin⁻ cells from region R4 are next analyzed for CD45 expression on CD45 vs. SSC dot plot, separating non-hematopoietic CD45⁻ fraction (R5) and

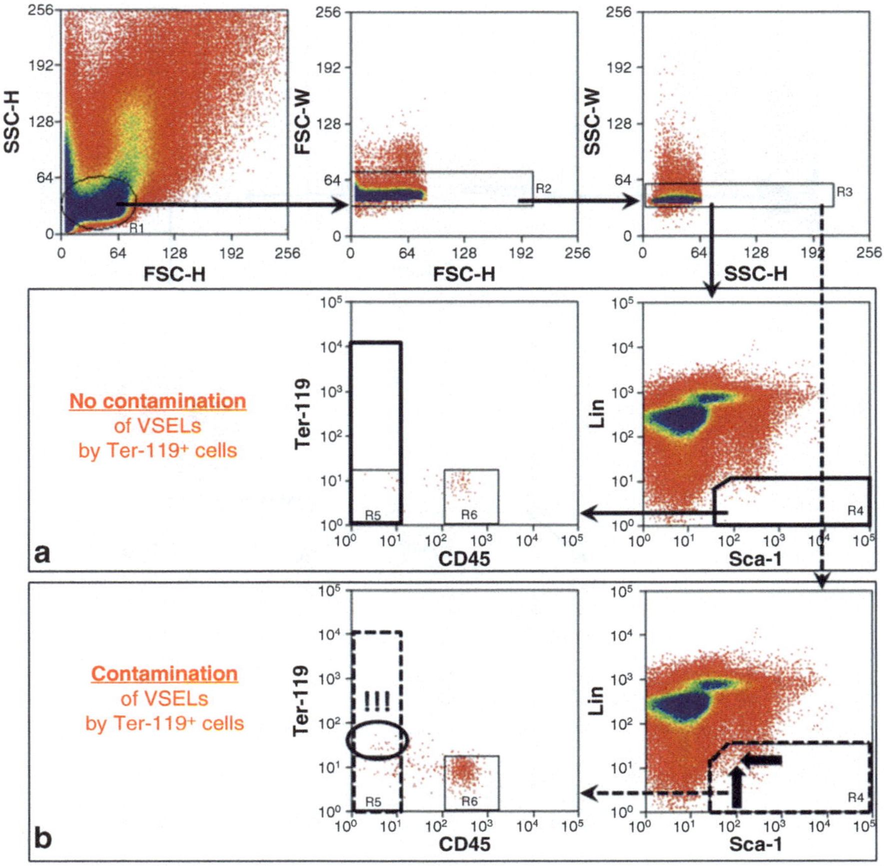


**Fig. 6.2** The effect of correct and incorrect gating of Sca-1$^+$/Lin$^-$ objects on final purity of sorted VSELs. The FSC$^{low}$ and SSC$^{low}$ single-cell fraction from gate R3 is visualized on Sca-1 vs. Lin expression dot plot. A gate including Sca-1$^+$/Lin$^-$ (R4) cells should be set up strictly for Lin$^-$ and Sca-1$^+$ objects (**a**). Expanding the gate into Lin$^{dim}$ objects to increase VSEL yield may enrich the sorted population of VSELs in unwanted population of Ter-119$^+$ erythoblasts (**b**)

hematopoietic CD45$^+$ fraction (R6). Avoid gating CD45$^{dim}$ objects to not contaminate non-hematopoietic CD45$^-$ fraction (VSELs) with hematopoietic-committed progenitor cells;

5. Events from R5 and R6 gates can be visualized again by back-gating on FSC vs. SSC dot plots. VSELs should be seen on FSC scale as smaller-in size population in comparison to HSCs. Using size gates including smaller events for VSELs (R7) and bigger events for HSCs (R8) on the last FSC vs. SSC dot plots may additionally increase VSELs and HSCs purity.

## 6.3 Hematopoietic Specification of Murine VSELs on OP-9 Supportive Feeder Layer

While freshly isolated VSELs do not exhibit *in vitro* and *in vivo* hematopoietic potential they may, after co-culture with OP9 stromal cells, differentiate along the hematopoietic lineage in a similar way as ESCs or iPSCs become specified into hematopoietic cells using comparable co-culture conditions (Ratajczak et al 2011a, b). We reported that "OP9-primed and subsequently expanded" VSEL-derived cells acquired expression of several hemato/lymphopoiesis-specific genes and markers, gave rise to hematopoietic colonies *in vitro*, and protected lethally irradiated mice in both primary and secondary transplant models upon transplantation (Ratajczak et al. 2011a).

Interestingly, in contrast to VSELs, neither ESCs nor iPSCs could be so far specified into long term repopulating hematopoietic stem cells (LT-HSCs). Based on these observations, we postulated that VSELs are the most primitive murine BM-residing population of stem cells, which have the potential to become specified into the hematopoietic lineage and thus may share some of the characteristics of LT-HSCs (Ratajczak et al. 2011a).

## 6.4 Ex-vivo Differentiation of VSELs into Hematopoietic Cells in Primary Co-cultures Over OP9 Stromal Cells

Freshly sorted $Lin^{-}/CD45^{-}/Sca\text{-}1^{+}$ VSELs ($1 \times 10^4$ cells) and in parallel as control, $Lin^{-}/CD45^{+}/Sca\text{-}1^{+}$ HSCs ($1 \times 10^4$ cells) isolated from e.g. green fluorescence protein transgenic mice ($GFP^{+}$ mice) are plated over OP9 cells in α-MEM (GIBCO, Cat. No. 11900-024) with 20 % FBS in glass bottom FluoroDishes (Tissue Culture Dish with Cover Glass Bottom, World Precision Instruments, Inc., Cat. No. FD35-100) without growth factors or cytokines. Five days later, some green cobblestone-area colonies (CAFCs) are formed by $Lin^{-}/CD45^{+}/Sca\text{-}1^{+}$ HSCs and single or only dividing once green $Lin^{-}/CD45^{-}/Sca\text{-}1^{+}$ VSELs are visible over OP-9 supportive feeder layer. Subsequently, these cultures are trypsinized, washed by centrifugation in α-MEM, and all cells from trypsinized dishes including $GFP^{+}$ HSCs and VSELs are plated into 6-well plates using methylcellulose-based medium (Human Methylcellulose Base Media R&D Systems; Cat. No. 390394) supplemented with cocktail of growth factors and cytokines, as described below.

## 6.5 Evaluation of the Clonogenic Potential of Sorted Cells in Methylcellulose Cultures

VSELs- or HSCs-derived cells harvested from OP9 cultures (primary cultures) are plated in methylcellulose-based medium supplemented with 5 ng/ml murine stem cell growth factor (mSCF, R&D System, Cat. No. 455-MC/CF), 10 ng/ml interleukin-3 (mIL-3, R&D System, Cat. No. 403-ML/CF), 5 ng/ml granulocyte-macrophage colony-stimulating factor (mGM-CSF, R&D System, Cat. No. 415 ML/CF), 5 ng/ml FLT3 (hFLT3, Stem Cell Tech., Cat. No. 308-FK/CF), 50 ng/ml thombopoietin (hTPO, Stem Cell Tech., Cat. No. 02522), 5 U/ml erythopoietin (hEpO, EPOGEN, Amgen Inc.), and 20 ng/ml insulin growth factor-2 (hIGF-2, Shenandoah Biotech. Inc., Cat. No. 50-990-667). Cells are subsequently cultured for 5 days and the colonies formed are scored under inverted microscope. Subsequently, methylcellulose cultures are solubilized by adding α-MEM and trypsinized for 5 min. Next, the cells are washed by centrifugation in α-MEM with 10 % FBS and plated into secondary methylcellulose cultures. Cells are grown again in the presence of the same growth factors and replated every 5 days into new methylcellulose cultures. The same procedure may be repeated for the next 2–3 passages. The clonogeneic potential of OP-9 primed VSELs-derived cells gradually increases and, 25 days after initial isolation, more and larger in size colonies are formed by VSELs initiated cells as compared to HSC-derived population. $GFP^+$ cells isolated from these VSELs-initiated cell cultures may be purified by FACS and employed for gene expression analysis, cell surface receptor staining or for *in vivo* transplant experiments.

## 6.6 Characterization of the Cells Obtained Using the Described Protocol

Sorted by FACS, small VSELs are evaluated for their purity by RQ-PCR analysis for Oct-4 expression. Small samples of cells may also be stained for expression of Oct-4 at protein level. HSCs specified from VSELs are tested in clonogenic assays, gene expression studies and by detection of surface specific hematopoietic markers (e.g. CD45, CD41). Of note, $CD45^-$ VSELs after hematopoietic specification give rise to a population of $CD45^+$ cells.

### *6.6.1 RQ-PCR Analysis of Sorted VSELs*

For real time-PCR analysis of Oct-4 expression, total mRNA is isolated from VSELs cells with the RNeasy Mini Kit (Qiagen, Cat. No. 74104), treated with DNase (DNA-free DNA Removal Kit, Ambion, Cat. No. AM1906) and mRNA is reverse-transcribed with TaqMan Reverse Transcription Reagents (Applied Biosystems, Cat. No. N8080234). Detection of Oct-4 and reference β2-microglobulin mRNA levels is performed by real-time RT-PCR using a 7500 Fast Real-Time PCR System (Applied Biosystems) in a 25 μl reaction mixture containing 12.5 μl Power SYBR® Green PCR Master Mix (Applied Biosystems, Cat. No. 4367660), 10 ng of cDNA template, and forward and reverse primers. The following primer pairs are used: Oct-4: 5′-ACA TCG CCA ATC AGC TTG G-3′ (forward), 5′-AGA ACC

ATA CTC GAA CCA CAT CC-3′ (reverse), β2-microglobin: 5′-CAT ACG CCT GCA GAG TTA AGC A-3′ (forward), 5′-GAT CAC ATG TCT CGA TCC CAG TAG-3′ (reverse). The threshold cycle (Ct), i.e., the cycle number at which the amount of amplified gene of interest reached a fixed threshold, is subsequently determined. The relative quantitation of target gene mRNA expression is calculated with the comparative Ct method. The relative quantitative value of the target, normalized to an endogenous β2-microglobulin gene control and relative to a calibrator, is expressed as $2^{-\Delta\Delta Ct}$ (fold difference), whereas $\Delta Ct = Ct$ of target genes *minus* Ct of an endogenous control gene (β2-microglobulin), and $\Delta\Delta Ct = \Delta Ct$ of samples measuring the target gene *minus* ΔCt of samples measuring the calibrator for the target gene. To avoid the possibility of amplifying contaminating DNA, (1) primers for real-time RT-PCR are designed to contain a DNA intron sequence for specific cDNA amplification, (2) reactions are performed with appropriate negative controls (template-free controls), (3) uniform amplification of the products is rechecked by analyzing the melting curves of the amplified products (dissociation graphs), (4) the melting temperature ($T_m$) is in the range 57–60 °C, with the product $T_m$ at least 10 °C higher than the primer $T_m$ and, (5) gel electrophoresis is performed to confirm the correct size of the amplified products and the absence of nonspecific bands.

### 6.6.2 *Detection of Oct-4 in VSELs by Immunostaining*

Samples of $Lin^-/CD45^-/Sca\text{-}1^+$ cells (VSELs) are sorted and plated on 22 mm diameter plates coated with poly-L-lysine (Sigma-Aldrich, Cat. No. P9155), and incubated for 24 h. Subsequently, cells are fixed in 3.5 % paraformaldehyde for 15 min, permeabilized by 0.1 % Triton X-100 (Bio-Rad, Cat. No. 1610407), washed in PBS, and pre-blocked with 2.5 % BSA for 2.5 h at RT to avoid nonspecific binding of antibodies. The cells are then stained at 4 °C, overnight, for Oct-4 (clone 7F9.2, mouse monoclonal IgG1, 1:150, Millipore, Cat. No. MAB4419). Appropriate secondary Alexa Fluor 488 goat anti-mouse IgG antibody is used (1:500, Molecular Probes, Cat. No. A11029) at 37 °C for 1 h. Nuclei are stained with DAPI (Invitrogen) at 37 °C for 20 min. Cells are evaluated under Olympus FV1000 confocal microscope.

## 6.7 The Potential Use of the Obtained Cells

Murine BM-purified VSELs could be employed in several murine models to regenerate damaged organs e.g. heart, liver, brain, skeletal muscles, retina or pancreas. As presented in this chapter VSELs-derived HSCs may be studied in hematopoietic transplant experiments in mice. More important, a model of hematopoietic specification of murine VSELs was helpful to develop a similar strategy for human BM- or umbilical cord blood (UCB)-purified VSELs (Ratajczak J. et al. 2011b). Since there are not available, at present time, efficient ex-vivo expansion protocols of HSCs, we believe that quiescent population of BM-residing VSELs is optimal as a starting cell population for efficient expansion of HSCs.

**Acknowledgments** This work was supported by NIH grants 2R01 DK074720 and R01HL112788, NCN Harmonia Grant UMO-2014/14/M/NZ3/00475 and the Stella and Henry Endowment to MZR.

## References

Anjos-Afonso F, Bonnet D (2007) Nonhematopoietic/endothelial SSEA-1+ cells define the most primitive progenitors in the adult murine bone marrow mesenchymal compartment. Blood 109:1298–1306

Asahara T, Murohara T, Sullivan A, Silver M, van der Zee R, Li T, Witzenbichler B, Schatteman G, Isner JM (1997) Isolation of putative progenitor endothelial cells for angiogenesis. Science 275:964–967

Beltrami AP, Cesselli D, Bergamin N, Marcon P, Rigo S, Puppato E, D'Aurizio F, Verardo R, Piazza S, Pignatelli A, Poz A, Baccarani U, Damiani D, Fanin R, Mariuzzi L, Finato N, Masolini P, Burelli S, Belluzzi O, Schneider C, Beltrami CA (2007) Multipotent cells can be generated in vitro from several adult human organs (heart, liver, and bone marrow). Blood 110:3438–3446

Chaganti RS, Ladanyi M, Samaniego F, Offit K, Reuter VE, Jhanwar SC, Bosl GJ (1989) Leukemic differentiation of a mediastinal germ cell tumor. Gene Chomosome Canc 1:83–87

D'Ippolito G, Diabira S, Howard GA, Menei P, Roos BA, Schiller PC (2004) Marrow-isolated adult multilineage inducible (MIAMI) cells, a unique population of postnatal young and old human cells with extensive expansion and differentiation potential. J Cell Sci 117:2971–2981

Jiang Y, Jahagirdar BN, Reinhardt RL, Schwartz RE, Keene CD, Ortiz-Gonzalez XR, Reyes M, Lenvik T, Lund T, Blackstad M, Du J, Aldrich S, Lisberg A, Low WC, Largaespada DA, Verfaillie CM (2002) Pluripotency of mesenchymal stem cells derived from adult marrow. Nature 418:41–49

Kogler G, Sensken S, Airey JA, Trapp T, Muschen M, Feldhahn N, Liedtke S, Sorg RV, Fischer J, Rosenbaum C, Greschat S, Knipper A, Bender J, Degistirici O, Gao J, Caplan AI, Colletti EJ, Almeida-Porada G, Muller HW, Zanjani E, Wernet P (2004) A new human somatic stem cell from placental cord blood with intrinsic pluripotent differentiation potential. J Exp Med 200:123–135

Kucia M, Reca R, Campbell FR, Zuba-Surma E, Majka M, Ratajczak J, Ratajczak MZ (2006) A population of very small embryonic-like (VSEL) CXCR4(+)SSEA-1(+)Oct-4+ stem cells identified in adult bone marrow. Leukemia 20:857–869

Kucia M, Halasa M, Wysoczynski M, Baskiewicz-Masiuk M, Moldenhawer S, Zuba-Surma E, Czajka R, Wojakowski W, Machalinski B, Ratajczak MZ (2007a) Morphological and molecular characterization of novel population of CXCR4+ SSEA-4+ Oct-4+ very small embryonic-like cells purified from human cord blood: preliminary report. Leukemia 21:297–303

Kucia M, Wu W, Ratajczak MZ (2007b) Bone marrow-derived very small embryonic-like stem cells: their developmental origin and biological significance. Dev Dyn 236:3309–3320

Kucia M, Wysoczynski M, Ratajczak J, Ratajczak MZ (2008) Identification of very small embryonic like (VSEL) stem cells in bone marrow. Cell Tissue Res 331:125–134

Li L, Clevers H (2010) Coexistence of quiescent and active adult stem cells in mammals. Science 327:542–545

Ling TY, Kuo MD, Li CL, Yu AL, Huang YH, Wu TJ, Lin YC, Chen SH, Yu J (2006) Identification of pulmonary Oct-4+ stem/progenitor cells and demonstration of their susceptibility to SARS coronavirus (SARS-CoV) infection in vitro. Proc Natl Acad Sci U S A 103:9530–9535

Mierzejewska K, Borkowska S, Suszynska E, Suszynska M, Poniewierska-Baran A, Maj M, Pedziwiatr D, Adamiak M, Abdel-Latif A, Kakar SS, Ratajczak J, Kucia M, Ratajczak MZ (2015) Hematopoietic stem/progenitor cells express several functional sex hormone receptors-novel evidence for a potential developmental link between hematopoiesis and primordial germ cells. Stem Cells Dev 24:927–937

Nichols CR, Hoffman R, Einhorn LH, Williams SD, Wheeler LA, Garnick MB (1985) Hematologic malignancies associated with primary mediastinal germ-cell tumors. Ann Intern Med 102:603–609

Ohtaka T, Matsui Y, Obinata M (1999) Hematopoietic development of primordial germ cell-derived mouse embryonic germ cells in culture. Biochem Biophys Res Commun 260:475–482

Petersen BE, Bowen WC, Patrene KD, Mars WM, Sullivan AK, Murase N, Boggs SS, Greenberger JS, Goff JP (1999) Bone marrow as a potential source of hepatic oval cells. Science 284:1168–1170

Prockop DJ (1997) Marrow stromal cells as stem cells for nonhematopoietic tissues. Science 276:71–74

Ratajczak J, Wysoczynski M, Zuba-Surma E, Wan W, Kucia M, Yoder MC, Ratajczak MZ (2011a) Adult murine bone marrow-derived very small embryonic-like stem cells differentiate into the hematopoietic lineage after coculture over OP9 stromal cells. Exp Hematol 39:225–237

Ratajczak J, Zuba-Surma E, Klich I, Liu R, Wysoczynski M, Greco N, Kucia M, Laughlin MJ, Ratajczak MZ (2011b) Hematopoietic differentiation of umbilical cord blood-derived very small embryonic/epiblast-like stem cells. Leukemia 25:1278–1285

Rich IN (1995) Primordial germ cells are capable of producing cells of the hematopoietic system in vitro. Blood 86:463–472

Serafini M, Dylla SJ, Oki M, Heremans Y, Tolar J, Jiang Y, Buckley SM, Pelacho B, Burns TC, Frommer S, Rossi DJ, Bryder D, Panoskaltsis-Mortari A, O'Shaughnessy MJ, Nelson-Holte M, Fine GC, Weissman IL, Blazar BR, Verfaillie CM (2007) Hematopoietic reconstitution by multipotent adult progenitor cells: precursors to long-term hematopoietic stem cells. J Exp Med 204:129–139

Shin DM, Liu R, Klich I, Ratajczak J, Kucia M, Ratajczak MZ (2010a) Molecular characterization of isolated from murine adult tissues very small embryonic/epiblast like stem cells (VSELs). Mol Cells 29:533–538

Shin DM, Liu R, Klich I, Wu W, Ratajczak J, Kucia M, Ratajczak MZ (2010b) Molecular signature of adult bone marrow-purified very small embryonic-like stem cells supports their developmental epiblast/germ line origin. Leukemia 24:1450–1461

Shin DM, Liu R, Wu W, Waigel SJ, Zacharias W, Ratajczak MZ, Kucia M (2012) Global gene expression analysis of very small embryonic-like stem cells reveals that the Ezh2-dependent bivalent domain mechanism contributes to their pluripotent state. Stem Cells Dev 21:1639–1652

Sovalat H, Scrofani M, Eidenschenk A, Pasquet S, Rimelen V, Henon P (2011) Identification and isolation from either adult human bone marrow or G-CSF-mobilized peripheral blood of CD34(+)/CD133(+)/CXCR4(+)/ Lin(-)CD45(-) cells, featuring morphological, molecular, and phenotypic characteristics of very small embryonic-like (VSEL) stem cells. Exp Hematol 39:495–505

Suszynska M, Poniewierska-Baran A, Gunjal P, Ratajczak J, Marycz K, Kakar SS, Kucia M, Ratajczak MZ (2014) Expression of the erythopoietin receptor by germline-derived cells - further support for a potential developmental link between the germline and hematopoiesis. J Ovarian Res 7:66

Wakao S, Kitada M, Kuroda Y, Shigemoto T, Matsuse D, Akashi H, Tanimura Y, Tsuchiyama K, Kikuchi T, Goda M, Nakahata T, Fujiyoshi Y, Dezawa M (2011) Multilineage-differentiating stress-enduring (Muse) cells are a primary source of induced pluripotent stem cells in human fibroblasts. Proc Natl Acad Sci U S A 108:9875–9880

Woodruff K, Wang N, May W, Adrone E, Denny C, Feig SA (1995) The clonal nature of mediastinal germ cell tumors and acute myelogenous leukemia. A case report and review of the literature. Cancer Genet Cytogenet 79:25–31
Yoshimoto M, Heike T, Chang H, Kanatsu-Shinohara M, Baba S, Varnau JT, Shinohara T, Yoder MC, Nakahata T (2009) Bone marrow engraftment but limited expansion of hematopoietic cells from multipotent germline stem cells derived from neonatal mouse testis. Exp Hematol 37:1400–1410
Yu H, Fang D, Kumar SM, Li L, Nguyen TK, Acs G, Herlyn M, Xu X (2006) Isolation of a novel population of multipotent adult stem cells from human hair follicles. Am J Pathol 168:1879–1888
Zuba-Surma EK, Kucia M, Wu W, Klich I, Lillard JW Jr, Ratajczak J, Ratajczak MZ (2008) Very small embryonic-like stem cells are present in adult murine organs: ImageStream-based morphological analysis and distribution studies. Cytometry A 73A:1116–1127
Zuba-Surma EK, Klich I, Greco N, Laughlin MJ, Ratajczak J, Ratajczak MZ (2010) Optimization of isolation and further characterization of umbilical-cord-blood-derived very small embryonic/epiblast-like stem cells (VSELs). Eur J Haematol 84:34–46

# Chapter 7
# Neural Crest Stem Cell Cultures: Establishment, Characterization and Potential Use

**Andréa Gonçalves Trentin, Ricardo Castilho Garcez, and Raul Bardini Bressan**

**Abstract** The neural crest (NC) is a remarkable structure of vertebrate embryos that continues to fascinate biologists due to its importance in development, evolution and disease. NC cells emerge in the neurulating embryo at the closing borders of the neural plate and, after a phase of extensive migration, settle in different parts of the body and contribute to the formation of a diversity of tissues and organs. Because of their broad potential, NC cells constitute an attractive system to investigate fundamental developmental processes, such as cell commitment and lineage diversification. Moreover, recent evidence for the persistence of multipotent NC-like stem cell in several adult locations opens up opportunities for use of human NC stem cells as a source of material for cell-based transplantation therapies. In this chapter, we describe procedures for the isolation and characterization of NC cells from mouse and avian embryos, as well as adult NC-derived stem cells from the mouse hair follicle. These cultures constitute an accessible *in vitro* system for modeling NC stem cell morphogenesis and have been extensively used to study intrinsic and environmental cues that regulate migration, self-renewal and fate choice.

**Keywords** Neural crest • NC-like stem cell • Transplantation • Therapies

## 7.1 Introduction

The neural crest (NC) corresponds to a transient, multi-fated embryonic cell population that emerges at the dorsal margins of the neural folds during development of vertebrates. Upon closure of the neural tube, NC cells undergo epithelial-to-mesenchymal

A.G. Trentin (✉) • R.C. Garcez
Department of Cell Biology, Embryology and Genetics, Federal University of Santa Catarina, Trindade, Florianopolis, SC 88040-900, Brazil
e-mail: andrea.trentin@ufsc.br

R.B. Bressan
MRC Centre for Regenerative Medicine, The University of Edinburgh, Edinburgh, UK


H. Ulrich, P. Davidson Negraes, *Working with Stem Cells*,
DOI 10.1007/978-3-319-30582-0_7

transition and, after a phase of extensive migration, become widely distributed within the embryo (Le Douarin and Kalcheim 1999).

The NC is remarkably multipotent giving rise to most of the peripheral nervous system and to a large variety of non-neural cell types, including pigmented cells of the skin, smooth muscle cells of the cardiovascular system, and craniofacial bones, cartilage and connective tissues (Baggiolini et al. 2015; Dupin and Sommer 2012; Le Douarin and Kalcheim 1999). Despite progressive restriction of developmental potential with time, some NC cells are capable of self-renewal—thereby fulfilling the stem cell criteria—and display a developmental repertoire only surpassed by embryonic stem (ES) cells (Baggiolini et al. 2015; Calloni et al. 2007; Douarin and Dupin 2003; Trentin et al. 2004).

Interestingly, NC stem cells are not only present in the developing embryo, but also in various NC-derived tissues in the adult body including gut, cornea, periodontal ligament and the skin (Coura et al. 2008; Fernandes et al. 2006; Kruger et al. 2002; Sieber-Blum et al. 2004; Yoshida et al. 2006). These persistent post-migratory NC stem cells molecularly resemble their embryonic counterparts and also have the ability to differentiate into a wide variety of cell types (reviewed in Dupin and Sommer 2012; Shakhova et al. 2010; Trentin and Calloni 2013).

Because of their broad differential potential, both embryonic NC and adult NC-derived stem cells (NCSCs) represent a valuable model system to study fundamental stem cell and developmental biology. Furthermore, their multilineage potential and easy accessibility in adult tissues open up opportunities for use of human NCSCs as a source of material for pharmaceutical screening, cell-based transplantation therapies and tissue engineering (Sieber-Blum 2014).

In this chapter, we describe protocols and troubleshooting advices for isolating and characterizing NC cells from mouse and avian embryos, as well as adult NCSCs from the mouse hair follicle. The technique involves microdissection of neural tube segments or mouse whisker hair follicles, followed by periods of primary culture in which NC cells migrate onto the culture dish surface. In a second step, tissue explants are removed and isolated NC cells passaged and grown as adherent monolayers. The cells can be further induced to differentiate into specific phenotypes upon appropriated culture conditions and are amenable to chemical screenings and population-based analysis of gene expression. These cultures have been extensively used to model NC morphogenesis and represent an accessible *in vitro* system to investigate cell intrinsic and environmental cues that regulate NC cell migration, self-renewal, lineage commitment and differentiation.

## 7.2 Embryonic NC Cultures

### *7.2.1 Reagents*

- Gestating mice or fertilized and incubated chicken/quail eggs.
- Alfa-Modified Eagle's Medium (αMEM, Gibco, Cat. No. 11.900-024).

- Phosphate buffer saline (PBS) without $Ca^{2+}$ and $Mg^{2+}$.
- Pancreatin solution at 6 mg/ml prepared in PBS (Sigma-Aldrich, Cat. No. P3292).
- Fetal calf serum (FCS, e.g. Vitrocell).
- Penicillin-streptomycin solution at 10,000 μ/ml (Invitrogen, Cat. No. 15140148).
- Hydrocortisone (Sigma-Aldrich, Cat. No. H0135).
- Transferrin (Sigma-Aldrich, Cat. No. T5391).
- 3,3,5-Thio-iodo-thyronine (T3, Sigma-Aldrich, Cat. No. T5516).
- Glucagon (Sigma-Aldrich, Cat. No. G3157).
- Epidermal growth factor (EGF, Invitrogen, Cat. No. PHG0314).
- Fibroblast growth factor 2 (FGF2, Invitrogen, Cat. No. PHG0024).
- Insulin (Sigma-Aldrich, Cat. No. I6634).
- Trypsin (0.25 %, wt/vol) 1 mM EDTA in PBS (Invitrogen, Cat. No. 25200-056).
- Rat-tail Collagen I (e.g. Corning, Cat. No. 354236).

### 7.2.2 *Equipment*

- Stereomicroscope (e.g., Leica MZ 7.5; Olympus SZ or equivalent).
- Swan-neck fiber optic illumination system.
- Sterile 35 and 100 mm tissue culture dishes.
- 4″ Curved scissor (Fig. 7.1).
- Vannas or Pascheff-Wolff spring scissors (Fig. 7.1).
- Pair of dissecting forceps (e.g. Dumont no. 5) (Fig. 7.1).
- Stainless steel holder with tungsten dissecting needle (Fig. 7.1).
- Perforated spoon (Fig. 7.1).
- Glass Pasteur pipettes.
- Sterile tissue culture hoods (stereomicroscope should be fit in).
- Tissue culture $CO_2$ incubator.
- Neubauer chamber.
- Inverted phase contrast microscope (e.g., Olympus CK2).

### 7.2.3 *Reagent and Equipment Setup*

1. **Embryo preparation**. For mouse NC cell cultures, gestating mice should be plugged 8–10 days prior to the procedure. For avian cultures, freshly laid quail eggs should be incubated at 38 °C with 60 % humidity for 26–55 h. Table 7.1 lists the optimal developmental stages for the preparation of avian and murine NC cell cultures.
2. **Equipment sterilization**. Bake microdissection tools and glass Pasteur pipettes in a dry oven at 150 °C for 2 h. Rinse stereomicroscope and illumination system with

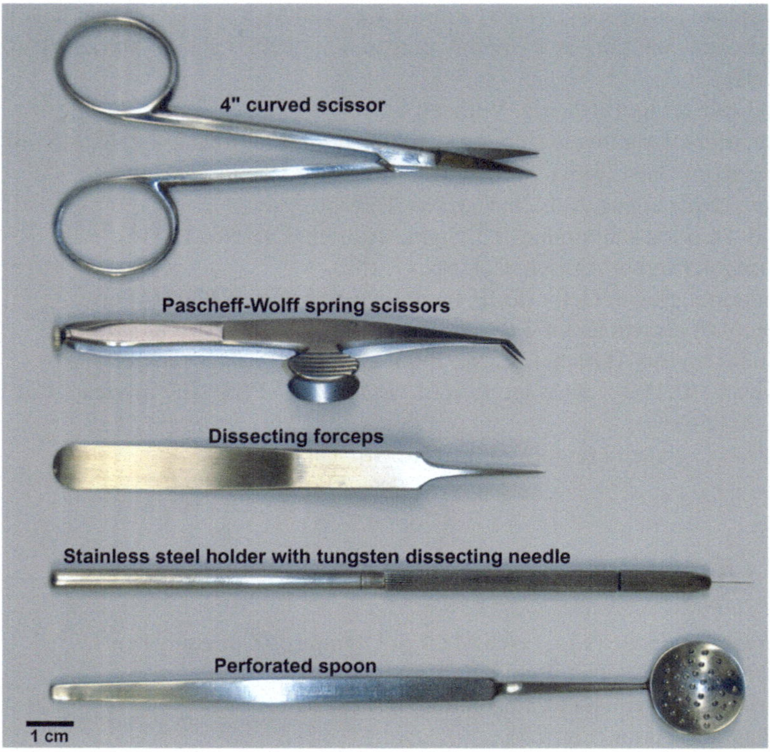


**Fig. 7.1** Microsurgery tools required for the protocol. Ideally, instruments should be made of stainless steel and sterilized by dry heat for 2 h at 150 °C before the procedures

**Table 7.1** Developmental stages of mouse and quail embryos for NC cell cultures

| | Mouse | Quail |
|---|---|---|
| Cephalic NC | 8.0–8.5 gestation days | 26–30 h[a] (5–7 somites) |
| Trunk NC | 9.0–10.0 gestation days | 36–46 h[a] (16–26 somites) |

Embryonic ages may vary slightly according to the researcher objectives
[a]Incubation time of quail eggs at 38 °C in 60 % humidity

70 % ethanol and fit inside the tissue culture hood. UV-C germicidal light can be used to further sterilize the equipment. Importantly, use sterile technique at all times.

3. **Media preparation.** Before the dissection procedure, prepare the complete NC cell culture medium by combining 9 ml α-MEM, 1 ml FCS and 200 μl chicken embryo extract. Filter the medium through a 0.22 μm filter and add the following components: 100 ng/ml hydrocortisone, 10 μg/ml transferrin, 1 ng/ml insulin, 400 pg/ml T3, 10 pg/ml glucagon, 100 pg/ml EGF, 200 pg/ml FGF2, 200 U/ml penicillin and 10 μg/ml streptomycin. For wash medium, combine 9 ml α-MEM with 1 ml FCS. Media can be stored for up to 2 weeks at 4 °C.
4. **Chicken embryo extract.** Incubate fertilized white chicken eggs for 11–14 days at 38 °C with 60 % humidity. Clean the shell eggs with 70 % ethanol. Break the eggshells and place embryos on 100 mm dishes with ice-cold PBS. Remove the

eyes with the aid of scissors and macerate embryos by passing them through a 50 ml syringe. Transfer material directly to a 50 ml sterile centrifuge tube (use around 10 embryos per tube). Add equal volume of α-MEM medium to the tube. Pass the entire tube contents again through the 50 ml syringe. Centrifuge at 2000 × *g* for 10 min at 4 °C. Filter the supernatant with 0.8 and 0.22 μm strainers sequentially. Aliquot and store at −20 °C.

5. **Culture dish coating with rat-tail collagen I**. Start by diluting cold solubilized collagen with sterile 17 μM acetic acid to 60 mg/ml. This can be stored at 4 °C for 1 week. Cover the bottom of the 35 mm tissue culture dishes with this solution and let sit for 2 h. Make sure the entire surface is covered. Remove the solution and allow plates to dry at room temperature. Wash once with PBS.

## 7.2.4 Procedure

### 7.2.4.1 Embryo Isolation

The first step of the protocol consists of isolating embryos from extra-embryonic tissues and other contaminants.

***Note:*** *Several embryos can be isolated and stored in ice-cold PBS for up to 2 h with no adverse effects before the fine dissection.*

Mouse Embryos

1. Sacrifice mice by cervical dislocation, spray with 70 % ethanol and, with fine scissors, open the peritoneal cavity. Remove both uteri of the animal and place them into a 100 mm Petri dish with ice-cold PBS;
2. Under the stereomicroscope, remove embryos from uterus with blunt forceps and scissors. Hold firmly between individual decidua against the bottom of dish with the forceps, and gently tease uterine tissue until embryo pops out. There are usually 2–4 embryos per uterus in mice;
3. Using a perforated spoon, transfer embryos to a fresh 100 mm dish with PBS. Remove decidua, amniotic and yolk sac membranes with the forceps and dissecting needles. Transfer embryos to a new dish with PBS for the dissection of neural tube.

Quail Embryos

1. Wipe eggs with 70 % ethanol and cut a small hole in the blunt end of eggshell with dissecting scissors;
2. Transfer carefully the content to a 100 mm dish with PBS. Grasp extra-embryonic membranes with forceps. Cut entirely around its perimeter with the Vannas or Pascheff-Wolff spring scissors;
3. Transfer the embryos to a fresh dish with ice-cold PBS. Use only embryos of age-appropriate stage and normal morphology.

#### 7.2.4.2 Neural Tube Isolation

This step consists of isolating the neural tube from surrounding tissues according to specific NC region of interest (Fig. 7.2). During this step, we recommend to place the dishes containing the embryos on a black background under a stereomicroscope. Use the swan-neck fiber optic illumination system to assist the visualization.

***Note:*** *Several isolated neural tube explants can be stored in washing medium on ice for up to 1 h with no adverse effects.*

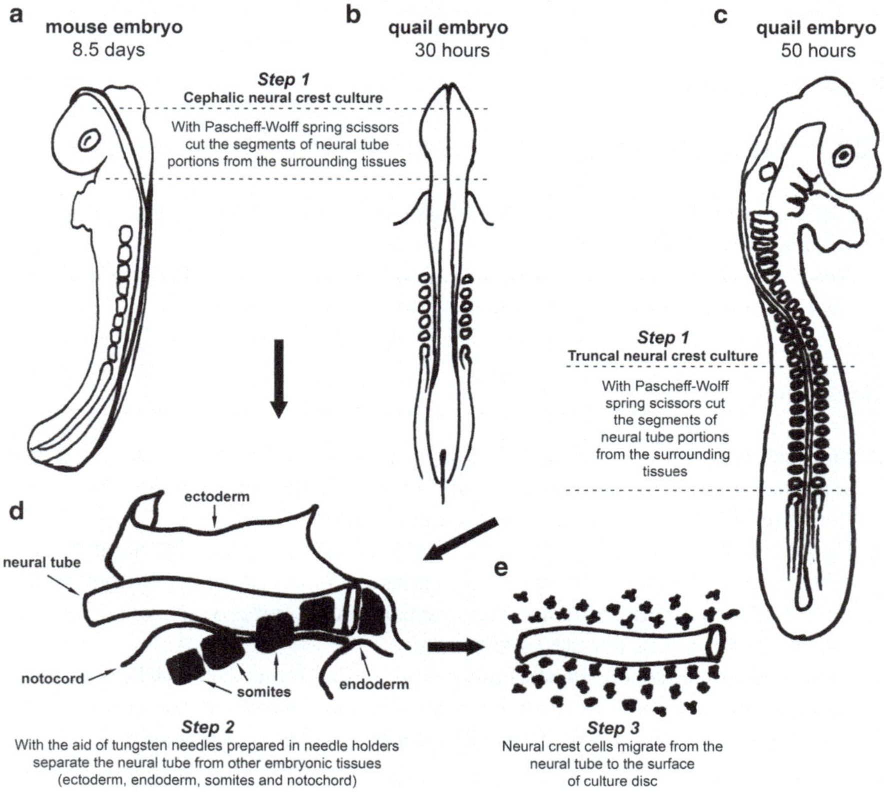


**Fig. 7.2** General scheme for embryonic NC cultures. Step 1 for mouse (**a**) and quail (**b**) cephalic NC cultures: cut the anterior region of embryos (*dashed lines* in **a** and **b**) as close as possible of neural tube. Step 2 for quail trunk NC culture: (**c**) cut the truncal region of embryo corresponding to the last 10 somites. Observe the appropriated developmental stage of embryos. Step 2 for mouse trunk NC culture use the same step shown in (**c**) (very similar in quail and mouse at this age). Step 2 for both cephalic and trunk NC cultures: incubate isolated sections in pancreatin solution for 15 min. Subsequently separate neural tube d from other embryonic tissues with aid of tungsten needles (**d**). Step 3: transfer isolated neural tube to culture dish in the desired culture condition and incubate at 37 °C for 24 h in humidified atmosphere of 5 % $CO_2$ and 10 % $O_2$ (**e**). NC cells migrate out of neural tubes toward the surface of culture recipient

1. Using Vannas or Pascheff-Wolff spring scissors, make two transversal cuts at the desired levels of neural tube (Fig. 7.2);
2. With a glass pipette, transfer tissue segments to a 35 mm dish containing pancreatin solution and incubate at room temperature for 15 min;
   ***Note:*** *Exact time length is empirical and can vary with embryo stage and specific activity of pancreatin. Explants should be removed from the pancreatin solution when the explants become slightly wavy and somites can be separated intact from the ectoderm and endoderm upon gentle agitation with the forceps.*
3. Transfer the tissue segments to a new 35 mm dish containing washing medium. With dissecting needles, gently remove the ectoderm, endoderm, somites and notochord away from the neural tube. Be careful not to damage the neural tube.
   ***Note:*** *Neural tube usually bends slightly with dorsal side convex, which helps to identify dorso-ventral polarity.*

#### 7.2.4.3 Primary NC Cell Culture (Migration Phase)

1. Carefully transfer neural tube explants with the glass Pasteur pipette to a 35 mm culture dish containing approximately 800 µl of NC complete medium. For best results, pre-coat culture dish with rat-tail collagen I;
2. Incubate at 37 °C in humidified atmosphere of 5 % $CO_2$ and 10 % $O_2$. Culture time may vary from 18 to 48 h according to the experimental design;

***Note:*** *Medium volume is critical. Use just enough so that the neural tubes explants are not exposed. Minimal medium volume increases neural tube contact with culture dish surface facilitating attachment and NC cell emigration.*

#### 7.2.4.4 Secondary NC Cell Cultures (Differentiation Phase)

After NC cells have migrated out of the neural tube explants, secondary cultures can be prepared by dislodging and replating emigrated NC cells at sparse density. At this point, NC cells can also be replated at single cell density, therefore making it possible to perform clonal analysis.

1. Under a stereomicroscope carefully remove neural tubes out of the culture dish surface. Use tungsten needles or aforceps to scrape away explanted neural tubes and neuroepithelial tissue;
2. Wash 3 times with washing medium to remove residual tissue fragments;
3. Incubate with trypsin/EDTA solution at 37 °C for 5 min. Control the reaction in an inverted phase contract microscope;
4. Block trypsin reaction with wash medium. Transfer cells to a 15 ml centrifuge tube and spin at room temperature at 1000 × *g* for 5 min.

5. Ressuspend the pelleted cells in NC complete medium and count them in a Neubauer chamber. Plate cells into a fresh culture dish. Cell density and media composition may vary according to the experimental design.
6. Incubate at 37 °C in humidified atmosphere of 5 % $CO_2$ and 10 % $O_{2.}$ NC cells may be maintained until 15 days under these conditions.

***Note:*** *Secondary cultures may not be required depending on the experimental design. Subculturing, however, allows cell expansion, which is particularly important for experiments that require larger amounts of cells (such as analysis of gene and protein expression). Secondary cultures are also recommended for cell differentiation experiments since it provides a more homogenous cell population. In addition, NC cells can also be plated at single-cell density for clonal analysis (for more details refer to Bittencourt et al.* 2013; *Trentin et al.* 2004; *Calloni et al.* 2007*). For these type of cultures, make sure a single cell solution has been obtained. This can be accomplished by serial of dilution in 96-well plates. Use the inverted phase-contrast microscope to verify the presence of single cells per well.*

## 7.3 Epidermal Neural Crest Stem Cell (Epi-NCSC) Cultures

Here, we also provide a detailed protocol for preparing mouse EPI-NCSCs cultures from whisker hair follicles, which can also be adapted for isolation of human EPI-NCSCs of scalp hair follicles (Sieber-Blum et al. 2004). Reagents and equipment necessary for this protocol are the same listed above for embryonic NC cell cultures.

### *7.3.1 Isolation of the Bulge Area of Mouse Whisker Hair Follicles*

1. Sacrifice animals by cervical dislocation. We normally use 8- to 12-weeks-old B57BL/6 mice. Other mouse strains can also be used;
2. Rinse facial region with 70 % ethanol;
3. Remove the whisker pads using forceps and dissecting scissors. Avoid cutting into the hair bulbs. Place the isolated pads on ice-cold PBS;
4. Under a stereomicroscope placed into the culture hood, dissect individual whisker follicles. Hold skin next to the follicles with a forceps and lift them out of the whisker pad with another forceps. Poll individual hair follicles in a new dish with fresh PBS;
5. Using Vannas or Pascheff-Wolff spring scissors, make a transverse cut to remove the dermal papilla/bulb of the follicles. Then make a second transverse cut right below the sebaceous gland. The isolated segment corresponds to the bulge area of whisker hair follicle, which is surrounded by a collagen capsule;

6. Grab an end of the capsule with the forceps and roll out the bulge with a bent dissecting needle. You would now be able to see the empty capsule and the isolated bulge;
7. Pool the isolated bulges in a new culture dish with α-MEM. These can be kept on ice for 2 h with no adverse effects.

### 7.3.2 *Culture of Bulge Explants*

1. Pre-incubate rat-tail collagen I-coated 35 mm dishes with 500 μl of culture medium in $CO_2$ incubator for approximately 1 h. Medium consists of 85 % α-MEM and 15 % FCS;
2. Plate 10–15 bulge explants per 35 mm dish. Alternatively, bulges can be plated individually into 24-well plates. Incubate for 2 h at 37 °C in humidified atmosphere of 5 % $CO_2$ and 10 % $O_2$. Explants should adhere to the collagen substratum;
3. Gently add 1 ml of fresh culture medium. After that, replace 50 % of the medium every 2–3 days;
4. Within 4–7 days, migratory cells will start emigrating from the bulge explants. These cells display spindle or stellate morphology and lack close cell-to-cell contact. Over the next few days, more cells will migrate out the explants and start to proliferate rapidly;

   ***Note:*** *Rare cells with flattened or cobblestone-like morphology might also emerge in some cultures. These cells correspond to putative epidermal stem/progenitor cells and/or follicular keratinocytes. Cultures mixed with these cells should be discarded.*
5. Remove the bulge explants 3–5 days after the onset of cell migration. Cells can be analyzed at this time point by different assays according to the research goal. Avoid keeping the cells for too long in primary cultures, as they tend to differentiate or undergo apoptosis at high cell density. Alternatively, cells can be harvested by tripsinization, replated and further grown as homogenous monolayers. For passaging the cells, we recommend a density of at least 50,000 cells/25 $cm^2$ flask.

## 7.4 Maintenance

Once migrated away from the neural tube explants, embryonic NC cells tend to differentiate quickly. Long-term NC cultures may therefore be challenging, as more specific culture conditions may be required for survival and self-renewal of the cells. Moreover, NC cells maintained in culture for longer periods may not faithfully represent their *in vivo* counterparts; therefore, extensive passaging should be avoided. Using the standard conditions described above, we recommend keeping

NC cells for not more than 15 days in secondary cultures. Media should be refreshed every 2–3 days. To avoid spontaneous differentiation and arrest of proliferation, cells must not be more than 70 % confluent.

## 7.5 Characterization

Embryonic NC cell cultures can be easily characterized by immunostaining for specific cell markers. Conventionally, the carbohydrate epitope HNK-1 is used to identify avian and human NC cells (Bronner-Fraser 1986; Coura et al. 2008; Etchevers 2011) (Fig. 7.3). In mice, the neurotrophin receptor p75 (p75NTR) and the intermediate filament protein nestin are used as general NC markers (Morrison et al. 1999). The transcription factor Sox10, which is required for early development of NC and play different roles throughout its development, is another commonly used NC marker (Sauka-Spengler and Bronner-Fraser 2008). Other transcription factors related to early events of NC morphogenesis include Sox9, Twist, Slug, Snail, FoxD3 and Pax3 (Sauka-Spengler and Bronner-Fraser 2008). The expression of these markers can be identified in NC cultures at the RNA level by reverse transcription (RT)-PCR or at protein level by immunostaining with specific antibodies. Similarly, to embryonic NC

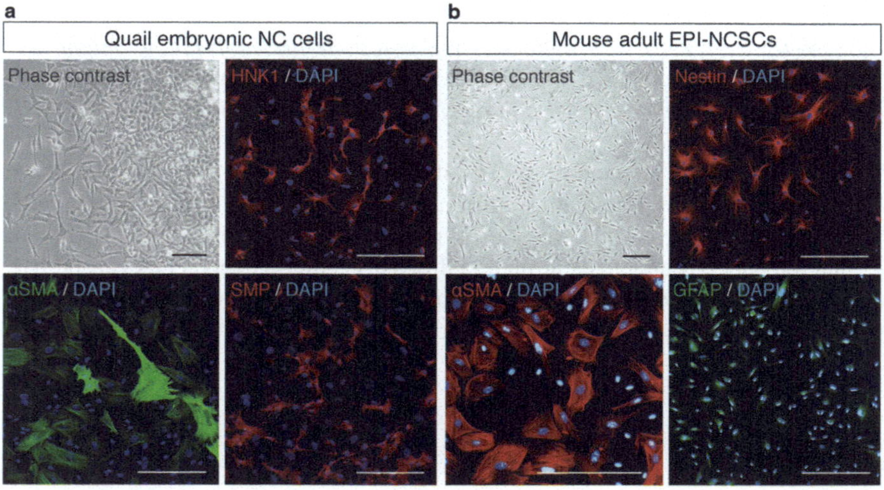


**Fig. 7.3** Characterization of NC cell cultures. (**a**) Quail embryonic NC cells and (**b**) adult mouse EPI-NCSCs. Top panel shows phase contrast images of NC cells after emigration from the tissue explants and staining for the common undifferentiated NC markers HNK-1 (quail) and Nestin (mouse). Bottom images show differentiated NC-derived cell types in secondary cultures. Smooth muscle cells are identified by staining for alpha smooth muscle actin (αSMA) in both avian and murine cultures. Glial cells are identified by staining for Schwann myelin protein (SMP) and glial fibrillary acidic protein (GFAP) in avian and murine cultures, respectively. Cell nuclei are stained with DAPI. Scale bar: 200 μm

cells, EPI-NCSC cultures can be characterized by the expression of Nestin (Fig. 7.3), p75NTR, Sox10 and other transcription factors involved in NC determination and migration *in vivo* (Bressan et al. 2014; Sieber-Blum et al. 2004). We recommend the following primary antibodies and dilutions for the characterization of NC cell cultures: HNK-1 (1:50, DHSB, Cat. No. 3H5), p75NTR (1:200, Millipore, Cat. No. Ab1554), Nestin (1:250, Abcam, Cat. No. Ab92391) and Sox10 (1:100, Abcam, Cat. No. Ab27655).

In addition to marker expression, NC cells can be further characterized by their differentiation potential *in vitro*. As the NC originates a wide range of differentiated cell types (e.g. neurons, glial cells, melanocytes, smooth muscle cells, bone, cartilage and adipose tissue), expression of differentiated cell markers are used for functional characterization of NC cultures (Calloni et al. 2007, 2009; Mayor and Theveneau 2013; Trentin et al. 2004) (Fig. 7.3). Furthermore, single cell cultures can be used to determine the differentiation potential and self-renewal capacity of NC cell clones (for further details see Trentin et al. 2004; Bittencourt et al. 2013). In some cases, differentiation into NC-derived cell types may require more specific culture conditions, such as inductive media and culture substratum.

Functionally, NC cells can be also characterized by their migratory behavior *in vitro*. Cultured NC cells display high motility, which is a reminiscent of their migratory ability *in vivo*. Various *in vitro* assays can be used to assess NC cell migration, including time-lapse cell tracking, transwell migration assay or simple measurement of the area occupied by NC cells after emigration from the NT explants in primary cultures (Costa-Silva et al. 2009). A comprehensive review on cell migration assay is provided by (Kramer et al. 2013).

## 7.6 Potential Use of NC Stem Cell Cultures

The NC cultures described in this chapter have been extensively used as model of NC morphogenesis *in vitro*. The relative ease and high fidelity manipulation render NC cultures an excellent *in vitro* system to study cellular and molecular mechanisms underlying epithelium-to-mesenchyme transition, cell migration, lineage commitment and differentiation (Dupin and Sommer 2012; Trentin and Calloni 2013).

NC cell behavior is highly influenced by environmental signals that these cells encounter along their migratory pathways during development (Le Douarin and Kalcheim 1999). We and others have employed the NC culture system to investigate the effects of environmental cues such as growth factors and extracellular matrix components on NC cell migration, proliferation and differentiation (Bittencourt et al. 2013; Costa-Silva et al. 2009; Garcez et al. 2008; Ito and Morita 1995; Kleber et al. 2005; Lahav et al. 1996).

In addition to the population-based analysis, clonal NC cultures can be used to assess the whole spectrum of developmental potential of single NC cells. By the use of this technique, the presence of several types of multipotent progenitor and stem cells has been reported in different regions of the avian NC (reviewed by Dupin et al. 2010; Dupin and Sommer 2012). The *in vitro* clonal approach also permits to challenge NC cell differentiation options in order to characterize the influences of extracellular factors on the behavior of specific sets of NC progenitors (Trentin and Calloni 2013). For instance, the influence of soluble factors such as endothelin-3, sonic hedgehog and FGF2 on self-renewal and fate choice have been investigated by series of cloning and subcloning of avian NC cells (Bittencourt et al. 2013; Trentin et al. 2004; Calloni et al. 2007).

Despite its importance to developmental biology, studying the NC is also relevant for the understanding of a diverse range of human pathologies, which are known to be associated with abnormal NC development. Collectively referred as neurocristopathies, these conditions include multiple neoplasia, skin pigmentation disorders and various craniofacial and cardiovascular malformations (Hall 2010). Embryonic NC cell cultures, thus, also represent an accessible system to investigate the etiopathogeny of NC-related human disorders. For instance, mouse and chicken primary NC cultures have been used to study cellular mechanisms involved in cephalic and cardiac malformations associated with folic acid deficiency and/or hyperhomocysteinemia as well as fetal alcohol syndrome (Boot et al. 2003; Jaurena et al. 2011; Rovasio and Battiato 2002).

In addition to the study of fundamental developmental biology questions, the recent identification of NC-derived stem/progenitors cells in several postnatal tissues opens up opportunities for the use of NC stem cells for pharmaceutical screenings, cell-based transplantation therapies and tissue engineering (Sieber-Blum 2014). The protocol described in this chapter allows the isolation of a relatively pure population of adult EPI-NCSCs from mouse whisker follicles (Sieber-Blum et al. 2004). This protocol can also be adapted for the isolation of human EPI-NCSCs from scalp hair follicles (Krejčí and Grim 2010; Clewes et al. 2011). Both mouse and human EPI-NCSCs are easily accessible by a minimally invasive procedure and can generate multiple cells types of clinical relevance *in vitro*, including neurons, glia, myofibroblast/smooth muscle cells, chondrocytes, and melanocytes (reviewed in Sieber-Blum 2014).

The easy isolation together with their multilineage potential render the possibility of using these cells for transplantation in variety of disorders very attractive. For instance, it has been shown in animal models that transplanted mouse EPI-NCSCs are able to integrate into the injured spinal cord tissue and elicit improvement of sensory connectivity and in touch perception. The intraspinal EPI-NCSC did not form tumours and were able to differentiate to some extend into GABAergic neurons and myelinating oligodendrocytes. Besides cell replacement, the transplanted EPI-NCSCs are through to exert other pertinent functions in the contused spinal cord, including neuroprotection, angiogenesis and modulation of scar formation (Sieber-Blum et al. 2006; Hu et al. 2010; Sieber-Blum 2010).

In addition, human EPI-NCSCs have recently been shown as a feasible autologous cell source for future applications in Parkinson's disease and peripheral nerve injury, as they can be used for the generation of highly pure populations of dopaminergic neurons and Schwann cells (Narytnyk et al. 2014; Sakaue and Sieber-

Blum 2015). The ability of isolate these cells together with the application of robust differentiation protocols and *in vivo* injury models should now help to further assess the potential of EPI-NCSCs in the field of regenerative medicine.

**Acknowledgments** We would like to thank Nicole Le Douarin and Elisabeth Dupin for constant encouragement and Diego Amarante da Silva for providing the images of quail NC cultures. RBB is supported by a fellowship from the Science Without Borders Program (CAPES, Brazil).

## References

Baggiolini A, Varum S, Mateos JM, Bettosini D, John N, Bonalli M, Ziegler U, Dimou L, Clevers H, Furrer R, Sommer L (2015) Premigratory and migratory neural crest cells are multipotent in vivo. Cell Stem Cell 16:314–322

Bittencourt DA, Costa MC, Calloni GW, Alvarez-Silva M, Trentin AG (2013) Fibroblast growth factor 2 promotes the self-renewal of bipotent glial smooth muscle neural crest progenitors. Stem Cells Dev 22:1241–1251

Boot MJ, Steegers-Theunissen RPM, Poelmann RE, Van Iperen L, Lindemans J, Gittenberger-de Groot AC (2003) Folic acid and homocysteine affect neural crest and neuroepithelial cell outgrowth and differentiation in vitro. Dev Dyn 227:301–308

Bressan RB, Melo FR, Almeida PA, Bittencourt DA, Visoni S, Jeremias TS, Costa AP, Leal RB, Trentin AG (2014) EGF-FGF2 stimulates the proliferation and improves the neuronal commitment of mouse epidermal neural crest stem cells (EPI-NCSCs). Exp Cell Res 327:37–47

Bronner-Fraser M (1986) Analysis of the early stages of trunk neural crest migration in avian embryos using monoclonal antibody HNK-1. Dev Biol 115:44–55

Calloni GW, Glavieux-Pardanaud C, Le Douarin NM, Dupin E (2007) Sonic hedgehog promotes the development of multipotent neural crest progenitors endowed with both mesenchymal and neural potentials. Proc Natl Acad Sci USA 104:19879–19884

Calloni GW, Le Douarin NM, Dupin E (2009) High frequency of cephalic neural crest cells shows coexistence of neurogenic, melanogenic, and osteogenic differentiation capacities. Proc Natl Acad Sci USA 106:8947–8952

Clewes O, Narytnyk A, Gillinder KR, Loughney AD, Murdoch AP, Sieber-Blum M (2011) Human epidermal neural crest stem cells (hEPI-NCSC)—characterization and directed differentiation into osteocytes and melanocytes. Stem Cell Rev 7:799–814

Costa-Silva B, Costa MC, Melo FR, Neves CM, Alvarez-Silva M, Calloni GW, Trentin AG (2009) Fibronectin promotes differentiation of neural crest progenitors endowed with smooth muscle cell potential. Exp Cell Res 315:955–967

Coura GS, Garcez RC, Mendes de Aguiar CB, Alvarez-Silva M, Magini RS, Trentin AG (2008) Human periodontal ligament: a niche of neural crest stem cells. J Period Res 43:531–536

Douarin NM, Dupin E (2003) Multipotentiality of the neural crest. Curr Opin Gen Dev 13:529–536

Dupin E, Sommer L (2012) Neural crest progenitors and stem cells: from early development to adulthood. Dev Biol 366:83–95

Dupin E, Calloni GW, Douarin NML (2010) The cephalic neural crest of amniote vertebrates is composed of a large majority of precursors endowed with neural, melanocytic, chondrogenic and osteogenic potentialities. Cell Cycle 15:238–249

Etchevers H (2011) Primary culture of chick, mouse or human neural crest cells. Nat Protoc 6:1568–1577

Fernandes KJL, Kobayashi NR, Gallagher CJ, Barnabé-Heider F, Aumont A, Kaplan DR, Miller FD (2006) Analysis of the neurogenic potential of multipotent skin-derived precursors. Exp Neurol 201:32–48

Garcez RC, Teixeira BL, Schmitt SS, Alvarez-Silva M, Trentin AG (2008) Epidermal growth factor (EGF) promotes the in vitro differentiation of neural crest cells to neurons and melanocytes. Cell Mol Neurobiol 29:1087–1091

Hall BK (2010) The neural crest and neural crest in vertebrate development and evolution. Springer, New York

Hu YF, Gourab K, Wells C, Clewes O, Schmit BD, Sieber-Blum M (2010) Epidermal neural crest stem cell (EPI-NCSC)—mediated recovery of sensory function in a mouse model of spinal cord injury. Stem Cell Rev 6:186–198

Ito K, Morita T (1995) Role of retinoic acid in mouse neural crest cell development in vitro. Dev Dyn 204:211–218

Jaurena MB, Carri NG, Battiato NL, Rovasio RA (2011) Trophic and proliferative perturbations of in vivo/in vitro cephalic neural crest cells after ethanol exposure are prevented by neurotrophin 3. Neurotoxicol Teratol 33:422–430

Kleber M, Lee HY, Wurdak H, Buchstaller J, Riccomagno MM, Ittner LM, Suter U, Epstein DJ, Sommer L (2005) Neural crest stem cell maintenance by combinatorial Wnt and BMP signaling. J Cell Biol 169:309–320

Kramer N, Walzl A, Unger C, Rosner M, Krupitza G, Hengstschlager M, Dolznig H (2013) In vitro cell migration and invasion assays. Mutat Res 752:10–24

Krejčí E, Grim M (2010) Isolation and characterization of neural crest stem cells from adult human hair follicles. Folia Biol (Praha) 56:149–157

Kruger GM, Mosher JT, Bixby S, Joseph N, Iwashita T, Morrison SJ (2002) Neural crest stem cells persist in the adult gut but undergo changes in self-renewal, neuronal subtype potential, and factor responsiveness. Neuron 35:657–669

Lahav R, Ziller C, Dupin E, Le Douarin NM (1996) Endothelin 3 promotes neural crest cell proliferation and mediates a vast increase in melanocyte number in culture. Proc Natl Acad Sci USA 93:3892–3897

Le Douarin NM, Kalcheim C (1999) The neural crest, 2nd edn. Cambridge Universtiy Press, Cambridge

Mayor R, Theveneau E (2013) The neural crest. Development 140:2247–5221

Morrison SJ, White PM, Zock C, Anderson DJ (1999) Prospective identification, isolation by flow cytometry, and in vivo self-renewal of multipotent mammalian neural crest stem cells. Cell 96:737–749

Narytnyk A, Verdon B, Loughney A, Sweeney M, Clewes O, Taggart MJ, Sieber-Blum M (2014) Differentiation of human epidermal neural crest stem cells (hEPI-NCSC) into virtually homogenous populations of dopaminergic neurons. Stem Cell Rev 10:316–326

Rovasio RA, Battiato NL (2002) Ethanol induces morphological and dynamic changes on in vivo and in vitro neural crest cells. Alcohol Clin Exp Res 26:1286–1298

Sakaue M, Sieber-Blum M (2015) Human epidermal neural crest stem cells as a source of Schwann cells. Development 142:1–10

Sauka-Spengler T, Bronner-Fraser M (2008) A gene regulatory network orchestrates neural crest formation. Nature Rev Mol Cell Biol 9:557–568

Shakhova O, Sommer L, Biology D (2010) Neural crest-derived stem cells. StemBook 1–20 (http://www.ncbi.nlm.nih.gov/books/NBK44752/)

Sieber-Blum M (2010) Epidermal neural crest stem cells and their use in mouse models of spinal cord injury. Brain Res Bull 83:189–193

Sieber-Blum M (2014) Human epidermal neural crest stem cells as candidates for cell-based therapies, disease modeling, and drug discovery. Birth Defects Res C Embryo Today 102:221–226

Sieber-Blum M, Grim M, Hu YF, Szeder V (2004) Pluripotent neural crest stem cells in the adult hair follicle. Dev Dyn 231:258–269

Sieber-Blum M, Schnell L, Grim M, Hu YF, Schneider R, Schwab ME (2006) Characterization of epidermal neural crest stem cell (EPI-NCSC) grafts in the lesioned spinal cord. Mol Cell Neurosci 32:67–81

Trentin AG, Calloni GW (2013) The stem cells of neural crest. In: Resident stem cells and regenerative therapy. Academic, Oxford, UK, pp 157–176

Trentin AG, Glavieux-Pardanaud C, Le Douarin NM, Dupin E (2004) Self-renewal capacity is a widespread property of various types of neural crest precursor cells. Proc Natl Acad Sci USA 101:4495–4500

Yoshida S, Shimmura S, Nagoshi N, Fukuda K, Matsuzaki Y, Okano H, Tsubota K (2006) Isolation of multipotent neural crest-derived stem cells from the adult mouse cornea. Stem Cells 24:2714–2722

# Chapter 8
# Cancer Stem Cells: Issues with In Vitro Expansion and Model Systems

**Khadidiatou Guiro, Garima Sinha, Oleta Sandiford, Treena L. Arinzeh, and Pranela Rameshwar**

**Abstract** Despite tremendous progress, breast cancer (BC) still remains a clinical issue. Reports have confirmed the presence of a subset of BC cells (BCCs) with self-renewal ability, which are referred as cancer stem cells (CSCs). The CSCs can remain dormant for decades following either incomplete removal of the primary tumor or through resisting current chemo-, radio- and endocrine treatments. The biology of breast CSCs is poorly understood. These cells can persist either by completely withdrawing from the cell cycle or by continuing to proliferate at a slower rate. Therefore, developing methods to characterize breast CSCs may help develop innovative-targeted therapies to control or eliminate the resistant cells. Traditional *in vitro* studies for breast CSCs usually use two-dimensional (2-D) monolayer cultures. Moreover, in order to mimic the architectural features of the *in vivo* environment, most recent reports use thee-dimensional (3-D) culture systems to address some of the limitations of the 2-D monolayer culture. This chapter discusses the current methodologies used for cancer stem cells (CSCs). The methods are also discussed in the context of specific applications.

**Keywords** Cancer stem cells • Breast cancer • Bone marrow

K. Guiro • O. Sandiford
Department of Medicine – Hematology/Oncology, Rutgers, New Jersey Medical School, Newark, NJ, USA

G. Sinha • P. Rameshwar (✉)
Department of Medicine – Hematology/Oncology, Rutgers, New Jersey Medical School, Newark, NJ, USA

Rutgers Graduate School of Biomedical Health Sciences at New Jersey Medical School, 185 South Orange Avenue, MSB E-585, Newark, NJ 07103, USA
e-mail: rameshwa@njms.rutgers.edu

T.L. Arinzeh
Department of Biomedical Engineering, New Jersey Institute of Technology, Newark, NJ, USA



H. Ulrich, P. Davidson Negraes, *Working with Stem Cells*,
DOI 10.1007/978-3-319-30582-0_8

## 8.1 Introduction

Cancer Stem Cells (CSCs) are a relatively small subset of cancer cells that resist radiation, chemotherapy and can initiate tumor (Rycaj and Tang 2015). CSCs can originate from mutation of normal stem cells or can be derived from mutation of differentiated cells (transit amplifying or progenitor), resulting in these cells adapting the property of stem cells. CSCs are generally characterized by relatively long doubling time, asymmetric division, and *in vitro* and *in vivo* serial passages (Patel et al. 2012). Most of the genes expressed by CSCs are also expressed by normal stem cells, such as Notch, Nanog and Oct4 (Yin et al. 2015). Therefore, it is important to combine multiple techniques when determining if a subset of cancer cells can be designated as CSCs. Among these techniques, functional studies are crucial before assigning the cancer cells as those with stem function. Specifically, it is advisable that <100 cells of the suspected population should be able to initiate the tumor *in vivo*.

The literature overwhelmingly reports on specific markers for CSCs with the reports of particular tissue-specific phenotype (Visvader and Lindeman 2008, 2012). Although the assignment of specific markers was based on sound scientific data, as more techniques are developed, it is generally accepted that the reported markers cannot adequately identify CSCs. One of the major issues that may be responsible for the confusion in the literature is based on the methods used to select the CSCs for experimental studies. During the early period of biologists taking note of CSCs, several markers were reported for CSCs in different types of tumors. Since these reports were presented to shown these markers as definitive, they were used in other reports on CSCs. Based on the extensive literature on cells selected with these markers, it is clear that the subsets originally reported as CSCs are in fact, a heterogeneous population. More importantly, the reported markers were applied in research studies to select CSCs from patients' tissues for expansion *in vitro*. The expanded cells were then used in multiple assays with the intent that the studies were performed with CSCs. The current data in the literature showed that CSCs cannot be expanded without these cells reverting into cells showing heterogeneity (Patel et al. 2012). The heterogeneity is a natural course when the stem cells differentiate and then proliferate to make the tumors into a mass of different malignant cells (Patel et al. 2012).

The definition of cancer dormancy varies depending on the author. However, it is accepted by most scientists that dormancy can be divided into two main categories: a tumor mass that lacks the appropriate microenvironmental cues and stop dividing, and those that are single cells, mostly found as disseminated tumor cells (Osisami and Keller 2013 ). These latter cells are generally the CSC subset. Dormant BCCs can be found in several organs, including the region around the vasculature (Ghajar et al. 2013).

The development of drugs to target CSC is a major challenge. This is especially a problem if the target is a stem cell-associated gene, such as the current Notch inhibitors (Jamdade et al. 2015). By selecting stem cell-linked genes to target the CSCs add to the challenge of eliminating the CSCs since this type of treatment could also target the normal stem cells, resulting in overt toxicity. Another challenge is how to develop methods to reverse the dormant phenotype adapted by CSCs. In dormancy, CSCs interact with cells of the microenvironment and are generally in cycling quiescence and, are chemoresistant (Ford et al. 2015; Mitra et al. 2015; Sosa et al. 2014).

This chapter focuses on methods to study breast BCCs and the complex interaction that allows the cancer cells to exist within the bone marrow (BM) niche (Patel et al. 2011). This organ is selected because it is the area where the BCCs seems to prefer at an early phase of dissemination (Pantel and Alix-Panabieres 2014). In BM, dormant BCCs exist within the stromal compartment among hematopoietic supporting cells (Bliss et al. 2014). This region is home to endogenous hematopoietic stem cells (Hanoun et al. 2015). Thus, this presents a scenario in which the dormant CSCs and the hematopoietic stem cells are in the same region. Since both are stem cells and shares long doubling time, it would be a challenge to target the CSCs without harm to the endogenous hematopoietic cells. To develop new treatment, it is important to first study the interaction between breast CSCs and cells of the microenvironment. This would provide regions of therapeutic targets to reverse dormancy for further chemotherapeutic intervention.

CSCs and cells of the BM microenvironment such as stroma and mesenchymal stem cells (MSCs) can communicate through channels formed by gap junction (GJIC) (Patel et al. 2014). In addition, the interaction between different cancer cells and cells of the BM microenvironment can occur through the release of small vesicles such as exosomes from one cell type to the other (Ono et al. 2014). The molecular consequence in the intercellular communication between BCCs and BM cells are a subject of intense investigation that will be discussed in this chapter. The studies reported on miR-222, -223,-127 being transported BM stromal cells to BCCs via GJIC to decrease the production of CXCL12 in the BCCs (Lim et al. 2011). As CXCL12 is decreased, connexin43 (Cx43) is activated to sustain GJIC (Park et al. 2013). Other studies have reported the complexity of a multi-step process in the ability of BCCs to adapt dormancy in the BM (Patel et al. 2011).

In order to understand the complex process by which BCCs and endogenous BM cells interact, the model and culture system have to be carefully selected. The methods will be established to assess how soluble factors such as cytokines and small vesicles such as exosomes, mediate intercellular complex to achieve dormancy or metastasis.

## 8.2 2-Dimensional (2-D) Culture System

Stem cells (SCs) are defined as multipotent cells that can both self-renew and differentiate long specific lineage into specialized cells. The increased interest in cell-based therapy has redirected research in every field to understand how SC property can be exploited in medicine. The is one of these SCs is one of the first step in achieving our goal.

This section will discuss 2-D cultures of adult BM-derived SCs as well as CSCs. The problems for prolonged culture are also discussed in this chapter. 2-D cultures refer to the use of tissue culture plates in which the cells are allowed to adhere to the plate's surface. The chapter will describe the isolation and when required, the expansion of BM-derived human mesenchymal stem cells (MSCs), human hematopoietic stem/progenitor cells (HSCs/HPSCs), and breast CSCs.

## 8.2.1 Culture of MSCs

The method described in this section use human BM aspirates that can be purchased from a commercial source, or can be taken by a hematologist from the iliac crest. However, it should be noted that human MSCs can be obtained from other sources such as adipose tissues, placenta, amniotic fluid, dental pulp and Wharton jelly of the umbilical cord. In addition, BM-derived MSCs can be obtained from several commercial sources such as American Type Culture Collection and Texas A&M Health Science Center.

If the BM is obtained from the research institution, the process needs an approval by the Institutional Review Board (IRB).

### 8.2.1.1 Reagents and Antibodies

- Culturing plasma treated dish (BD Falcon, Cat. No. 353003).
- High glucose DMEM (Sigma-Aldrich, Cat. No. D6546).
- Low glucose DMEM (Sigma-Aldrich, Cat. No. D5921).
- Fetal Bovine Serum, FBS (Sigma-Aldrich, Cat. No. F2442).
- L-glutamine (Sigma-Aldrich, Cat. No. G8540).
- Antibiotic penicillin/streptomycin (Sigma-Aldrich, Cat. No. P0781).
- Histopaque, for gradient separation (Sigma-Aldrich, Cat. No. 10771).
- Heparin sodium salt (Sigma-Aldrich, Cat. No. H3393).
- 0.05 % Trypsin (Sigma-Aldrich, Cat. No. 59417C).
- Phosphate Saline Buffer, PBS (Sigma-Aldrich, Cat. No. D8537).
- α-MEM with glutamax (Thermo Fisher, Gibco, Cat. No. 32571).
- pEGFP1-Oct3/4, used for differentiating between the different subset, was provided by Dr. Wei Cui (Imperial College London, UK).
- anti-human antibodies: CD45-PerCP-Cy™5.5, CD34-FITC, CD105-PE, CD44-APC, CD29-APC, CD-38-PE (BD Biosciences).

### 8.2.1.2 Protocol for Culture

1. Aspirates are obtained in a syringe containing 1/10 volume of 50 U sodium heparin diluted in any tissue culture medium with fetal calf sera (FCS);
2. Transfer 1 ml of the aspirate to plasma-treated tissue culture flasks or dishes with four volumes of Dulbecco's Modified Eagles media (DMEM) containing 10% FCS, L-glutamine and if needed, antibiotics (Penicillin-Streptomycin) (D10 media);
3. Incubated plates for 3 days at 37 °C with 5 % $CO_2$. At day 3, the red blood cells and neutrophils are removed by density gradient;

***Gradient separation using Ficoll-Hypaque***

4. Mix the media in the plates or dishes and then layer it on top of the Ficoll Hypaque at a 1:1 ratio in 15 or 50 ml tissue culture tubes;
5. Centrifuged at 800×*g* at room temperature for 30 min. During this time, add fresh D10 media to the flasks or dishes. The culture flasks/plates can be kept under the hood until the completion of density gradient separation, or they can be replaced in the incubator;
6. Immediately after the centrifugation with the Ficoll Hypaque, carefully aspirated the upper layer, which would be media mixed with plasma and sera. Next, carefully remove the white buffy coat layer on top of the Ficoll and then transfer to a clear tissue culture tube;

   ***Note:*** *To ensure that all of the mononuclear cells within the buffy coat are acquired, it is acceptable to leave some of the upper layer during the initial aspiration and to collect as well as some of the Ficoll immediately below the buffy coat.*
7. Fill the tubes with serum-free tissue culture media. The elimination of serum is important to dilute residual Ficoll and sera. This dilution will allow for efficient pelleting of the cells. After this first pellet, you can then add sufficient media to the tubes so that 3 ml of the cell suspension are re-added to the flasks/plates.
8. If preferred, the pelleted cells can undergo one cell wash with tissue culture media. It should be noted that multiple washes would lead to the loss of the recovered mononuclear cells.
9. Incubate the flasks/plates at 37 °C with 5 % $CO_2$ until the cells are 70–80 % confluent. During this time, 50 % of the media is replaced with fresh D10 media every 5 days.

   ***Note:*** *It is important that the cells do not reach 100 %, otherwise, cells will begin to differentiate.*

#### 8.2.1.3 Passaging of Cells

1. During cell passages , collect the media and save in a large tube;
2. Wash adherent cells with tissue-culture grade phosphate buffered saline (PBS, pH 7.4) and trypsinize them using enough amounts that barely cover the surface area of the flasks/plates;
3. Add half of the trypsin to a new plate/flask. Dilute the D10 media at 1:1 ratio using the saved media and then add this mixture to both sets of plates and flasks. *This method will allow the cultures to use factors that were released by the cultured MSCs.* At this time, there will be twice the amounts of flasks/plates.
4. Repeat the above process three to five times before and then, the MSCs are ready to be used in further assays.

Before the cells are selected for assays, they should be phenotyped with the following antibodies: CD45 (–), CD34 (–), CD105 (+), CD44 (+), CD29 (+). The morphology should be symmetrical at confluence and should the cells be able to undergo multi-lineage differentiation. The least cumbersome method to assess the latter is to study the ability of the cells to undergo osteogenic, adipogenic and/or chondrogenic differentiation. These methods can be easily performed with commercial kits.

The above-explained protocol is the method performed by our research group using BM aspirates. Other labs have some slight variation in the method, such as the use of low glucose in the D10 media and the use of α-minimum essential media (α-MEM) as well as increased FCS. In our lab, the use of α-MEM caused the MSCs to differentiate into fibroblasts. We also compared high and low glucose DMEM and found that the reduced glucose caused a high level of cell senescence.

#### 8.2.1.4 Selection of HSCs/HPSCs

The culture of HSCs or their progenitors (HPSCs) is included in this chapter because to study CSC dormancy or metastasis, there may be a need to study the process in the context of the hematopoietic system. The HSCs cannot be expanded *in vitro*, but can be isolated by undergoing positive/negative selection with a cocktail of antibodies, which can be purchased from commercial sources. Otherwise, the lineage positive cells are depleted. The remaining cells are selected based on CD34 expression; the cells that co-express CD38 are selected as the progenitors.

The hematopoietic cells can be selected from BM aspirates or cord blood. In both cases, the cells are subjected to Ficoll Hypaque density gradient as described above. Since the frequency of HSCs is low, it is important that the mononuclear fraction is recovered with high efficiency. This can be accomplished by diluting the aspirates or cord blood with equal volume of serum-free tissue culture media.

#### 8.2.1.5 Expanding BCCs

The cell lines are obtained from the American Type Culture Collection (Manassas, VA, USA). In the case of MDA-MB-231, cells are maintained at 37 °C, 5% $CO_2$ humidified incubator in Dulbecco's Modified Eagle's Medium (Invitrogen, Carlsbad, CA, USA) containing 10% fetal bovine serum (Fisher Scientific, Pittsburg, PA, USA). If needed, antibiotic-antimycotic solution (Invitrogen, Carlsbad, CA, USA) is added to the media.

#### 8.2.1.6 Selection of CSCs

As discussed in the introduction section, CSCs represent a small subset of the tumor cells that resist chemotherapy and can survive at distant organs as dormant cells, ready to initiate a tumor during cancer recurrence (Naume et al. 2007). We reported on the ease by which breast CSCs adapt and survive as dormant cells in the BM (Patel et al. 2012). Therefore, isolation, characterization and analyses are important to understand how the mechanisms of their survival occur. More importantly, it is necessary to ensure that the cells selected as CSCs are indeed the tumor initiating cells.

As previously mentioned, the current markers have not been proved to be specific for CSCs. We took the approach to select different subsets of BCCs based on the expression of the stem cell gene, *Octamer4A* (*Oct4A*) (Patel et al. 2012). This poses a current problem on how to isolate CSCs from primary breast tissue. In the meantime, the different Oct4 subset has to form the basis to isolate specific genes. In the following sections, we will discuss the method used to isolate different subsets of Oct4-expressing BCCs.

BCCs are stably transfected with a vector with the Oct4A promoter that controls a heterologous gene, for cell selection. If the intent is to use flow cytometry to select a subset of CSCs, the heterologous gene could be green fluorescence protein (GFP). If the cells are intended for *in vivo* imaging, the heterologous gene could be luciferase. We have transfected BC cells with pEGFP1-Oct3/4 and then used the intensity of GFP to select different subsets (Patel et al. 2012). These high Oct4 expressing cells have been shown to be the CSCs using several methods: time lapse microscopy to study cell division, noble agar cultures with expansion and secondary noble agar assay, tumor sphere assay on low attachment plates used for serial tumor sphere. The *in vivo* approaches used different amounts of Oct4 (hi) BCCs to study serial passaging in nude mice.

The Oct4 (hi) BCCs can be placed in culture for up to 3 days or even for 5 days. However, by 2 weeks, the CSCs (Oct4 high BCCs) expand as heterogeneous BCCs in culture, indicating that there is no known method to expand CSCs *in vitro*. Until a method is developed to expand CSCs, studies with CSCs are limited. Thus, when data are described, the limitation of the methods needs to be discussed. The challenge with expanded CSCs is similar to what is now known for HSCs, as discussed above. The difference is, HSCs once taken from their microenvironment, will differentiate whereas CSCs will undergo self-renewal but will differentiate into multiple subtypes.

## 8.3 3-D Culture Systems: Overview

Several studies have aimed to target CSCs for eradication. However, as discussed above, there is a need to better understand the CSCs, including their identification. A few of these studies are cited in this section (Al-Hajj et al. 2003; Bonnet and Dick 1997; Braun et al. 2009; Diehn et al. 2009; Gonzalez-Angulo et al. 2007; Li et al. 2008; Morris and Carey 2007; Ponti et al. 2005; Sorokin 2010). Adding to the limitations of studying CSCs is the use of 2-D culture methods, which are generally considered as the gold standard for BC *in vitro* studies. However, these models to recapitulate metastasis do not always represent their occurrence in an endogenous 3-D microenvironment. The 2-D cultures lack the spatial cues, what makes it difficult to study *in vivo* cancer progression and propagation of breast CSCs (Nyga et al. 2011).

In this section, we will discuss why a 3-D culture method is needed instead of the traditional 2-D BC *in vitro* BC model. There are fundamental limitations with regards to reproducibility and flexibility of design. For instance, several 2-D models and *in vivo* animal models have been used to investigate BCCs within specific milieu, for instance the bone milieu (Kaemmerer et al. 2014; Khanna and Hunter 2005). Nonetheless, the biological complexity of human bone is difficult to recapitulate and control effectively in these models (Weigelt and Bissell 2008; Yoneda and Hiraga 2005). Moreover, 2-D cell–cell *in vitro* model often fails to mimic cellular interaction with native extracellular matrices (ECM). This lack of cell–matrix interactions can affect gene expression of both normal and malignant cells (Birgersdotter et al. 2005), potentially affecting the design of drugs for targeted therapy. To this end, scientists have implemented the integration of cancer biology and tissue engineering for the development of novel *in vitro* approaches to study CSCs behavior, and assess how the 3-D microenvironment can modulate the cells' behavior. The intent is to build a bridge to close the gap currently seen with 2-D monolayer cultures.

As one of the pioneers to use 3-D model systems, Bissell et al. confirmed that the design criteria to adequately model breast cancer *in vitro* in 3-D are similar to those used in tissue engineering (Nelson and Bissell 2005). 3-D *in vitro* culture system models for BC can be classified into the following: cells cultured as multicellular aggregates (spheroids/mammosphere), and cells embedded in constructs made of natural or synthetic matrices (scaffolds) (Ricci et al. 2013).

### 8.3.1 3-D Spheroid Culture

Aggregates of BCCs have been referred to by different names such as mammosphere, micromass, spheroids and microfabricated tissues (Talmadge 2007). Spheroids have been widely used in the cancer field as a model system in several studies involving 3-D cell culture for drug screening predominantly for high-throughput applications (Kelm et al. 2003). Spheroids are 20–1000 μm-diameter clusters of cells that self-aggregate when cultured. Unlike cells seeded on 2-D monolayer cultures, spheroid cultures have been described to generate heterogeneous cell populations that vary in response to diffusion limitations similar to the *in vivo* environment (Kunz-Schughart 1999; Lin and Chang 2008; Ghosh et al. 2005; Hauptmann et al. 1998). Spheroids have been shown to respond to chemotherapy and radiation therapy, and to maintain the physiological shape of the tumor (Hirschhaeuser et al. 2010; Ghajar et al. 2013; Pampaloni et al. 2007) while sustaining tissue-specific properties of the primary tissue (Kobayashi et al. 1993; Kunz-Schughart 1999; Sutherland and Durand 1976).

Several techniques have been presented in the literature for culturing spheroids. The most widely used techniques involve growing cells on plates coated with low attachment substrates to prevent cell-substrate interaction (Carlsson and Yuhas 1984; Ivascu and Kubbies 2006), in hanging drops (Kelm et al. 2003), or in a continually rotated suspension such as spinner flasks (Sutherland et al. 1971).

#### 8.3.1.1 Protocol: Sphere Formation Assay

Several authors have adopted the sphere formation assay generally performed as follows:

1. First, grow BCCs as non-adherent spherical clusters of cells (usually named "spheres" or "mammospheres"). Every 3 days, these spheres are enzymatically dissociated by incubation in a trypsin-EDTA solution (Invitrogen at 37 °C) for 3 min;
2. After the primary spheres dissociation, plate 100 cells/well in 96-well culture dishes in growth medium. The number of spheres for each well is then evaluated after 7 days of culture;
3. Next, downstream assays are performed to characterize the cells by their phenotype. More importantly, it is necessary to determine if the spheres contain self-renewal cells. This could be done by disassociating one sphere and then plating at one cell/well in a 96-well plate to determine if the spheres can be passaged. This method can continue up to five times.

***Note:*** *The spheroid cultures can be used to study anti-cancer drugs. There are several reports on methods to successfully use the spheroid assays to isolate, enrich, maintain or expand the tumor initiating cells/CSC subpopulations from various tumor sites (Grimshaw et al.* 2008; *Ponti et al.* 2005; *Liu and Wicha* 2010; *Wicha* 2008*).*

### 8.3.2 3-D Fibrous Scaffolds

Currently, study of CSCs is hindered by the inability to propagate these cells in tissue culture without promoting their differentiation (Patel et al. 2012). CSCs maintenance and self-renewal are also highly dependent on the tumor microenvironment Lately, it was reported that the proliferation and differentiation can be modified by culturing these cells on electrospun nanofibrous and 3-D microfibrous scaffolds that are intended to mimic the architecture of the tumor microenvironment. Tissue engineered 3-D culture techniques usually include adding cell suspension to matrices such as collagen type I or Matrigel, or culturing cells on biomaterial scaffolds that can be fabricated into various desired architectures from different materials (Hutmacher et al. 2010; Lee et al. 2008; Burdett et al. 2010; Hutmacher 2010; Ghajar and Bissell 2010). Hydrogel matrices, such as collagen type I, Matrigel and synthetic matrices have been extensively used to examine how ECM topography and chemistry affect tumor cell behavior (Levental et al. 2009; Sabeh et al. 2009). Usually in these studies, cancer cells are uniformly seeded inside a 3-D ECM and their phenotypic characteristics are monitored over given culture periods (Levental et al. 2009; Kim and Peyton 2012).

Many groups have shown the advantages of using 3-D scaffolds over 2-D tissue culture polystyrene to closely mimic the *in vivo* environment (Yamada and Cukierman 2007; Griffith and Swartz 2006; Horning et al. 2008; Kim et al. 2004; Hutmacher et al. 2009). Overall, studies comparing 3-D biomaterials models to 2-D monolayer cultures using cell lines across a range of cancer types have established *in vitro* proliferation rates closer to those found *in vivo* (Fischbach et al. 2007; Leung et al. 2010), increased gene expression especially upregulation of angiogenic factors (Fischbach et al. 2007; Kievit et al. 2010; Leung et al. 2010; Szot et al. 2011), and enhanced drug resistance (Horning et al. 2008; Fischbach et al. 2007; Leung et al. 2010; Dhiman et al. 2005) and enhanced stem cell properties. In other studies, it appears that CSCs could be selected by seeding them on electrospun nanofibrous scaffolds made of polycaprolactone and chitosan (Sims-Mourtada et al. 2014).

To culture the cells on the scaffolds, you are expected to optimize the seeding density according to your question. The seeded cells are kept on nanofiber scaffolds or polystyrene culture dishes for the required time period. In order to analyze the cells growing on the fibers, the fiber ring and the attached cells are removed and transferred to a fresh well, rinsed with phosphate-buffered saline, and the cells are removed by trypsinization. Cell viability and size are determined using a Countess automated cell counter (Invitrogen) after staining with Trypan blue (Invitrogen).

#### 8.3.2.1 Culture of CSCs on PCL-Chitosan Nanofibrous Scaffolds

First, we describe how the fibers are prepared. Polycaprolactone (PCL) and Chitosan (CS) (both from Sigma-Aldrich) are used without further purification.

**Scaffold fabrication**

1. Prepare 10% percent (w/v [weight/volume]) PCL solution in 1,1,1,3,3,3-hexafluoro-2-propanol;
2. Prepare a 0.5% (w/v) CS solution in 1,1,1,3,3,3-hexafluoro-2-propanol;
3. Sonicate both solutions for 3 h, leave overnight, and sonicated for an additional hour to ensure dissolution of all solid materials;
4. Prepare PCL-CS solutions by mixing the prepared PCL and CS solutions in equal ratio; vortex for 1 min before use;
5. Fabricate PCL-CS fibers using an electrospinning apparatus containing a stationary plate (outer dimensions 12 cm × 8 cm; inner gap dimensions 9.5 cm × 2.5 cm) covered with aluminum foil at a distance of 10.5 cm from the syringe tip;
6. Next, load 1 ml of solution into a 5 ml syringe equipped with a 21-gauge needle.
7. Apply voltage of 15.0 kV with a flow rate of 0.4 ml/h.

**Culture of CSCs on Nanofiber Scaffolds**

After fabricating electrospun nanofiber scaffolds from 10% (w/v) polycaprolactone (PCL) electrospun in equal ratio with 0.5% (w/v) chitosan (CS), these constructs are characterized for their mechanical properties. A 1:1 ratio of PCL to CS has been reported to merge the high mechanical strength properties of PCL with the biocompatibility of CS polymers. Sims-Mourtada et al. examined the growth of cells on plastic culture dishes and PCL-CS nanofibers by microscopy, focusing initially on cell morphology (Sims-Mourtada et al. 2014). Unsorted BCCs (MCF-7 and T47D) were seeded on fibers at a density of $5 \times 10^4$ cells/sample and then studied for morphology, sphere formation and chemoresistance.

#### 8.3.2.2 Culture of CSCs on PCL Microfibrous Scaffolds

Similarly, we have been able to maintain CSCs on electrospun PCL random and aligned oriented microfibrous scaffolds after 7 days, as described in the next paragraph. As depicted in Fig. 8.1, random and aligned microfibrous scaffolds can modulate BCC behavior *in vitro*. Random and aligned fibers mimic, respectively the random and highly organized collagen fibers found in the bone extracellular matrix.

**BCCs Culture**

MDA-MB-231 (highly invasive, basal-like) and T47D (low-invasive, luminal) breast cancer cell lines were obtained from the American (ATCC). The cells were propagated following ATCC instructions. MDA-MB-231 cells are cultured at 37 °C, 5 % $CO_2$ in a humidified incubator in DMEM containing 10 % fetal bovine serum and 1 % penicillin and streptomyosin. T47D breast cancer cell lines are cultured in RPMI

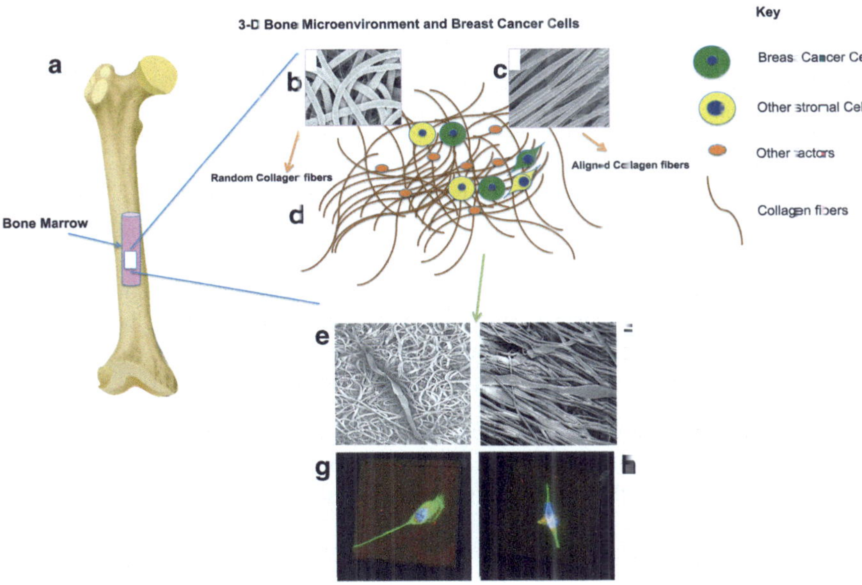


**Fig. 8.1** Breast cancer cell-bone microenvironment interactions are depicted with the breast cancer (BC) cells (BCCs) interacting with the supporting stromal cells. (**a**) Looking closer into the bone/bone marrow microenvironment, it is known to be composed of a collagen fibrous extracellular matrix (ECM). In order to mimic the bone ECM fiber architecture which is composed of random and aligned fibers, we fabricated scaffolds with (**b**) random and (**c**) aligned oriented microfibrous scaffolds using the electrospinning method. This 3-D *in vitro* model can also incorporate other key factors of (**d**) the bone microenvironment such as stromal cells and growth factors. The model can be used to study breast cancer recurrence by investigating how the microenvironment alters the different subsets of BCCs and breast cancer stem cells (CSCs). Our findings suggested that breast CSCs can respond to the different fiber orientation. Scanning Electron Microscopy of (**e**) random fibrous scaffolds and (**f**) aligned oriented fibrous scaffolds. Confocal Microscopy images of BCCs on (**g**) random fibrous scaffolds and (**h**) aligned fibrous scaffolds. Blue indicates nuclei (DAPI); green indicates F-actin (Alexa 488)

1640 supplemented with 10 % FBS and 0.2 Units/ml of bovine insulin. Change media every 2–3 days; split the cells at 80 % confluence using 0.25 % EDTA-trypsin.

**Scaffold Fabrication**

Fabricate polycaprolactone (PCL) scaffold using the electrospinning method, as previously described (Lee et al. 2011). Briefly,

1. Dissolve 15 wt% PCL in methylene chloride.
2. The electrospinning setup used for the fabrication of scaffolds consists of
   - a syringe pump (Cole Parmer, Vernon Hills, IL)
   - a syringe containing a polymer solution
   - a needle attached to the syringe
   - a grounded collector (aluminum plate) and
   - a high-voltage power supply (Gamma high voltage, Florida).
3. The parameters used for the fabrication of the electrospun mats are
   - a 10-ml plastic disposable syringe
   - a 20-gauge needle
   - 20 kV voltage; an infusion rate of 3 ml/h and distance of 40 cm between the syringe needle and grounded collector.
4. For aligned fiber electrospinning, instead of a collector plate, a rotating drum is used for aligned fiber collection. Rotate the drum at 1000–1300 rpm during the process. The electrospinning parameters to collect aligned fibrous scaffolds from the drum are similar to the random fiber collection;
5. Air dry the electrospun mats for 1 day to remove any residual solvents, and stored in vacuum desiccators. At this point, scaffolds are ready to be sterilized and used for downstream application in subsequent studies such as morphology, proliferation, viability, cell cycle protein and gene expressions.

Further studies of nanofiber and microfiber features including alignment, density, stiffness mineral component and porosity should be further investigated to optimize this culture method for drug screening applications aiming the design of potential novel therapies.

## 8.4 Potential Use of CSCs

Understanding the behavior of CSCs in different microenvironments can facilitate the development of novel targeted therapies to control or eliminate these resistant cancer cells within a given niche. For instance, within a milieu with specific signaling pathways, CSCs can be targeted via cell surface molecules by blocking their interactions within this environment, thus causing the CSCs to differentiate and become targetable. Moreover, these generated CSCs can be used in the development of predictive assays to test and screen drug responses in order to successfully design therapy treatment regimens for patients. In addition, interactions between generated

CSCs and other cells of the tumor microenvironment using a co-culture system can be further investigated to mimic closely the *in vivo* environment.

## 8.5 Future Expectations

CSCs have been shown to have similar properties as healthy stem cells, and usually sit near HSCs within the bone marrow. Furthermore, MSCs line the central sinus and essentially are encountered when CSCs come into the bone marrow. Thus, the challenge becomes how do we target the CSCs without harming HSCs and MSCs? The answer is, to understand how CSCs integrate within this microenvironment and try to get them out of this environment to target them instead of giving chemotherapy alone.

## References


Al-Hajj M, Wicha MS, Benito-Hernandez A, Morrison SJ, Clarke MF (2003) Prospective identification of tumorigenic breast cancer cells. Proc Natl Acad Sci 100:3983–3988

Birgersdotter A, Sandberg R, Ernberg I (2005) Gene expression perturbation in vitro—a growing case for thee-dimensional (3D) culture systems. Sem Cancer Biol 15:405–412

Bliss SA, Greco SJ, Rameshwar P (2014) Hierarchy of breast cancer cells: key to reverse dormancy for therapeutic intervention. Stem Cells Transl Med 3:782–786

Bonnet D, Dick JE (1997) Human acute myeloid leukemia is organized as a hierarchy that originates from a primitive hematopoietic cell. Nat Med 3:730–737

Braun S, Auer D, Marth C (2009) The prognostic impact of bone marrow micrometastases in women with breast cancer. Cancer Invest 27:598–603

Burdett E, Kasper FK, Mikos AG, Ludwig JA (2010) Engineering tumors: a tissue engineering perspective in cancer biology. Tissue Eng Part B 16:351–359

Carlsson J, Yuhas JM (1984) Liquid-overlay culture of cellular spheroids. Recent results in cancer research. Progres dans les recherches sur le cancer 95:1–23

Dhiman HK, Ray AR, Panda AK (2005) Thee-dimensional chitosan scaffold-based MCF-7 cell culture for the determination of the cytotoxicity of tamoxifen. Biomaterials 26:979–985

Diehn M, Cho RW, Clarke MF (2009) Therapeutic implications of the cancer stem cell hypothesis. Sem Radiation Oncol 19:78–86

Fischbach C, Chen R, Matsumoto T, Schmelzle T, Brugge JS, Polverini PJ, Mooney DJ (2007) Engineering tumors with 3D scaffolds. Nat Methods 4:855–860

Ford AM, Mansur MB, Furness CL, van Delft FW, Okamura J, Suzuki T, Kobayashi H, Kaneko Y, Greaves M (2015) Protracted dormancy of pre-leukemic stem cells. Leukemia 29:2202–2207

Ghajar CM, Bissell MJ (2010) Tumor engineering: the other face of tissue engineering. Tissue Eng Part A 16:2153–2156

Ghajar CM, Peinado H, Mori H, Matei IR, Evason KJ, Brazier H, Almeida D, Koller A, Hajjar KA, Stainier DYR, Chen EI, Lyden D, Bissell MJ (2013) The perivascular niche regulates breast tumour dormancy. Nat Cell Biol 15:807–817

Ghosh S, Spagnoli GC, Martin I, Ploegert S, Demougin P, Heberer M, Reschner A (2005) Thee-dimensional culture of melanoma cells profoundly affects gene expression profile: a high density oligonucleotide array study. J Cell Physiol 204:522–531

Gonzalez-Angulo AM, Morales-Vasquez F, Hortobagyi GN (2007) Overview of resistance to systemic therapy in patients with breast cancer. Adv Exp Med Biol 608:1–22

Griffith LG, Swartz MA (2006) Capturing complex 3D tissue physiology in vitro. Nat Rev Mol Cell Biol 7:211–224
Grimshaw MJ, Cooper L, Papazisis K, Coleman JA, Bohnenkamp H, Chiapero-Stanke L, Taylor-Papadimitriou J, Burchell JM (2008) Mammosphere culture of metastatic breast cancer cells enriches for tumorigenic breast cancer cells. Breast Cancer Res 10:R52
Hanoun M, Maryanovich M, Arnal-Estap A, Frenette P (2015) Neural regulation of hematopoiesis, inflammation, and cancer. Neuron 86:360–373
Hauptmann S, Gebauer-Hartung P, Leclere A, Denkert C, Pest S, Klosterhalfen B, Dietel M (1998) Induction of apoptosis in the centre of multicellular tumour spheroids of colorectal adenocarcinomas—involvement of CD95 pathway and differentiation. Apoptosis 3:267–279
Hirschhaeuser F, Menne H, Dittfeld C, West J, Mueller-Klieser W, Kunz-Schughart LA (2010) Multicellular tumor spheroids: an underestimated tool is catching up again. J Biotechnol 148:3–15
Horning JL, Sahoo SK, Vijayaraghavalu S, Dimitrijevic S, Vasir JK, Jain TK, Panda AK, Labhasetwar V (2008) 3-D tumor model for in vitro evaluation of anticancer drugs. Mol Pharm 5:849–862
Hutmacher DW (2010) Biomaterials offer cancer research the third dimension. Nat Mater 9:90–93
Hutmacher DW, Horch RE, Loessner D, Rizzi S, Sieh S, Reichert JC, Clements JA, Beier JP, Arkudas A, Bleiziffer O, Kneser U (2009) Translating tissue engineering technology platforms into cancer research. J Cell Mol Med 13:1417–1427
Hutmacher DW, Loessner D, Rizzi S, Kaplan DL, Mooney DJ, Clements JA (2010) Can tissue engineering concepts advance tumor biology research? Trends Biotechnol 28:125–133
Ivascu A, Kubbies M (2006) Rapid generation of single-tumor spheroids for high-throughput cell function and toxicity analysis. J Biomol Screen 11:922–932
Jamdade VS, Sethi N, Mundhe NA, Kumar P, Lahkar M, Sinha N (2015) Therapeutic targets of triple-negative breast cancer: a review. Br J Pharmacol 172:4228–4237
Kaemmerer E, Melchels FP, Holzapfel BM, Meckel T, Hutmacher DW, Loessner D (2014) Gelatine methacrylamide-based hydrogels: an alternative thee-dimensional cancer cell culture system. Acta Biomater 10:2551–2562
Kelm JM, Timmins NE, Brown CJ, Fussenegger M, Nielsen LK (2003) Method for generation of homogeneous multicellular tumor spheroids applicable to a wide variety of cell types. Biotechnol Bioeng 83:173–180
Khanna C, Hunter K (2005) Modeling metastasis in vivo. Carcinogenesis 26:513–523
Kievit FM, Florczyk SJ, Leung MC, Veiseh O, Park JO, Disis ML, Zhang M (2010) Chitosan-alginate 3D scaffolds as a mimic of the glioma tumor microenvironment. Biomaterials 31:5903–5910
Kim HD, Peyton SR (2012) Bio-inspired materials for parsing matrix physicochemical control of cell migration: a review. Integr Biol 4:37–52
Kim JB, Stein R, O'Hare MJ (2004) Thee-dimensional in vitro tissue culture models of breast cancer—a review. Breast Cancer Res Treatment 85:281–291
Kobayashi H, Man S, Graham CH, Kapitain SJ, Teicher BA, Kerbel RS (1993) Acquired multicellular-mediated resistance to alkylating agents in cancer. Proc Natl Acad Sci 90:3294–3298
Kunz-Schughart LA (1999) Multicellular tumor spheroids: intermediates between monolayer culture and in vivo tumor. Cell Biol Intl 23:157–161
Lee J, Cuddihy MJ, Kotov NA (2008) Thee-dimensional cell culture matrices: state of the art. Tissue Eng Part B 14:61–86
Lee YS, Collins G, Arinzeh TL (2011) Neurite extension of primary neurons on electrospun piezoelectric scaffolds. Acta Biomater 7:3877–3886
Leung M, Kievit FM, Florczyk SJ, Veiseh O, Wu J, Park JO, Zhang M (2010) Chitosan-alginate scaffold culture system for hepatocellular carcinoma increases malignancy and drug resistance. Pharm Res 27:1939–1948

Levental KR, Yu H, Kass L, Lakins JN, Egeblad M, Erler JT, Fong SF, Csiszar K, Giaccia A, Weninger W, Yamauchi M, Gasser DL, Weaver VM (2009) Matrix crosslinking forces tumor progression by enhancing integrin signaling. Cell 139:891–906

Li X, Lewis MT, Huang J, Gutierrez C, Osborne CK, Wu MF, Hilsenbeck SG, Pavlick A, Zhang X, Chamness GC, Wong H, Rosen J, Chang JC (2008) Intrinsic resistance of tumorigenic breast cancer cells to chemotherapy. J Natl Cancer Instit 100:672–679

Lim PK, Bliss SA, Patel SA, Taborga M, Dave MA, Gregory LA, Greco SJ, Bryan M, Patel PS, Rameshwar P (2011) Gap Junction-mediated import of MicroRNA from bone marrow stromal cells can elicit cell cycle quiescence in breast cancer cells. Cancer Res 71:1550–1560

Lin RZ, Chang HY (2008) Recent advances in thee-dimensional multicellular spheroid culture for biomedical research. Biotechnol J 3:1172–1184

Liu S, Wicha MS (2010) Targeting breast cancer stem cells. J Clin Oncol 28:18115–18120

Mitra A, Misha L, Li S (2015) EMT, CTCs and CSCs in tumor relapse and drug-resistance. Oncotarget 6:10697–10711

Morris SR, Carey LA (2007) Molecular profiling in breast cancer. Rev Endocr Metab Disord 8:185–198

Naume B, Zhao X, Synnestvedt M, Borgen E, Russnes HG, Lingjaerde OC, Stromberg M, Wiedswang G, Kvalheim G, Karesen R, Nesland JM, Borresen-Dale AL, Sorlie T (2007) Presence of bone marrow micrometastasis is associated with different recurrence risk within molecular subtypes of breast cancer. Mol Oncol 1:160–171

Nelson CM, Bissell MJ (2005) Modeling dynamic reciprocity: engineering thee-dimensional culture models of breast architecture, function, and neoplastic transformation. Sem Cancer Biol 15:342–352

Nyga A, Cheema U, Loizidou M (2011) 3D tumour models: novel in vitro approaches to cancer studies. J Cell Commun Signal 5:239–248

Ono M, Kosaka N, Tominaga N, Yoshioka Y, Takeshita F, Takahashi R, Yoshida M, Tsuda H, Tamura K, Ochiya T (2014) Exosomes from bone marrow mesenchymal stem cells contain a microRNA that promotes dormancy in metastatic breast cancer cells. Sci Signal 7:ra63

Osisami M, Keller ET (2013) Mechanisms of metastatic tumor dormancy. J Clin Med 2:136–150

Pampaloni F, Reynaud EG, Stelzer EH (2007) The third dimension bridges the gap between cell culture and live tissue. Nat Rev Mol Cell Biol 8:839–845

Pantel K, Alix-Panabieres C (2014) Bone marrow as a reservoir for disseminated tumor cells: a special source for liquid biopsy in cancer patients. BoneKEy Rep 3:584

Park JM, Munoz JL, Won BW, Bliss SA, Greco SJ, Patel SA, Kandouz M, Rameshwar P (2013) Exogenous CXCL12 activates protein kinase C to phosphorylate connexin 43 for gap junctional intercellular communication among confluent breast cancer cells. Cancer Lett 331:84–91

Patel LR, Camacho DF, Shiozawa Y, Pienta KJ, Taichman RS (2011) Mechanisms of cancer cell metastasis to the bone: a multistep process. Future Oncol 7:1285–1297

Patel SA, Ramkissoon SH, Bryan M, Pliner LF, Dontu G, Patel PS, Amiri S, Pine SR, Rameshwar P (2012) Delineation of breast cancer cell hierarchy identifies the subset responsible for dormancy. Sci Rep 2:906

Patel SA, Dave MA, Bliss SA, Giec-Ujda AB, Bryan M, Pliner LF, Rameshwar P (2014) T/Th17 polarization by distinct subsets of breast cancer cells is dictated by the interaction with mesenchymal stem cells. J Cancer Stem Cell Res 2:e1003

Ponti D, Costa A, Zaffaroni N, Pratesi G, Petrangolini G, Coradini D, Pilotti S, Pierotti MA, Daidone MG (2005) Isolation and in vitro propagation of tumorigenic breast cancer cells with stem/progenitor cell properties. Cancer Res 65:5506–5511

Ricci C, Moroni L, Danti S (2013) Cancer tissue engineering-new perspectives in understanding the biology of solid tumours-a critical review. Tissue Eng 1:4

Rycaj K, Tang DG (2015) Cell-of-origin of cancer versus cancer stem cells: assays and interpretations. Cancer Res 75:4003–4011

Sabeh F, Shimizu-Hirota R, Weiss SJ (2009) Protease-dependent versus -independent cancer cell invasion programs: thee-dimensional amoeboid movement revisited. J Cell Biol 185:11–19

Sims-Mourtada J, Niamat RA, Samuel S, Eskridge C, Kmiec EB (2014) Enrichment of breast cancer stem-like cells by growth on electrospun polycaprolactone-chitosan nanofiber scaffolds. Intl J Nanomed 9:995–1003

Sorokin L (2010) The impact of the extracellular matrix on inflammation. Nat Rev Immunol 10:712–723

Sosa MS, Bragado P, Aguirre-Ghiso JA (2014) Mechanisms of disseminated cancer cell dormancy: an awakening field. Nat Rev Cancer 14:611–622

Sutherland RM, Durand RE (1976) Radiation response of multicell spheroids—an in vitro tumour model. Curr Topics Rad Res 11:87–139

Sutherland RM, McCredie JA, Inch WR (1971) Growth of multicell spheroids in tissue culture as a model of nodular carcinomas. J Natl Cancer Inst 46:113–120

Szot CS, Buchanan CF, Freeman JW, Rylander MN (2011) 3D in vitro bioengineered tumors based on collagen I hydrogels. Biomaterials 32:7905–7912

Talmadge JE (2007) Clonal selection of metastasis within the life history of a tumor. Cancer Res 67:11471–11475

Visvader JE, Lindeman GJ (2008) Cancer stem cells in solid tumours: accumulating evidence and unresolved questions. Nat Rev Cancer 8:755–768

Visvader JE, Lindeman GJ (2012) Cancer stem cells: current status and evolving complexities. Cell Stem Cell 10:717–728

Weigelt B, Bissell MJ (2008) Unraveling the microenvironmental influences on the normal mammary gland and breast cancer. Sem Cancer Biol 18:311–321

Wicha MS (2008) Cancer stem cell heterogeneity in hereditary breast cancer. Breast Cancer Res 10:105

Yamada KM, Cukierman E (2007) Modeling tissue morphogenesis and cancer in 3D. Cell 130:601–610

Yin X, Zhang BH, Zheng SS, Gao DM, Qiu SJ, Wu WZ, Ren ZG (2015) Coexpression of gene Oct4 and Nanog initiates stem cell characteristics in hepatocellular carcinoma and promotes epithelial-mesenchymal transition through activation of Stat3/Snail signaling. J Hematol Oncol 8:23

Yoneda T, Hiraga T (2005) Crosstalk between cancer cells and bone microenvironment in bone metastasis. Biochem Biophys Res Commun 328:679–687

# Chapter 9
# Spontaneous Generation of Patient-Specific Retinal Pigment Epithelial Cells Using Induced Pluripotent Stem Cell Technology

**David A. Carter, Britta Nommiste, Pete J. Coffey, and Amanda-Jayne F. Carr**

**Abstract** Stem cell technology has a number potential uses when it comes to the eye, particularly disease and developmental modelling, and as potential therapeutic source. A variety of protocols have been developed that facilitate the generation of the different cell types found within the eye, as well as those that produce a facsimile of the developing eye *in vitro*. This chapter introduces the importance of the Retinal Pigment Epithelium (RPE) in maintaining visual function. We then focus on methods developed by our group to produce RPE from patient skin samples using human induced pluripotent stem cell technology (iPSC).

**Keywords** Human induced pluripotent stem cells • Retinal pigment epithelium • Transplantation

## 9.1 Introduction

The Retinal pigment epithelium (RPE) is the monolayer of epithelium found at the back of the eye, behind the retina. The RPE performs a support and maintenance role for the retina and is involved in many processes crucial to the health of the overlying photoreceptive cells (Strauss 2005). The transport of nutrients, water and ions (Reichhart and Strauss 2014) from the blood supply to photoreceptor cells is

D.A. Carter • B. Nommiste • A.-J. F. Carr (✉)
Division for Ocular Biology and Therapeutics, UCL Institute of Ophthalmology,
11-43 Bath Street, London EC1V 9EL, UK
e-mail: a.carr@ucl.ac.uk

P.J. Coffey
Division for Ocular Biology and Therapeutics, UCL Institute of Ophthalmology,
11-43 Bath Street, London EC1V 9EL, UK

BMRC (NIH), Moorfields Eye Hospital, 162 City Road, London EC1V 2PD, UK

Center for Stem Cell Biology and Engineering, Neuroscience Research Institute,
University of California, Santa Barbara, CA, USA

 
H. Ulrich, P. Davidson Negraes, *Working with Stem Cells*,
DOI 10.1007/978-3-319-30582-0_9

regulated by the RPE monolayer. The RPE is involved in the phagocytosis of photoreceptor outer segments (POS) discarded daily by the retina; being responsible for binding (mediated by Integrins and CD36 and 81), engulfing (FAK and MERTK) and breaking down the shed POS (Kevany and Palczewski 2010). The RPE is also important role for the visual cycle and produces various proteins, including RPE65 and LRAT, which play crucial roles in retinoid recycling (Kiser et al. 2012; Wright et al. 2015).

The RPE is a highly pigmented monolayer, with each cell containing many melanosomes that aid vision by absorbing any stray light within the eye (Boulton 1998). Within the monolayer, individual RPE cells are tightly packed together to give a 'cobblestone'-like morphology. The formation of tight junctions between cells is crucial for the role of the RPE as a component of the blood:retina barrier (Rizzolo 2014). The development of tight junctions enables the establishment of cell polarity, with concentration gradients maintained through ion pumps/channels such as $Na^+/K^+$-ATPase and Bestrophin at the apical and basolateral surface respectively (Lehmann et al. 2014), and the polarized secretion of growth factors, such as Pigment epithelium-derived growth factor (PEDF) and Vasoactive Endothelial Growth Factor (VEGF) (Sonoda et al. 2009).

The RPE is vital for the maintenance of vision, therefore the deterioration of, or deficiencies in the functional performance of the RPE can result in various forms of blindness (Sparrow et al. 2010). RPE-specific rare inherited genetic disorders (Table 9.1) and a number of degenerative conditions, such as age-related macular degeneration (AMD) arise when the RPE is lost or dysfunctional. AMD is the leading cause of sight loss in the developed world (de Jong 2006); there are over half a million people diagnosed with late-stage AMD in the UK, and around half of these are registered as visually impaired. Clinically, there are two forms of AMD, the slowly progressing non-exudative form, known as dry AMD, and the rapidly developing exudative or "wet" AMD, which results from the infiltration of new, leaky blood vessels through the RPE barrier. A number of pharmaceutical products are available to treat wet AMD, namely Anti-VEGF drugs e.g. Ranibizumab (Lucentis) and Bevacizumab (Avastin) (Martin et al. 2011). New drugs in the anti-VEGF class are being developed but these treatments tend only to arrest decline in vision rather than inducing any significant improvement. Anti-VEGFs are ineffective in some patients with wet AMD and are of no benefit to the patient population with dry AMD. Over recent years, focus has turned to using a cell therapy for the treatment of AMD, however, although it is possible to utilise RPE from an extra-macula area of a patients own eye (Chen et al. 2009), it is technically quite difficult to do so, and the cells/tissue yielded are fragile and limited in amount. Human primary RPE is also difficult to obtain due to reliance on post-mortem tissues. These sources are inconsistent and limited, which makes planning studies for transplantation therapeutics very difficult. This makes non-donor derived RPE a highly desirable research and clinical resource for cell transplantation.

With the advent of the regenerative medicine era, much effort has been directed to using stem cells as a potential source to generate new RPE (Klimanskaya et al. 2004, Vugler et al. 2008, Carr et al. 2013). These cell therapies have the potential to

**Table 9.1** Commonly used markers to identify retinal pigment epithelial cells

| Gene | Description | Role | Notes | Reference |
|---|---|---|---|---|
| RPE65 | RPE specific protein 65kDA | Visual cycle | Mutations lead to RP, LCA, FA | Wright et al. (2015), den Hollander et al. (2008) |
| LRAT | Lecithin retinol acyltransferase | Visual cycle | Mutations lead to early onset RD, LCA | Den Borman et al. (2012), den Hollander et al. (2008) |
| RLBP1 | Retinaldehyde binding protein 1 | Visual cycle | Mutations lead to severe RCD, FA | Maw et al. (1997), Eichers et al (2002) |
| FAK | Focal adhesion kinase | Phagocytosis | POS binding and engulfment | Finnemann (2003) |
| MerTK | Mer tyrosine protein kinase | Phagocytosis | Mutations leads to RP, LCA | den Hollander et al. (2008) |
| ITGAV | Alpha V integrin | Phagocytosis | Extracellular interactions with POS | Finnemann et al (1997) |
| ITGB5 | Integrin beta 5 | Phagocytosis | Extracellular interactions with POS | Finnemann et al. (1997) |
| CD36 | Cluster of differentiation 36 | Phagocytosis | Extracellular interactions with POS | Sun et al. (2006) |
| CD81 | Cluster of differentiation 81 | Phagocytosis | Extracellular interactions with POS | Chang and Finnemann (2007) |
| CTSD | Cathepsin D | Phagocytosis | Lysosomal protease—degrades internalised POS | Bosch et al. (1993) |
| GAS6 | Growth arrest specific 6 | Phagocytosis | MerTK interactions | Hall et al. (2001), (2005) |
| MFGE8 | Milk fat globule EGF factor 8 | Phagocytosis | Interacts with integrins | Nandrot et al. (2007) |
| BEST1 | Bestrophin 1 | Ion transport | Expressed basolaterally, mutations lead to BVMD | Marmorstein et al. (2000) |
| ATP1A | Na+/K+-ATPase | Ion Transport | Expressed apically | Hu and Bok (2001) |
| TYR | Tyrosinase | Melanogenesis | Pigment production | Murisier and Beermann (2006) |
| PMEL17 | Pre-melanosome protein 17 | Melanogenesis | Biogenesis of pre-melanosomes | Vugler et al. (2008) |
| MITF | Micropthalmia-associated transcription factor | RPE development | Transcription factor for RPE specification | Martínez-Morales et al. (2004) |

(continued)

**Table 9.1** (continued)

| Gene | Description | Role | Notes | Reference |
|---|---|---|---|---|
| OTX2 | Orthodenticle homolog 2 | RPE development | Transcription factor for RPE specification | Martínez-Morales et al. (2004) |
| PAX6 | Paired box protein 6 | Transcription factor | Early eye field marker—associated with eye defects | Martínez-Morales et al. (2004) |
| COL4A1 | Collagen IV | Extracellular matrix | Secreted component of RPE extracellular matrix | Campochiaro et al. (1986) |
| PEDF | Pigment epithelium derived factor | Growth factor | Anti-angiogenic factor | Dawson et al. (1999) |
| VEGF | Vascular epithelium growth factor | Growth factor | Pro-angiogenic factor—associated with AMD | Witmer et al. (2003) |
| KRT8 | Keratin 8 | Intermediate filament | Marker of proliferative RPE | Vugler et al. (2008) |

*RPE* retinal pigment epithelium, *RP* retinitis pigmentosa, *LCA* Leber's congenital amaurosis, *RD* retinal dystrophy, *RCD* rod-cone dystrophy, *FA* Fundus albipunctatus, *POS* photoreceptor outer segments, *AMD* age-related macular degeneration, *BVMD* Best's vitteliform macular dystrophy

replace lost or damaged cells and therefore improve the recipients' visual acuity. Human embryonic stem cells (hESC) have been used to generate RPE for treatment of AMD and Stargardt's Macular dystrophy in on-going clinical trials (Schwartz et al. 2012, 2014), however, due to the origin of these cells, there may be concerns with immune rejection. Induced pluripotent stem cells (iPSC) offer an alternative source of tissue for transplantation therapies. Somatic cells isolated from patient tissues, such as skin and blood, can be reprogrammed into iPSCs, which can then be used to generate any tissue within the body (Takahashi et al. 2007; Yu et al. 2007). Development of new methods to reprogram cells, e.g. using episomal vectors, rather than the original retroviral constructs, make generation of iPSC safer for use in humans (Okita et al. 2011; Goh et al. 2014). Many groups have described methods to differentiate RPE from iPSC, these range from protocols describing spontaneous differentiation (Carr et al. 2009a), differentiation via embryoid body/neural differentiation (Meyer et al. 2009) and directed differentiation (Westenskow et al. 2012). Studies have shown that these iPSC-derived RPE are morphologically similar to human RPE, perform many of the functions required to maintain the health of retinal cells (Carr et al. 2009a; Meyer et al. 2011; Vaajasaari et al. 2011; Kokkinaki et al. 2011; Westenskow et al. 2012) and do not form teratomas upon transplantation in the subretinal space (Kanemura et al. 2014, Kamao et al 2014). The possibility of producing stem cells from a patient's own tissue, and differentiating these cells into an autologous source of functional RPE, may alleviate rejection issues associated with HESC-derived tissues.

The generation of RPE from induced pluripotent cells also offers other areas of possibility for the novel treatment of diseases. Genetic disorders affecting the RPE may be rare in terms of overall incidence, but still affect thousands of people worldwide. In many of these conditions there is no sufficiently analogous animal model available. Using iPSCs derived from patients to generate the required tissue in the laboratory presents an opportunity to study these conditions at the cellular and molecular levels (Yvon et al. 2015). Furthermore, the generation of RPE from a patients own cells provides a unique opportunity to investigate new medicinal products, using the patients own cells as a model platform to screen for novel therapeutics in human diseased cells (Meyer et al. 2011; Schwarz et al. 2015). Additionally, the generation of RPE from pluripotent stem cells provides a new arena in which to understand the development of the RPE.

## 9.2 Materials

### *9.2.1 Solutions and Chemicals*

- 70 % Alcohol solution.
- Fibroblast cell media: Remove 55 ml of DMEM:F12 medium (Life Technologies, Cat. No. 31331-028) from the bottle. Add 50 ml of fetal bovine serum (Life Technologies, Cat. No 16000-044) and 5 ml penicil in/streptomycin (Life Technologies, Cat. No. 15140-122). Store at 4 °C for up to 4 weeks.
- Sodium butyrate (Tocris, Cat. No. 3850): Dilute 50 mg in 4.54 ml $H_2O$ and store aliquots at −20 °C. Dilute 1:200 in cell culture medium for a final concentration of 0.5 mM.
- DPBS, no calcium no magnesium (Life Technologies, Cat. No. 14190-94).
- TrypLE Select Solution (Life Technologies, Cat. No. 12563-011).
- Cell freezing solution: For 50 ml of solution, combine 30 ml DMEM (Life Technologies, Cat. No 41966-029) with 15 ml fetal bovine serum and 5 ml DMSO (Life Technologies, Cat. No. D12345). Store at 4 °C.
- Gelatin-coated plates: Add 1 g gelatin (Sigma-Aldrich, Cat. No. G9136) to 500 ml of $ddH_2O$. Autoclave the solution to dissolve and sterilise. Add 5 ml of the solution to a 10 $cm^2$ tissue culture treated dish in the laminar flow hood. Incubate at room temperature for 1 h. Remove the excess gelatin prior to use. Plates can be wrapped in parafilm and stored at 4 °C for up to a week.
- Amaxa Cell Line Nucleofector® Kit R (Lonza, Cat. No. VCA-1001).
- Yamanaka episomal reprogramming plasmids (Okita et al. 2011—Addgene, pCXLE-hOCT3/4-shp53-F (Cat. No.27077); pCXLE-hUL (Cat. No. 27080); pCXLE-hSK (Cat. No. 270778); pCXLE-GFP (Cat. No. 27082)).
- HESC qualified Matrigel-coated plates: All tissue cultureware, plastics and media should be ice-cold to prevent gelling of the Matrigel. Defrost the BD Matrigel hESC-qualified matrix solution (BD Biosciences Cat. No. 354277) overnight in an ice bucket within a fridge (4 °C). Dilute 1:1 with DMEM, prepare 1 ml aliquots in 1.5 ml microcentrifuge tubes and store at −20 °C. Slowly thaw aliquots overnight as described above and dilute in cold DMEM to the final ratio suggested according manufacturers certificate of analysis. Coat tissue culture

plastic (1 ml/well of a 6-well plate, 3 ml/T25 flask) and incubate at room temperature for at least 1 h. Aspirate off excess Matrigel immediately prior to use.

- mTeSR1 Media Complete Kit (Stem Cell Technologies, Cat. No. 05850). Prepare according to the manufacturers instructions.
- Stainalive Tra-1-60 Antibody (DyLight 488), mouse anti-human (STEMGENT, Cat. No. 09-0068).
- Stainalive Tra-1-81 Antibody (DyLight™ 488), mouse anti-human (STEMGENT, Cat. No. 09-0069).
- iPSC-RPE differentiation media: All components are supplied from Life Technologies. Remove 111.2 ml of media from a bottle of Knockout DMEM (Cat. No. 10829-018), then add 100 ml of KnockOut Serum Replacement (Cat. No. 10828010), 5 ml of 200 mM L-Glutamine (Cat. No. 25030-081), 5 ml of 100× MEM Non-essential amino acids (Cat. No. 11140-035), 909 μl of 55 mM β-mercaptoethanol (31350-010) and 300 μl of 50 mg/ml Gentamicin (Cat. No. 15750-037) to the bottle. Store at 4 °C for up to 4 weeks.
- Accutase® Solution (Sigma-Aldrich, Cat. No. A6964).
- BD Matrigel Growth Factor reduced (GFR) matrix (BD Biosciences, Cat. No. 356230): Thaw, dilute 1:1 with DMEM, aliquot and store as described above. Dilute 1:15 with DMEM to coat plates for a final dilution of 1:30, incubate at 37 °C for at least 2 h, place at room temperature for 1 h and aspirate excess Matrigel immediately prior to use.
- 30 % sucrose cryopreservation solution: Dissolve 30 g of sucrose (Sigma, Cat. No. S0389) in 0.01 M PBS (Sigma, Cat. No. P4417) to a final volume of 100 ml.
- OCT embedding compound (CellPath, Cat. No. KMA-0100-00A).
- Immunocytochemistry cell permeabilization solution: For a 50 ml solution, dilute 1.5 ml of 10 % Triton-X solution (Sigma-Aldrich, Cat. No. 93443) in 0.01 M PBS.
- Immunocytochemistry blocking solution: Add 0.5 ml of normal donkey serum (Jackson ImmunoResearch Laboratories Ltd.) to 0.3 g of Bovine serum albumin and make up to 10 ml with PBS.
- RPE specific antibodies: Pmel17 Mouse monoclonal antibody (Dako, Cat. No. M0634, 1:500 final dilution), MerTK Rabbit monoclonal antibody (Abcam, Cat. No. ab52968, 1:50 final dilution), Bestrophin Mouse monoclonal antibody (Abcam, Cat. No. a2182, 1:1000 final dilution).
- Secondary antibodies: Donkey Anti-mouse IgG Alexa Fluor® 488 (Abcam, Cat. No. ab150105) and Donkey Anti-Rabbit IgG Alexa Fluor® 555 (Abcam, Cat. No. ab150074).
- VECTASHEILD anti-fade mounting medium with DAPI (Vector Laboratories, Cat. No. H-1200).

### *9.2.2 Disposables*

- 7 ml Bijou tube.
- Conical tubes (15 and 50 ml).
- 1.5 ml microcentrifuge tube.

- Cryovial cryogenic preservation tubes.
- 6-Well tissue culture-treated plate.
- 10 cm$^2$ tissue culture-treated plastic dish.
- 25 cm$^2$ (T25) tissue culture-treated plastic flask.
- 160 cm$^2$ (T60) tissue culture-treated plastic flask.
- Extended fine tip Pasteur pipette (Alpha Labs, Cat. No. LW4636).
- Sterile serological pipette (5, 10 and 25 ml).
- 40 μm Cell Strainer (Corning, Cat. No. 352340).
- 22×22 mm glass cover slip (VWR, Cat. No. 631-0124). Sterilise the coverslips overnight by immersing in 100 % ethanol. Allow to dry thoroughly before placing on cells.
- 22×50 mm glass coverslip (VWR, Cat. No. 631-0094).
- Sterile cell scraper (Greiner Bio-one, Cat. No. 541070).
- 11 mm Crescent blade microknife (Interfocus, Cat. No 10317-14).
- Superfrost Plus glass slides (VWR, Cat. No. 631-0447).

### *9.2.3 Equipment*

- Tissue culture incubator (humidified to 95 % with 5 % $CO_2$ maintained at 37 °C).
- Laminar flow hood (Class I and II).
- Centrifuge.
- Media Aspirator.
- Sterile dissection forceps.
- Sterile scalpel blades.
- Dissection teasing needle.
- Water bath.
- Inverted Microscope with LCD screen (2, 4, 10 and 20× objectives) contained inside a laminar flow hood
- Cell counter (automated or haemocytometer).
- CoolCell SV2 cell freezing container (Biocision Cat. No. BCS-172).
- Liquid nitrogen storage tank with cryovial storage racking.
- Amaxa Nucleofector device.

## 9.3 Methods

### *9.3.1 Generation of Fibroblast Cells from a Patient Dermal Skin Biopsy*

1. A 5-mm skin biopsy should be obtained from a patient with informed consent under aseptic conditions by a trained physician. The protocol should be approved by the appropriate research ethics committee and review board.

2. Collect the biopsy and transfer to a sterile 7 ml Bijou tube containing fibroblast cell media and transport on ice to the laboratory;
3. Remove the biopsy sample from the tube using sterile forceps and place into a 10 $cm^2$ sterile culture dish containing DPBS to wash the sample;
4. Remove the epidermal layer of the skin using a sterile scalpel blade;
5. Place the biopsy into a fresh 10 $cm^2$ dish and dissect the biopsy into small pieces (approx. 1 mm) using sterile scalpel blades;
6. Transfer the biopsy fragments to two wells of a 6-well tissue culture-treated plate and overlay with a sterile 22×22 mm glass coverslip;
7. Carefully add 500 μl of fibroblast media to the well so that sufficient media is drawn underneath the coverslip;
8. Culture the sample overnight in a humidified incubator at 37 °C, 5 % $CO_2$;
9. The following day add 1.5 ml of fibroblast media and culture the tissue for 2–3 weeks to permit primary fibroblast cell emergence (Fig. 9.1a);
10. When sufficient fibroblast cell outgrowth has occurred (approx. 70 % confluence) the cells are ready to be passaged;
11. Aspirate the media and transfer the coverslip to a fresh well, inverting the coverslip so that the side which has been in contact with cells is uppermost (see ***Note 1***);
12. Wash the coverslip and cells with sterile DPBS and add 500 μl of TrypLE Select solution to each well. Incubate at room temperature until cells have detached (approx. 10 min);
13. Add 2 ml of fibroblast medium to each well. Collect the media and pool in a 15 ml conical tube. Remove clumps of tissue by placing the media through a cell strainer (see ***Note 2***);
14. Resuspend the cells in an appropriate volume of media (2 ml per well of a 6-well plate or 6 ml per T25) and passage at a ratio of 1:3 (see ***Note 3***);
15. Replace the fibroblast medium twice weekly and passage the cells using TrypLE solution as described above when the cells reach approx. 75 % confluency, maintaining a 1:3 split ratio (see ***Note 4***);
16. Patient fibroblast cells can be cryopreserved long-term in liquid nitrogen. Dissociate cells using TrypLE solution as above and resuspend cells in fibroblast media. Centrifuge the cells at 250×*g* for 5 min and aspirate the supernatant leaving a small miniscus of media (approx. 50 μl) in which to resuspend the cells. Add 1 ml of Cell Freezing solution and ensure cells are fully dispersed. Aliquot the cells into cryo-preservation vials (cells from 1 well of a 6-well plate aliquoted into one vial, from a T25 into thee vials). Place the cryovials inside a CoolCell SV2 controlled rate freezing container and freeze overnight in a −80 °C freezer. Vials should be transferred into liquid nitrogen storage the following day.
17. Cells can be recovered from cryopreservation prior to reprogramming. Warm the vial in a 37 °C water bath until most of the tube has thawed. Wipe the tube with 70 % alcohol before adding 1 ml of prewarmed fibroblast cell media.

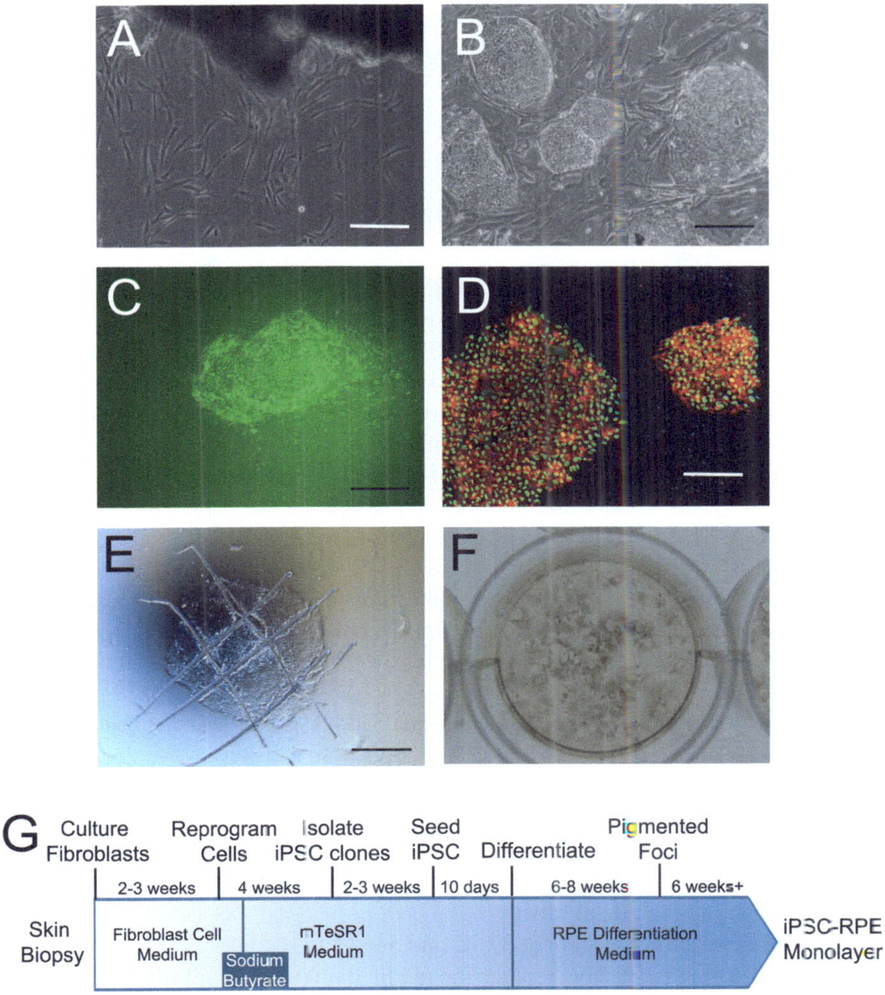


**Fig. 9.1** Generation of iPSC-RPE from a patient skin biopsy. (**a**) Fibroblast outgrowth from a skin biopsy. (**b**) Emergence of iPSC colonies in reprogrammed cultures (**c**) stained with Stainalive Tra-1-60 antibody. (**d**) Oct4 (*green*) and Tra-1-81 (*red*) staining of iPSCs. (**e**) Dissection of iPSC colony for clonal expansion. (**f**) Appearance of pigmented foci from patient iPSC cultures. (**g**) Timeline of iPSC-RPE generation from a patient skin biopsy. Scale bars—All 200 μm

Carefully transfer the contents of the cryovial to a 15 ml conical tube containing 5 ml of warm fibroblast media. Centrifuge the cells at 250 × *g* for 3 min and aspirate the supernatant. Resuspend the cells in 2 ml of fibroblast media and transfer to a well of a 6-well plate. The cells should be cultured as described above.

### 9.3.2 *Reprogramming of Patient Fibroblast Cells into Induced Pluripotent Stem Cell Lines*

1. On the day of the transfection prepare gelatin-coated 10 $cm^2$ culture dishes;
2. Pre-warm the Nucleofector reagent and DPBS to room temperature. Warm the fibroblast media to 37 °C;
3. Dissociate fibroblast cells from a T160 using TrypLE solution as described above (see ***Note 5***);
4. Resuspend in the cells in 1 ml of DPBS and centrifuge at $250 \times g$ for 3 min;
5. Count cells using an automated cell counter or haemocytometer;
6. Aliquot $1 \times 10^6$ cells into a sterile 1.5 ml microcentrifuge tube, centrifuge at $250 \times g$ for 3 min. Remove the supernatant and resuspend the cells in 100 μl of Nucleofector solution (prepared according to the manufacturers instructions). Add 1 μg of each of the thee Yamanaka episomal reprogramming plasmids (pCXLE-hUL, pCXLE-hOCT3/4-shp53-F and pCXLE-hSK). A control transfection reaction to examine transfection efficiency can be prepared by adding 3 μg of pCXLE-GFP to cells to electroporate in parallel with the reprogramming cells;
7. Pipette the reaction gently to mix, then transfer the reaction into an Amaxa electroporation cuvette ensuring minimal bubble formation;
8. Electroporate the cells using an Amaxa *Nucleofector* device using program U-023;
9. Remove the cuvette from the device and add 1 ml of warm fibroblast cell medium;
10. Remove the gelatin solution from the 10 $cm^2$ plates and add 8 ml of warmed MEF media;
11. Transfer the cell suspension from the cuvette into the prepared 10 $cm^2$ dish;
12. Add an additional 1 ml of medium to the cuvette to collect any remaining cells and transfer to the 10 $cm^2$ dish;
13. Incubate the cells overnight at 37 °C and 5 % $CO_2$;
14. Over the following 6 days replace the media daily with 10 ml fibroblast media containing 0.5 mM Sodium butyrate (see ***Note 6***);
15. Prepare HESC qualified Matrigel-coated plates on the 7th day following the transfection;
16. Dissociate the transfected patient fibroblast cells using TrypLE solution as previously described and resuspend in 5 ml of fibroblast media;
17. Count the cells and plate at a concentration of $2 \times 10^5$ cells/well of the coated 6-well plate in a final volume of 2 ml media per well. Culture the cells the overnight 37 °C and 5 % $CO_2$;
18. Remove the medium from each well on the following day and replace with 2 ml of mTeSR1 medium supplemented with 0.5 mM Sodium butyrate. The cell media should be replaced daily, supplementing with sodium butyrate for the first 5 days only;

19. The appearance of pluripotent colonies in reprogrammed cultures can vary between patients however ESC-like colonies should begin to emerge within 3 weeks of the initial transfection. Under a x2 microscope objective the cells grow within tightly compacted colonies that have a phase positive appearance with defined edges (Fig. 9.1b). Cells have a high nucleus:cytoplasm ratio.

### *9.3.3 Isolation of Clonal iPSC Colonies*

1. Pluripotent colonies can also be identified using StainAlive stem cell antibodies for the stem cell surface proteins Tra-1-60 and Tra-1-81. Aspirate the medium from the cells and add fresh mTeSR1 medium containing 3 μg/ml StainAlive antibody. Incubate the cells at 37 °C, 5 % $CO_2$ for 30 min Remove the medium, and wash twice with cell medium. Add fresh mTeSR1 and image the cells using a fluorescent microscope (Fig. 9.1c, see ***Note 7***).
2. On the day of picking, prepare the required number of 6-well plates by coating wells with HESC-qualified Matrigel Basement membrane matrix (1 ml of solution per well). For clonal iPSC lines prepare 1 well per colony;
3. Transfer the plate to laminar flow hood containing an EVOS XL or equivalent LCD display microscope;
4. To pick clonal iPSCs, gently score criss-cross over an iPSC colony with the edge of a fine tip plastic Pasteur pipette (See Fig. 9.1E). Use the tip of the pipette to gently lift the colony fragments away from the surrounding cells (see ***Note 8***). Collect the free-floating colony fragments using the Pasteur pipette and transfer to a well of the Matrigel coated 6-well plate and add 1 ml of medium;
5. Replace the medium on the original plate containing the reprogrammed cells with fresh mTeSR1 medium cell. Repeat the picking procedure for any remaining iPSC colonies;
6. Incubate the cells at 37 °C, 5 % $CO_2$, taking care not to disturb the plate for 24 h in order to permit colony attachment. Next, cells should be fed daily with fresh mTeSR1 medium. Pluripotency can be confirmed by immunostaining for pluripotent stem cell markers e.g. Oct4 and Tra-1-81 (Fig. 9.1d);
7. iPSC colonies will be ready to passage within 4–7 days. To maintain and expand the iPSC line it is important to removal any differentiated cells prior to passaging; this can be achieved by scrapping across any areas of differentiation with the tip of a Pasteur pipette to detach cells. The well should then be washed with media before adding fresh mTeSR1 medium prior to passaging. Fragment and detach the cells by scoring across the clonal colonies within a well and as described above (Fig. 9.1e) and replate onto HESC qualified Matrigel-coated tissue culture plastic at a split ration of 1:3–1:6.

### 9.3.4 Spontaneous Differentiation of Patient iPSCs into RPE

1. Passage the iPSCs onto Growth Factor reduced Matrigel-coated plates or flasks as described above using mTeSR1 medium, incubate the cells at 37 °C, 5 % $CO_2$;
2. Replace the medium daily with mTeSR1 until the individual iPSC colonies merge and become confluent over the dish/flask; this will occur approximately 8–10 days post-seeding;
3. Once cells have become confluent, replace the medium with a 1:1 mix of mTeSR1:iPSC-RPE Differentiation medium;
4. The following day replace the medium with Differentiation medium only and culture the cells at 37 °C, 5 % $CO_2$. From this point change the media twice weekly;
5. Within a 4–6-week period, pigmented foci of retinal pigment epithelial cells, which are detectable by eye, should begin to appear within the cultures (Fig. 9.1f).

### 9.3.5 Isolation of Pigmented Foci from iPSC Cultures and Culture as a Monolayer

1. Once pigmented foci have reached a sufficient size (>1 mm diameter) dissect around the foci using a crescent blade. Place the foci into a fresh dish containing differentiation medium and carefully remove any non-pigmented tissue using sterile forceps and scalpel blades;
2. Place the pigmented foci into a 15 ml conical tube containing Accutase solution and incubate at 37 °C for 2–3 h to dissociate cells;
3. Pellet the cells by centrifugation at 250 × $g$ for 3 min and resuspend in differentiation medium;
4. Pass the cells through a 40 μm cell strainer to create a single cell suspension of retinal pigment epithelial cells;
5. Count the cells using a haemocytometer;
6. Seed out the pigmented cells in differentiation medium (See ***Note 9***) at a minimum density of 50,000 cells/$cm^2$ on Growth factor reduced Matrigel-coated tissue culture plastic dishes. Incubate cells at 37 °C, 5 % $CO_2$;
7. The media should be replaced with differentiation media twice weekly for approximately 6 weeks, until a pigmented monolayer of cells forms (see ***Note 10***);
8. iPSC-derived RPE can be maintained in culture for several months to permit further maturation of the cell monolayer (see ***Note 11***).

A schematic overview of the whole process is shown in Fig. 9.1g.

### 9.3.6 Immunocytochemical Staining of iPSC-Derived RPE

1. Remove the medium from the cells and wash twice with PBS;
2. Fix the cells in 4 % paraformaldehyde in 0.01 M phosphate buffer at 4 °C for 30 min. Wash twice with PBS;

3. Using a cell scraper, slowly but firmly scrape off the RPE cells from the dish in one sweep, so that an intact sheet of cells is lifted;
4. Carefully transfer the sheet to a bijou tube containing 30 % sucrose using a teasing needle. Cryopreserve the sheet overnight at 4 °C;
5. The following day, cut the RPE sheet into smaller pieces (approx. 1 $cm^2$) for embedding. Fill a Cryomold with OCT compound being careful to exclude large air bubbles. Transfer the sheets into the OCT cryomolds and gently orientate the sheet into a vertical, on-edge position, using teasing needles;
6. Place the bottom of the cryomold into a dry ice/acetone slurry bath and slowly freeze the block, ensuring the sheet maintains its vertical orientation;
7. Store the blocks at −80 °C;
8. Section the tissue at 14 μm on a cryostat and collect tissue sections on warm Superfrost Plus slides;
9. Air dry the sections and use immediately or store at −80 °C;
10. Permeabilise the tissue in 0.3 % Triton in PBS for 10 min at 4 °C;
11. Remove the permeabilization solution and incubate in blocking solution for a minimum of 30 min;
12. Place the slides inside an immunostaining moisture chamber, add RPE specific antibodies diluted in blocking solution to the sections and incubate at 4 °C overnight;
13. The following day wash the slides five times in PBS;
14. Pipette the secondary antibodies diluted in blocking solution onto the sections and incubate for 1 h inside a moisture chamber;
15. Wash the slides five times in PBS;
16. Mount the cells by adding a drop of VECTASHIELD anti-fade mounting medium with DAPI over the tissue sections. Cover with a clean 22×50 mm coverslip and seal with nail polish.

### 9.3.7 Characterization of the Patient iPSC-Derived RPE

1. iPSC-RPE cells should be characterised to ensure their similarity to human RPE. iPSC-derived RPE cells can be examined initially by their morphological appearance. By eye, an even layer of brown/black cells should be visible on the tissue culture plastic (Fig. 9.2a). The RPE form a monolayer of hexagonal cells arranged in a regular pattern (cobblestone-like appearance Fig. 9.2b), are highly pigmented and appear brown/black under a standard microscope (Fig. 9.2c);
2. The intricate ultrastructure of the RPE cell can be analysed by electron microscopy (Fig. 9.2d). RPE cells are highly polarised with prominent microvilli and coated pits apparent on the apical surface, densely packed melanosomes within the apical cytoplasm and a basal nucleus. The RPE possess cell-cell adhesion structures including adherins junctions, tight junctions and desmosomes, and secrete their own basement lamina;
3. Gene and protein expression can be examined for a panel of RPE markers (Table 9.1) using RT-PCR, immunostaining (see ***Note 12***) and Western blot.

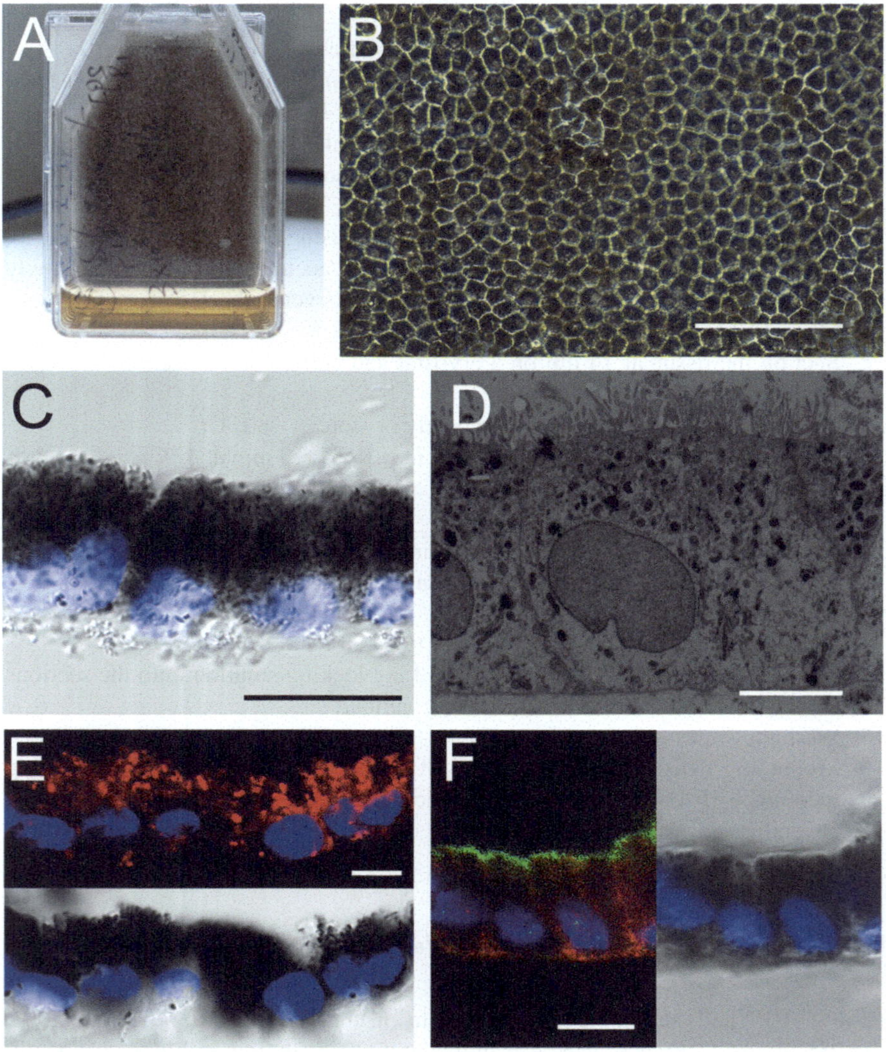


**Fig. 9.2** Characterization of patient iPSC-derived RPE. (**a**) A T25 tissue culture flask containing purified RPE. (**b**) Cobblestone-morphology of RPE monolayer. RPE cells are highly polarised epithelial cells with (**c**) a basal nucleus (*blue*) and are packed with melanosomes (*black* granules). (**d**) Electron microscopy of the RPE cell ultrastructure. Immunostaining of cells with (**e**) the pre-melanosome marker Pmel17 (*red*) and (**f**) apical expression of MerTK (*green*) and basolateral localisation of Best1 (*red*). Scale bars—(**b**) 100 μm, (**c**) 20 μm, (**d**) 5 μm, (**e**) 10 μm, (**f**) 10 μm

RPE cell markers should be expressed in the correct cellular compartment. For example, Pmel17 is observed with punctate staining of the cytoplasm (Fig. 9.2e), MerTK should be expressed on the apical surface, whilst Bestrophin should be localised basolaterally (Fig. 9.2f);

4. RPE cells secrete a number of factors crucial for the survival and maintenance of photoreceptor cells, including Vasoactive Endothelial Growth Factor (VEGF) and Pigment epithelium-derived growth factor (PEDF). To analyse secretion of these factors iPSC-RPE should be grown on permeable culture membrane inserts. Apical and basal media can then be collected for cells and growth factor secretion measured using an ELISA kit;
5. The RPE is a tight epithelium and is a component of the blood:retina barrier separating the retina from the choriocapillaris in vivo. The barrier function of iPSC-RPE cells can be assessed by measuring the transepithelial resistance. iPSC-RPE cells should be seeded onto permeable culture membrane insert and development of a functional barrier measured weekly over the course of RPE cell maturation using an epithelial voltohmmeter;
6. An important function of the RPE in the eye is the phagocytosis of the outer segment debris shed daily by the retinal photoreceptor cells. Functional phagocytosis can be examined in patient iPSC-RPE by exposing cells to fluorescently labelled photoreceptor outer segments isolated from porcine, bovine or ovine eyes. Alternatively, an isolated sheet of retina can be co-cultured above the RPE monolayer; RPE cells should then be assessed for ingestion of rhodopsin positive material by immunostaining;
7. The RPE cell is vital for the recycling of retinoids in the visual cycle. In the photoreceptor cells the detection of light results in the isomerisation of the 11-cis-retinal chromophore into all-trans-retinal, after further reduction the by-product of the visual cycle is released from the photoreceptor cell as all-trans-retinol. The RPE is responsible for the recycling of all-trans-retinol into 11-cis-retinal, which can then be transported back to the photoreceptor cells to form a new light-absorbing photopigment molecule. Retinol derivatives of the visual cycle can be measured in cell lysates and cell media by High Performance Liquid Chomatography (HPLC) or Liquid Chomatography-Mass Spectrometry (LC-MS);
8. The functional properties of iPSC-RPE cells can be assessed by transplantion of cells into the subretinal space of a retinal degenerate animal e.g. the Royal College of Surgeons (RCS) rat. Preservation of visual responses in these animals over time is the ultimate test of cell function.

## 9.4 Notes

1. Patient fibroblast cells can grow on the underside of the coverslip; these can easily be collected for alongside those on the plate for expansion.
2. Any clumps of tissue from the patient skin biopsy can be reseeded onto fresh tissue culture plates as described above for further cell growth if required.
3. Cells from a single well of a 6-well plate can be passaged into one T25 flask for expansion. With fibroblast cells it is important not to split for passage at a high ratio (>1:3) as the seeding density will be too low for the cells to survive.

4. For reprogramming $1 \times 10^6$ fibroblast cells are required. A sufficient number of fibroblast cells can be achieved by expanding the cells and preparing a culture for the reprogramming in a T160 flask.
5. For reprogramming cells should be at low passage (<p15) and still be in a proliferative phase, approximately 2–3 days following passage.
6. To establish the efficiency of transformation examine the ratio of GFP expressing:non expressing cells on the control GFP transfected plates the day after electroporation.
7. For ease of picking, the position of Stainalive™ verified pluripotent colonies can be marked on the underside of the plate using an objective stamp attached to the fluorescent microscope.
8. When picking the colonies for the first time it is important to avoid disturbing any untransformed fibroblast cells surrounding each colony as these will be carried over in the sub-culture.
9. A variety of media can be used to culture iPSC-RPE cells; commonly used media include RPE medium (Klimanskaya et al. 2004) and Human Fetal RPE medium (Maminishkis et al. 2006).
10. The iPSC-derived retinal pigment epithelial cells will initially de-pigment after passage, undergoing an EMT-like transformation, re-pigmentation and development of the epithelial cell morphology will occur over the 6-week culture period.
11. iPSC-RPE derived cells can be expanded for a limited number of passages by dissociation using Accutase® solution. It is not advisable to continue passage of cells beyond p4 as the EMT phenotype prevails and cells fail to re-establish the RPE cell morphology.
12. Immunostaining of iPSC-derived RPE cells can be masked due to the high levels of pigmentation in cells. We recommend using the protocol described in Sect. 9.3.7 to prepare cell sections for staining.

## 9.5 Potential Use of iPSC-Derived RPE

There has been considerable success in generating human RPE *in vitro* from a number of human pluripotent stem cell sources including HESC (Klimanskaya et al. 2004, Vugler et al. 2008) and iPSC (Carr et al. 2009b; Vaajasaari et al. 2011; Kokkinaki et al. 2011). As a single layer of cells responsible for a number of eye diseases, the RPE is an ideal target for regenerative medicine, using stem cell derived RPE as a potential clinical source of replacement tissue. HESC-derived RPE are currently being assessed as a treatment for RPE-based diseases such as AMD and Stargardt's macular dystrophy (Schwartz et al. 2012) and in other clinical trials worldwide. More recently, as part of a first-in-human iPSC clinical study, autologous iPSC-derived-RPE have been made from, and transplanted back into a patient with AMD in Japan. However, the long-term success of these transplants the eye has yet to be evaluated. iPSC-derived RPE cells also offer a powerful new source of tissue for the modelling of RPE disorders, replacing animal models as a disease-in-a-dish tool to examine the development and pathogenesis of inherited and degenerative RPE disease (Yvon et al. 2015). As patient iPSC-derived RPE also contain the genetic background responsible for RPE disease, the cells will also be a viable tissue source for investigating potential therapeutics in high throughput drug discovery screens (Schwartz et al. 2014; Meyer et al. 2009).

## References


Bosch E, Horwitz J, Bok D (1993) Phagocytosis of outer segments by retinal pigment epithelium: phagosome-lysosome interaction. J Histochem Cytochem 41:253–63

Boulton M (1998) The role of melanin in the RPE. In: Marmor MF Wolfensberger TJ (eds) The retinal pigment epithelium. Oxford University Press, New York, pp 68–85

Campochiaro PA, Jerdon JA, Glaser BM (1986) The extracellular matrix of human retinal pigment epithelial cells in vivo and its synthesis in vitro. Invest Ophthalmol Vis Sci 27:1615–21

Carr AJ, Vugler AA, Hikita ST, Lawrence JM, Gias C, Chen LL, Buchholz DE, Ahmado A, Semo M, Smart MJK, Hasan S, da Cruz L, Johnson LV, Clegg DO, Coffey PJ (2009a) Protective effects of human iPS-derived retinal pigment epithelium cell transplantation in the retinal dystrophic rat. PLoS One 12:e8152

Carr AJ, Vugler AA, Lawrence JM, Chen LL, Ahmado A, Chen FK, Semo M, Gias C, da Cruz L, Moore HD, Walsh J, Coffey PJ (2009b) Molecular characterization and functional analysis of phagocytosis by human embryonic stem cell-derived RPE cells using a novel human retinal assay. Mol Vis 15:283–95

Carr A-JF, Smart MJK, Ramsden CM, Powner MB, da Cruz L, Coffey PJ (2013) Development of human embryonic stem cell therapies for age-related macular degeneration. Trends Neurosci 36:385–95

Chang Y, Finnemann SC (2007) Tetraspanin CD81 is required for the alpha v beta5-integrin-dependent particle-binding step of RPE phagocytosis. J Cell Sci 120:3053–63

Chen FK, Patel PJ, Uppal GS, Rubin GS, Coffey PJ, Aylward GW, Da Cruz L (2009) A comparison of macular translocation with patch graft in neovascular age-related macular degeneration. Invest Ophthalmol Vis Sci 50:1848–55

Dawson DW, Volpert OV, Gillis P, Crawford SE, Xu H-J, Benedict W Bouck NP (1999) Pigment epithelium-derived factor: a potent inhibitor of angiogenesis. Science 285:245–8

de Jong PTVM (2006) Age-related macular degeneration. N Engl J Med 355:1474–85

den Hollander AI, Roepman R, Koenekoop RK, Cremers FP (2008) Leber congenital amaurosis: genes, proteins and disease mechanisms. Prog Retin Eye Res 27:391–419

Dev Borman A, Ocaka LA, Mackay DS, Ripamonti C, Henderson RH, Moradi P, Hall G, Black GC, Robson AG, Holder GE, Webster AR, Fitzke F, Stockman A, Moore AT (2012) Early onset retinal dystrophy due to mutations in LRAT: molecular analysis and detailed phenotypic study. Invest Ophthalmol 53:3927–38

Eichers ER, Green JS, Stockton DW, Jackman CS, Whelan J, McNamara JA, Johnson GJ, Lupski JR, Katsanis N (2002) Newfoundland rod-cone dystrophy, an early-onset retinal dystrophy, is caused by splice-junction mutations in RLBP1. Am J Hum Genet 70:955–64

Finnemann SC (2003) Focal adhesion kinase signaling promotes phagocytosis of integrin-bound photoreceptors. EMBO J 22:4143–54

Finnemann SC, Bonilha VL, Marmorstein AD, Rodriguez-Boulan E (1997) Phagocytosis of rod outer segments by retinal pigment epithelial cells requires αvβ5 integrin for binding but not for internalization. Proc Natl Acad Sci U S A 94:12932–7

Goh PA, Caxaria S, Casper C, Rosales C, Warner TT, Coffey PJ, Nathwani AC (2014) A systematic evaluation of integration free reprogramming methods for deriving clinically relevant patient specific induced pluripotent stem (iPS) cells. PLoS One 8:e81622

Hall MO, Prieto AL, Obin MS, Abrams TA, Burgess BL, Heeb MJ, Agnew BJ (2001) Outer segment phagocytosis by cultured retinal pigment epithelial cells requires Gas6. Exp Eye Res 73:509–20

Hall MO, Obin MS, Heeb MJ, Burgess BL, Abrams TA (2005) Both protein S and Gas6 stimulate outer segment phagocytosis by cultured rat retinal pigment epithelial cells. Exp Eye Res 81:581–91

Hu J, Bok D (2001) A cell culture medium that supports the differentiation of human retinal pigment epithelium into functionally polarized monolayers. Mol Vis 7:14–9

Kamao H, Mandai M, Okamoto S, Sakai N, Suga A, Sugita S, Kiryu J, Takahashi M (2014) Characterization of human induced pluripotent stem cell-derived retinal pigment epithelium cell sheets aiming for clinical application. Stem Cell Rep 2:205–18

Kanemura H, Go MJ, Shikamura M, Nishishita N, Sakai N, Kamao H, Madai M, Morinaga M, Takahashi M, Kawamata S (2014) Tumorigenicity studies of induced pluripotent stem cell (iPSC)-derived retinal pigment epithelium (RPE) for the treatment of agre-related macular degeneration. PLoS One 9:e85336

Kevany BM, Palczewski K (2010) Phagocytosis of retinal rod and cone photoreceptors. Physiology (Bethesda) 25:8–15

Kiser PD, Golczak M, Maeda A, Palczewski K (2012) Key enzymes of the retinoid (visual) cycle in vertebrate retina. Biochim Biophys Acta 1821:137–51

Klimanskaya I, Hipp J, Rezai KA, West M, Atala A, Lanza R (2004) Derivation and comparative assessment of retinal pigment epithelium from human embryonic stem cells using transcriptomics. Cloning Stem Cells 6:217–45

Kokkinaki M, Sahibzada N, Golestaneh N (2011) Human induced pluripotent stem-derived retinal pigment epithelium (RPE) cells exhibit ion transport, membrane potential, polarized vascular endothelial growth factor secretion, and gene expression pattern similar to native RPE. Stem Cells 29:825–35

Lehmann GL, Benedicto I, Philp NJ, Rodriguez-Boulan E (2014) Plasma membrane protein polarity and trafficking in RPE cells: past, present and future. Exp Eye Res 126:5–15

Maminishkis A, Chen S, Jalickee S, Banzon T, Shi G, Wang FE, Ehalt T, Hammer JA, Miller SS (2006) Confluent monolayers of cultured human fetal retinal pigment epithelium exhibit morphology and physiology of native tissue. Invest Ophthalmol Vis Sci 47:2612–24

Marmorstein AD, Marmorstein LY, Rayborn M, Wang X, Hollyfield JG, Petrukhin K (2000) Bestrophin, the product of the best vitelliform macular dystrophy gene (VMD2), localizes to the basolateral plasma membrane of the retinal pigment epithelium. Proc Natl Acad Sci U S A 97:12758–63

Martin DF, Maguire MG, Ying GS, Grunwald JE, Fine SL, Jaffe GJ (2011) Ranibizumab and bevacizumab for neovascular age-related macular degeneration. N Engl J Med 364(20):1897–908

Martínez-Morales JR, Rodrigo I, Bovolenta P (2004) Eye development: a view from the retina pigmented epithelium. Bioessays 26:766–77

Maw MA, Kennedy B, Knight A, Bridges R, Roth KE, Mani EJ, Mukkadan JK, Nancarrow D, Crabb JW, Denton MJ (1997) Mutation of the gene encoding cellular retinaldehyde-binding protein in autosomal recessive retinitis pigmentosa. Na Genet 17:198–200

Meyer JS, Shearer RL, Capowski EE, Wright LS, Wallace KA, McMillan EL, Zhanga S-C, Gamm DM (2009) Modeling early retinal development with human embryonic and induced pluripotent stem cells. Proc Natl Acad Sci U S A 106:16698–703

Meyer JS, Howden SE, Wallace KA, Verhoeven AD, Wright LS, Capowski EE, Pinilla I, Martin JM, Tian S, Stewart R, Pattnaik B, Thomson J, Gamm DM (2011) Optic vesicle-like structures derived from human pluripotent stem cells facilitate a customized approach to retinal disease treatment. Stem Cells 29:1206–18

Murisier F, Beermann F (2006) Genetics of pigment cells: lessons from the tyrosinase gene family. Histol Histopathol 21:567–78

Nandrot EF, Anand M, Almeida D, Atabai K, Sheppard D, Finnemann SC (2007) Essential role for MFG-E8 as ligand for alphavbeta5 integrin in diurnal retinal phagocytosis. Proc Natl Acad Sci U S A 104:12005–10

Okita K, Matsumura Y, Sato Y, Okada A, Morizane A, Okamoto S, Hong H, Nakagawa M, Tanabe K, Tezuka K-I, Shibata T, Kunisada T, Takahashi M, Takahashi J, Saji H, Yamanaka S (2011) A more efficient method to generate integration-free human iPS cells. Nat Methods 8(5):409–14

Reichhart N, Strauss O (2014) Ion channels and transporters of the retinal pigment epithelium. Exp Eye Res 126:27–37

Rizzolo LJ (2014) Barrier properties of cultured retinal pigment epithelium. Exp Eye Res 126:16–26

Schwartz SD, Hubschman J-P, Heilwell G, Franco-Cardenas V, Pan CK, Ostrick RM, Mickunas E, Gay R, Klimanskaya I, Lanza R (2012) Embryonic stem cell trials for macular degeneration: a preliminary report. Lancet 379:713–20

Schwartz SD, Regillo CD, Lam BL, Eliott D, Rosenfeld PJ, Gregori NZ, Hubschman J-P, Davis JL, Heilwell G, Spirn M, Maguire J, Gay R, Bateman J, Ostrick RM, Morris D, Vincent M, Anglade E, Del Priore LV, Lanza R (2014) Human embryonic stem cell-derived retinal pigment epithelium in patients with age-related macular degeneration and Stargardt's macular dystrophy: follow-up of two open-label phase 1/2 studies. Lancet 385 509–16

Schwarz N, Carr A-JF, Lane A, Moeller F, Chen LL, Aguila M, Nommiste B, Muthiah MN, Kanuga N, Wolfrum U, Nagel-Wolfrum K, da Cruz L, Coffey PJ, Cheetham ME, Hardcastle AJ (2015) Translational read-through of the RP2 Arg120stop mutation in patient iPSC-derived retinal pigment epithelium cells. Hum Mol Genet 24:972–86

Sonoda S, Sreekumar PG, Kase S, Spee C, Ryan SJ, Kannan R, Hinton DR (2009) Attainment of polarity promotes growth factor secretion by retinal pigment epithelial cells: relevance to age-related macular degeneration. Aging (Albany NY) 2:28–42

Sparrow JR, Hicks D, Hamel CP (2010) The retinal pigment epithelium in health and disease. Curr Mol Med 10:802–23

Strauss O (2005) The retinal pigment epithelium in visual function. Pysiol Rev 85:845–81

Sun M, Finnemann SC, Febbraio M, Shan L, Annangudi SP, Podrez EA, Hoppe G, Darrow R, Organisciak DT, Salomon RG, Silverstein RL, Hazen SL (2006) Light-induced oxidation of photoreceptor outer segment phospholipids generates ligands for CD36-mediated phagocytosis by retinal pigment epithelium: a potential mechanism for modulating outer segment phagocytosis under oxidant stress conditions. J Biol Chem 281:4222–30

Takahashi K, Tanabe K, Ohnuki M, Narita M, Ichisaka T, Tomoda K, Yamanaka S (2007) Induction of pluripotent stem cells from adult human fibroblasts by defined factors. Cell 131:861–72

Vaajasaari H, Ilmarinen T, Juuti-Uusitalo K, Rajala K, Onnela N, Narkilahti S, Suuronen R, Hyttinen J, Uusitalo H, Skottman H (2011) Toward the defined and xeno-free differentiation of functional human pluripotent stem cell–derived retinal pigment epithelial cells. Mol Vision 17:558–75

Vugler AA, Carr AJ, Lawrence JM, Chen LL, Burrell K, Wright A, Lundh P, Semo M, Ahmado A, Gias C, da Cruz L, Moore H, Andrews P, Walsh J, Coffey PJ (2008) Elucidating the phenomenon of HESC-derived RPE: anatomy of cell genesis, expansion and retinal transplantation. Exp Neurol 214:347–61

Westenskow PD, Moreno SK, Krohne TU, Kurihara T, Zhu S, Zhang Z, Zhao T, Xu Y, Ding S, Friedlander M (2012) Using flow cytometry to compare the dynamics of photoreceptor outer segment phagocytosis in iPS-derived RPE cells. Invest Ophthalmol Vis Sci 53:6282–90

Witmer AN, Vrensen GF, Van Noorden CJ, Schlingemann RO (2003) Vascular endothelial growth factors and angiogenesis in eye disease. Prog Retin Eye Res 22:1–29

Wright CB, Redmond TM, Nickerson JM (2015) A history of the classical visual cycle. Prog Mol Biol Transl Sci 134:433–48

Yu J, Vodyanik MA, Smuga-Otto K, Antosiewicz-Bourget J, Frane JL, Tian S, Nie J, Jonsdottir GA, Ruotti V, Stewart R, Slukvin II, Thomson JA (2007) Induced pluripotent stem cell lines derived from human somatic cells. Science 318:1917–21

Yvon C, Ramsden CM, Lane A, Powner MB, da Cruz L, Coffey PJ, Carr AJ (2015) Using Stem Cells to Model Diseases of the Outer Retina. Comput Struct Biotechnol J 13:382–9

# Chapter 10
# Differentiation of Human Pluripotent Stem Cells into Cortical Neurons

**Cassiano Carromeu, Alexandre Vessoni, Ana Paula Diniz Mendes, and Patricia Cristina Baleeiro Beltrão-Braga**

**Abstract** Human neurologic and psychiatric disorders comprise a wide range of diseases, characterized by a strong genetic influence. Mutations can affect a broad spectrum of molecular pathways in neural cells leading to disturbances in the fine tune development of the nervous system. While transgenic animal models have led to important insights into neurological disorders, the complex circuitry of the human brain is difficult to fully recapitulate in these models, often producing misleading phenotypic correlations. Moreover, the translation of therapeutic compounds that were successful in animal models to human patients revealed to be a daunting task, with very low rates of success. With the recent advent of human Pluripotent Stem Cells (PSCs) field, researchers have now the possibility of modeling the human brain development *in vitro*. Moreover, using cellular reprogramming, which allows conversion of patient adult cells to ESCs-like stages, it is now possible to obtain neurons from patients with any genetic background of interest. In addition, the use of PSCs to model neurons *in vitro* provides a drug-screening platform and also personalized medicine, bearing in mind the cell donor. Considering this, several protocols were developed to differentiate human PSCs into neurons. The present chapter aims to introduce a detailed protocol to acquire human cortical neurons *in vitro*. Finally, it also discusses the potential use of these cells and their impact on the field of neuroscience.

---

C. Carromeu (✉)
University of California San Diego, La Jolla, CA 92093, USA
e-mail: carromeu@gmail.com

A. Vessoni
University of São Paulo, São Paulo, SP 05508-000, Brazil

A.P.D. Mendes
Salk Institute for Biological Studies, La Jolla, CA 92037, USA

P.C.B. Beltrão-Braga (✉)
Faculdade de Medicina Veterinária e Zootecnia (FMVZ), Universidade de São Paulo, Av. Prof. Orlando Marquês Paiva 87, São Paulo, SP 05508-270, Brazil
e-mail: patriciacbbbraga@usp.br

 
H. Ulrich, P. Davidson Negraes, *Working with Stem Cells*,
DOI 10.1007/978-3-319-30582-0_10

**Keywords** Human neurons • Cortical neurons • Neurodevelopment • Neural progenitor cells • Neuroectoderm • Neuronal differentiation

## 10.1 Introduction

Neurons are the basic functional unit of an intricate network inside the brain and the cells responsible for the coordination of all duties in the body. Probably because of this complex responsibility, the central nervous system (CNS) is the first one to be differentiated in embryos.

Coming from the embryonic ectoderm, the neural tube, prime of CNS, is differentiated and closed within 30 days in human embryos. Later, layers and cells from the early developing nervous system originate from the four fundamental zones: ventricular (VZ), subventricular (SVZ), intermediate (I), and marginal zones (MZ) (The Boulder Committee 1970). All cells of the CNS are derived from these developmental zones, on which neuroepithelial stem cells, after increasing thickness and surface of the VZ through mitosis, start an asymmetrical cell division, where one cell remains as a neuroprogenitor and the sister turns into glia or neuron (Takahashi et al. 1995; Bystron et al. 2006). This differentiation process is directed by a gradient of factors, like transforming growth factor β (TGF-β), activin and fibroblast growth factors (FGFs) (O'Rahilly 1999).

Even after birth, neurons and glial cells can continue to be differentiated from neuroprogenitor niches, like the subventricular zone and the dentate gyrus of the hippocampus in the adult brain. The pioneer for the discovery of neurogenesis was Joseph Altman, in 1962. He detected neurons that had incorporated the nucleotide thymidine-$H^3$ weeks later after its injection into the mouse brain. He then suggested the existence of undifferentiated cells that were dividing mitotically when this radiolabeled nucleotide was administered (Altman 1962).

Despite the many types of neurons, a typical neuron is composed by cell body, dendrites, axon and synaptic terminals. All these parts are important to connect and transmit information to other neurons in the form of electrical signals (Kandel et al. 2000). In the intricate environment of the CNS, and emerging from the release of a complex soup of neurotransmitters in a coordinated sequence, sensations are experienced, throughts came to live and we became aware of the world. It is not surprising that such complex tasks are only possible if all the different components of the brain are working harmonically. Any disturbance in its homeostasis can unbalance this system, leading to pathological conditions.

Until recently, most of the studies on how the brain works, especially in altered conditions, relied on animal models and human post-mortem brains (Dolmetsch and Geschwind 2011). It is known that the cerebral cortex, a laminated and complex structure filled with a huge amount and diversity of neurons, is the responsible for many assignments such as cognition, intelligence, emotions, and it is also the part of the brain that most turns human brain singularized from other species (Bystron et al. 2006). Therefore, while unquestionably progress was achieved using these

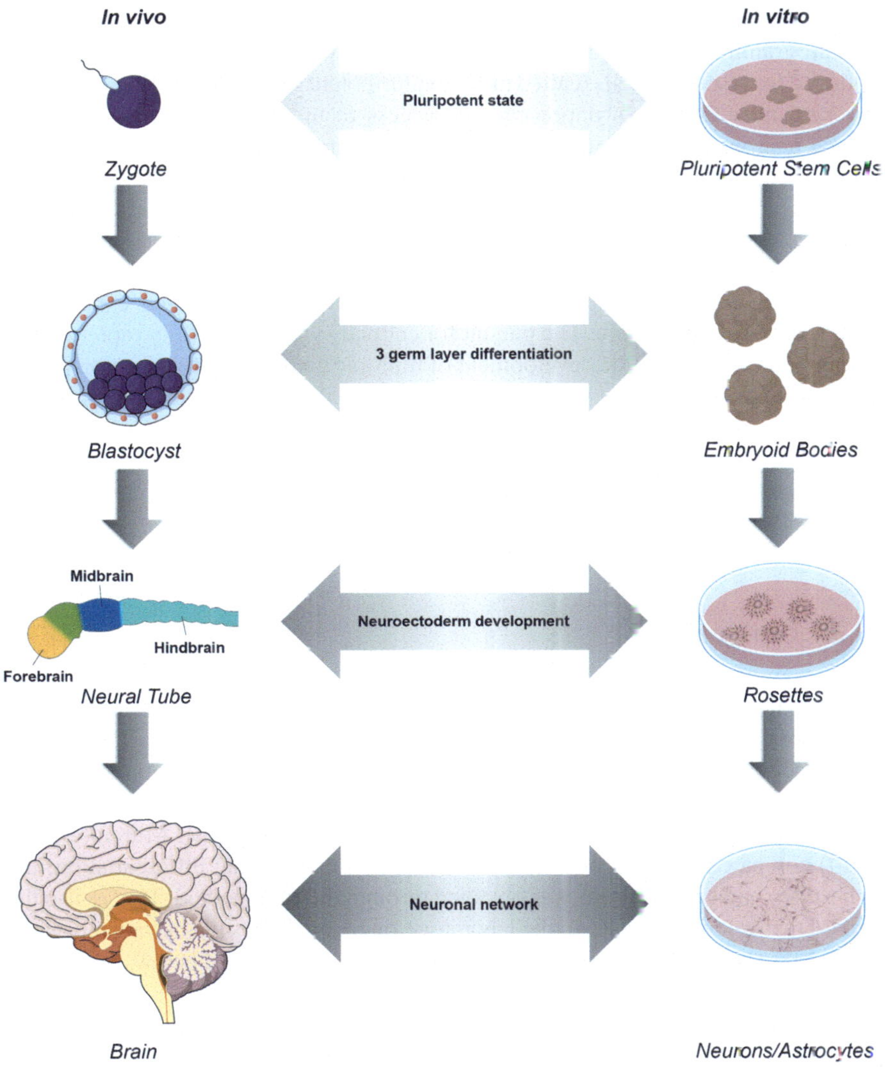


**Fig. 10.1** Schematic comparison of *in vivo* versus *in vitro* neuronal development

models, they have intrinsic limitations. The evolutionary gap between human and rodent species makes efforts to recapitulate the human phenotype *in vivo* very challenging (Dragunow 2008). Additionally, for neurodegenerative disorders studies, although post-mortem tissue fully represents the human disease, it offers just a snapshot, usually of an end-stage of the disorder (Dolmetsch and Geschwind 2011).

Bearing in mind that stem cells could be cultured and differentiated *in vitro*, soon they became targets for studies in developmental biology field and to turn powerful tools to give insights about the human brain functioning (Fig. 10.1). Taking advantage of the knowledge about the factors used for neurogenesis, embryonic stem cells

(ESCs) became interesting targets to produce neuronal cells. However, when cellular reprogramming techniques were developed and pluripotent stem cells could be produced from somatic cells (called induced pluripotent stem cells, iPSC) (Takahashi and Yamanaka 2006), an unprecedented access to unlimited amount of patient-derived neurons *in vitro* from various nervous system conditions started to be produced (Beltrão-Braga et al. 2013), reproducing very well the disorder *in vivo* and providing a powerful platform for drug screening tests (Marchetto et al. 2010a, b). Besides the use of cells to model diseases agree with the 3'Rs principle in research, iPSC carries the human genetic background involved on the target neurological disorder, thus recapitulating more closely the disorder and allowing exponential accumulation of knowledge. The pursuit for cellular systems that fully represent the human brain stimulated the development of multiple neuronal differentiation protocols. Different labs adopted varying approaches, usually consisting of optimizations of previous published versions (Muratore et al. 2014). This process of refining previous versions is still ongoing, with the ultimate goal of having a more representative modeling system *in vitro*. The protocol presented in this chapter represents the state-of-art for neuronal differentiation, focusing on human PSCs.

## 10.2 Protocol

Although the overwhelming amount of available techniques can be intimidating, the vast majority of the protocols to obtain neural cells are variations from only two different approaches: Embryoid Bodies (EBs) or monolayer differentiation. While the EBs protocol relies on tridimensional free-floating spheres, the monolayer method induces the differentiation of PSCs growing attached to the plate. Muratore et al. (2014) compared diverse commonly used methods of neuronal differentiation; the highest percentage of neurons was obtained through the EBs method. This is also the method that we routinely use in the laboratory, once it provides the best results when compared to other methods in our hands, and it will be described below.

### *10.2.1 Reagents*

Reagent preparation should be done in accordance with the manufacturer's recommendation. Additionally, unless otherwise started, stock solutions are kept at 1000× concentrated. Avoid repetitive cycles of freezing/thawing of the stock solutions.

- N2 media: 1× N-2 supplement (Thermo Fisher, Cat. No. 17502-048) in DMEM/F12 50/50 mix media (Cellgro, Cat. No. 10-090-CV). No antibiotics are added to this media. The media is stable at 4 °C for 1 month. Aliquots can be frozen (40 ml of media in 50 ml conical tubes) and stored at −20 °C.

- NBF media: 0.5× N-2 supplement (Thermo Fisher, Cat. No. 17502-048), 0.5× B-27 supplement (Thermo Fisher, Cat. No. 17504-044) and 1× Penicillin-Streptomycin (Thermo Fisher, Cat. No. 15140-122) in DMEM/F12 50/50 mix media (Cellgro, Cat. No. 10-090-CV). The media is stable at 4 °C for 1 month. Aliquots can be frozen (40 ml of media in 50 ml conical tubes) and stored at −20 °C. Add 20 ng/ml of fresh FGF2 (Thermo Fisher, Cat. No. PHG0023), before adding the media to the cells. FGF2 stock solution (100 μg/ml) aliquots should be kept at −80 °C for long-term storage, or at −20 °C for short term (<1 month). A thawed aliquot of FGF2 can be kept at 4 °C and is stable for 1 week (leftovers should be discarded after 1 week at 4 °C).
- Neuro media: 0.5× N-2 supplement (Thermo Fisher, Cat. No. 17502-048), 1× B-27 supplement (Thermo Fisher, Cat. No. 17504-044) and 1× Penicillin-Streptomycin (Thermo Fisher, Cat. No. 15140-122) in DMEM/F12 50/50 mix media (Cellgro, Cat. No. 10-090-CV). No FGF2 is added to this media. The media is stable at 4 °C for 1 month. Aliquots can be frozen (40 ml of media in 50 ml conical tubes) and stored at −20 °C.
- Dorsomorphin (EMD Millipore, Cat. No. 171261). Stock solutions should be kept at −20 °C.
- SB431542 (Tocris, Cat. No. 1614). Stock solutions should be kept at −20 °C.
- PBS without Calcium and Magnesium (Thermo Fisher, Cat. No. 14190-144).
- Accutase Cell Dissociation Reagent (Thermo Fisher Cat. No. A11105-01). Aliquots (10 ml) should be kept at −20 °C for long-term storage, or at 4 °C for short term (<2 weeks).
- Matrigel (BD Biosciences, Cat. No. CB 40230). Stock solutions should be kept at −20 °C. Make aliquots with enough volume to be thawed just once in each coating preparation. Follow manufacturer recommendations in the preparation of the plates.
- Poly-L-Ornithine (PLO, Sigma, Cat. No. P3655). Stock solutions should be kept at −20 °C.
- Laminin (Invitrogen, Cat. No. 23017-015). Stock solutions should be kept at −80 °C for long-term storage or at −20 °C for short term (<1 month).
- Rock Inhibitor Y27632 (Axxora, Cat. No. ALX-270-333-M025). Stock solutions should be kept at −20 °C. A thawed aliquot of Rock Inhibitor can be kept at 4 °C and is stable for 2 weeks.
- Brain-derived neurotrophic factor (BDNF) (R&D Systems, Cat. No. 248-BD). Stock solutions should be kept at −20 °C. A thawed aliquot of BDNF can be kept at 4 °C and is stable for 2 weeks.
- Glial cell-derived neurotrophic factor (GDNF) (R&D Systems, Cat. No. 212-GD). Stock solutions should be kept at −20 °C. A thawed aliquot of GDNF can be kept at 4 °C and is stable for 2 weeks.
- Blocking Solution for immunostainings: 5 % donkey serum in PBS. Prepare fresh.
- Triton X-100 (Sigma, Cat. No. X100-500ML).
- DAPI (4′,6-Diamidino-2-Phenylindole, Dihydrochloride) (Thermo Fisher, Cat. No. D1306).

### *10.2.2 Protocol 1: From Pluripotent Stem Cells (PSCs) to Neural Progenitor Cells (NPCs)*

To a successful cortical neural differentiation, it is crucial to start with good quality PSCs: no or only a few differentiation in the plate; passages between 10 (to erase any remaining epigenetic markers) and 40 (older than that have a higher chance of presenting chomosomal abnormalities); cells adapted and growing under feeder-free conditions (for example: growing on plates coated with Matrigel and using mTeSR media); and, free of mycoplasma contamination. Although a slight variability between different batches of differentiation of the same cell type can be noted, no significant differences are observed between ESC and iPSC, once the criteria above are matched. The protocol described here is highly efficient in generating a mix population of cortical neurons (for a schematic representation, see Fig. 10.2a, c). Although not a topic of this chapter, specific neuron subtypes can be obtained by adapting the protocol described herein to the many others available in this book or online (Maroof et al. 2013; Nicholas et al. 2013; Kriks et al. 2011; Kirkeby et al. 2012; Doi et al. 2014).

- **Day 1:** Starting from a confluent (***Note 1***) 10 cm or 6 cm plate of PSC, change the media to N2 media supplemented with 1 μM of Dorsomorphin and 10 nM of SB431542. This allows the cells to adapt to the new media before growing them in suspension as embryoid bodies (EBs). Also, once it still consists of a 2D culture system and not a multilayered tridimensional cell system, the concentration and availability of the inhibitors are optimal;
- **Day 2:** 24 h after starting the protocol, change the media again with N2 media supplemented with 1 μM of Dorsomorphin and 10 nM of SB431542. The cells are still attached to the plate and little (if any) morphological changes should be noted;
- **Day 3:** After 48 h growing in the presence of inhibitors as monolayer, the PSC are ready to differentiate in 3D as EBs. The interval between the first media change (Day 1) and the EB formation (Day 3) is important to efficiently: (1) prime the cells to a neural fate; and, (2) adapt the cells to the new media, avoiding severe cell death when transferring to suspension. Start cutting the colonies into smaller pieces using an 18G needle. Gently apply pressure to cut the cells, but avoid damaging the plastic of the plate. It is recommended to cut horizontally and vertically, forming squares. The goal is to decrease the size of the colonies, resulting in smaller EBs. After cutting the whole plate, add PBS without Calcium and Magnesium (8 ml in 10 cm plates, 4 ml in 6 cm dishes) and incubate at 37 °C for 5 min. The PBS will act as a passive chelating agent of Calcium and Magnesium, which are important ions in the cell-to-substrate adhesion. After the incubation at 37 °C, add N2 media supplemented with 1 μM of Dorsomorphin and 10 nM of SB431542 (6 ml of media per 10 cm plate and 3 ml per 6 cm). Gently scrap off the cells using a cell lifter. Make sure to suspend all the colonies of the plate. If necessary, do gently up and downs with a 5 ml pipette to decrease the size of big

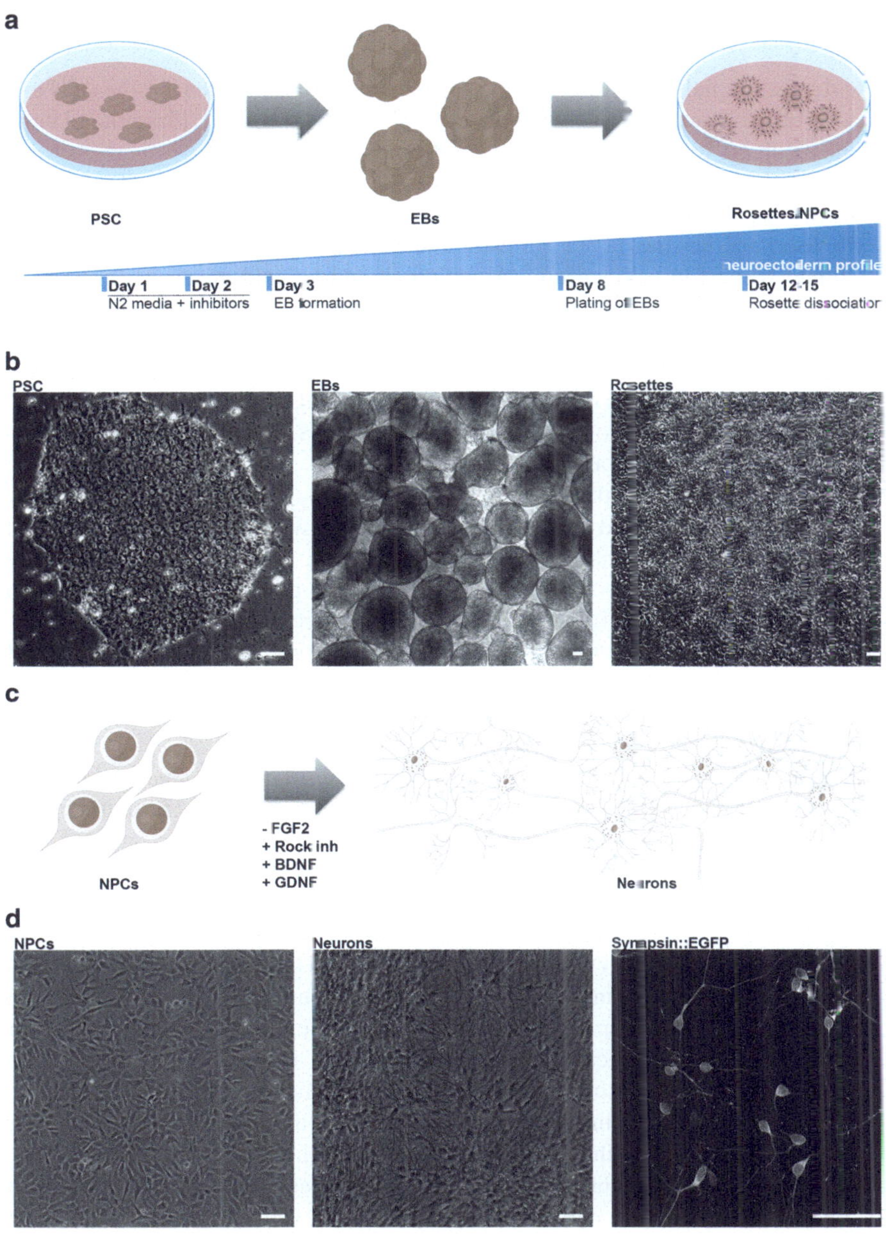


**Fig. 10.2** Obtaining human neurons from Pluripotent Stem Cells (PSCs). (**a**) Schematic representation of the derivation of Neural Progenitor Cells (NPCs) from PSCs. (**b**) Bright field microscopy showing (from *left* to *right*): a colony of pluripotent cells growing under feeder-free conditions; embryoid bodies (EBs) in suspension; rosettes structures after plating the dissociated EBs on matrigel plates. (**c**) Microscopy images showing (from *left* to *right*): bright field imaging of NPCs growing on Poly-L-Ornithine/Laminin coated tissue culture plates; bright field imaging of neurons after 4 weeks of differentiation; fluorescence imaging of 4 weeks neurons transduced with virus expressing EGFP under the promoter of *SYNAPSIN*. Only 1 % of the cells were transduced with the virus, to turn evident the cellular features of the neurons in a compact neuronal network. Bar represents 50 μm

colonies to small clumps. When the desired size is reached, transfer the cells to a well of a low attachment 6-well plate (3 ml of cells per well). Place the plate in the incubator (37 °C) under agitation (90 rpm);

- **Day 4:** 24 h after suspending the cells to form EBs, change the media with N2 media supplemented with 1 μM of Dorsomorphin and 10 nM of SB431542. Perform another media change using the same media on day 6. Media change should be performed carefully. Attention should be paid not to discarding the EBs when changing the media. To facilitate this task, it is possible to slowly perform a circular rotation of the plate inside the tissue culture hood, in order to concentrate the EBs at the center of the well. After that, carefully aspirate the media from the borders of the well, avoiding the EBs. They should remain concentrated at the center of the plate, while the dead (and smaller) cells are scattered over the well and aspirated;
- **Day 8:** Let the EBs sink in a 15 ml conical tube by gravity (the EBs normally sink really fast). Discard the supernatant (N2 media with inhibitors) and add 4 ml of NBF media (add fresh FGF2 to the media before adding to the cells). Gently break the EBs with ups and downs using a P1000 pipette. The goal here is to decrease the size of the EBs to smaller clumps (dissociation to single cell suspension will drastically decrease the efficiency of the protocol). Transfer the small clumps of cells in the NBF media to 6 cm Matrigel-coated plates (follow manufacturer's recommendation to perform the coating). Place the cells in the incubator making sure to evenly distribute them in plate;
- **Day 9:** Change the media with new NBF media. From now on, change the media every other day (48 h) to keep constant concentration levels of FGF2 in the culture, given the short half-life of the compound at 37 °C;
- **Days 12–15:** Closely monitor the emergence of rosettes structures *in vitro* (Fig. 10.2b). Usually, they are ready to passage between 4 and 7 days after plating the EBs. Before proceeding, change the media to new NBF. Manually select the rosettes under an EVOS Cell Imaging system (Thermo Fisher) with a P1000 tip, keeping them floating in the new media. Continue until you have collected all the rosettes in the plate (if too many are present, you can save some for another round of collection in the next day). Try to be very selective in the rosettes selection, picking only the flower-like structures (Fig. 10.2c). Neural crest cells are present in the surrounding areas of the rosettes and, given their proliferative potential, these cells can contaminate the neural progenitor culture. Move the rosette containing media to a 15 ml conical tube and dissociate them with a P1000 pipette by performing up and down movements, and transfer all the volume to Poly-L-Ornithine/Laminin plates (see *Auxiliary Protocol 1*) using NBF media.

***Note 1:*** *Good PSC grow in monolayer as colonies. PSC are confluent when the colonies have reached a large size and there are many colonies on the plate. If too many colonies are present, they can start touching each other before they are ready* *to start the differentiation protocol. Just a few colonies, on the other hand, will lead to few EBs and low efficiency. In both cases, it is recommended to passage the cells until the ideal confluence is reached to perform the protocol.*

### 10.2.3 Protocol 2: Maintenance of NPCs

After the dissociation and passage of the rosettes, the derived neural progenitor cells (NPCs) can be kept and passage in Poly-L-Ornithine/Laminin coated plates with NBF media. The media change is accomplished every 48 h and is very important to supplement it with FGF2 before adding the media to the cells. The earlier the passage number, the greater the neurogenic potential of the progenitor population. However, there are situations where passaging the NPCs is desirable, like in obtaining cells for a large screening platform or doing a cell stock for the lab, allowing the use of same batches of differentiation in future experiments. Additionally, early passages NPCs can be contaminated with other cell types and the passage helps to dilute this effect by the fast growing characteristic of the progenitors, leading to a more homogenous culture. We have noticed that NPCs up to passage 10 are good neurogenic sources and adopted the following practice in the lab:

1. Isolate the progenitors (NPCs) from rosettes (*Protocol 1*);
2. Expand these progenitors for a few passages (up to passage 2);
3. Access the quality of the NPCs. The most common ways are: (a) morphology under the microscope (achieved with experience, this is the easiest and quickest way to access the quality of the culture); (b) immunostaining of canonical NPCs markers (refer to *Immunostaining Protocol*) (c) real-time PCR comparing PSCs to NPCs using typical progenitor markers; and, (d) differentiation of NCPs to neurons and later quantification of the neuronal population by immunostaining. We routinely access the quality of the progenitors visually under the microscope and with immunostainings of the progenitor and the derived neuronal population. Generated NPCs approved on the criteria above are considered excellent neuronal sources and expanded to make frozen stocks of cells;
4. Finally, all the experiments for a particular study using these cells are performed using NPCs at similar passage number (to reduce variability between the assays).

   The passage of NPCs is achieved enzymatically, dissociating the cells to single cells before expanding to new culture dishes. Cells at this stage are growing as monolayers and should be passaged when they are 85–95 % confluent. It is not recommended to keep the cells at 100 % confluence for long periods of time, once the confluency can induce some premature differentiation. The passaging protocol is described below:

Starting with a confluent plate (85–95 % confluent), discard the media and wash the cells once with PBS;

1. Discard the PBS and add enough prewarmed (37 °C) Accutase (***Note 2***) to cover the monolayer of cells;
2. Transfer the cells to 37 °C and allow the enzyme to act for approximately 2 min;
3. Visually inspect the cells under the microscope to check the enzyme activity. When the cells start to round up, it is time to proceed to the next step. Otherwise, incubate 3 more min at 37 °C (do not allow the cells to stay longer than 5 min at the 37 °C with Accutase);
4. Add PBS and detach the remaining attached cells with gentle up and down pipetting. Avoid bubble formation at this stage;
5. Transfer the suspended cells to a conical tube and centrifuge at 200 × *g* for 5 min at room temperature;
6. Discard the supernatant;
7. Resuspend the cell pellet with fresh NBF media (remember to add fresh FGF2);
8. Transfer the resuspended cells to new plates coated with Poly-L-Ornithine/ Laminin. The cells can be split in a 1:3 ratio (for example: from 1× 10 cm, plate 3× 10 cm plate) during the maintenance stage;
9. Media change should be performed every 48 h and cells should be passaged again once an 85–95 % confluence is reached.

***Note 2:*** *Accutase enzyme is temperature sensitive. Over repetitive cycles of cooling (4 °C storage)/heating 37 °C), the activity of the enzyme suffers a significant decay. In order to have always the best activity, it is recommended to warm up just an aliquot of the amount to be used in the experiment. Also, the enzyme should not be kept longer than 5 min at 37 °C before transfered to the cells. These steps assure a highly active enzyme, with a short action period of time to dissociate the cells.*

## *10.2.4 Protocol 3: Differentiation of NPCs into Neurons*

Neuronal induction can be accomplished solely by removing the FGF2 from the media. The use of neurotrophic proteins as Brain-derived neurotrophic factor (BDNF) and Glial cell-derived neurotrophic factor (GDNF) helps in the survival and maturation of newly differentiated neurons. Their use, although advised, should also be cautioned especially when studying diseases that have a known deficiency of these factors, as some neurodevelopmental disorders. In these cases, the use of such factors may influence the result of the experimental assays, rescuing some of the potential phenotypes that would exist in physiological conditions. An alternative is to use these factors for a neuronal induction period (2 weeks, for example) and withdraw for a few days/weeks before the assays (but this has to be optimized for each specific case). The following protocol describes a standard neuronal differentiation starting from NPCs (Fig. 10.2c, d).

- **Day 1:** Passage the cells on the day before the differentiation. Plate the cells with NBF media supplemented with BDNF (20 ng/ml) and GDNF (20 ng/ml). They should be at 70–80 % confluence in the following day, before starting the differentiation;
- **Day 2:** induce the differentiation by changing the media to Neuro media supplemented with BDNF (20 ng/ml), GDNF (20 ng/ml) and Rock Inhibitor Y-27632 (10 μM). Withdraw of FGF2 will induce a cell cycle exit of the progenitor, forcing them to differentiate. Rock Inhibitor promotes neurite outgrowing and stimulates the differentiation (Gu et al. 2013). The neurotrophic factors BDNF and GDNF help in the survival and maturation during the differentiation process;
- **Day 4:** Change the media to Neuro media (48 h after the induction). Withdraw the Rock Inhibitor, but keep the neurotrophic factors BDNF and GDNF (both at 20 ng/ml);
- Perform regular media change (Neuro media + BDNF + GDNF) every 3–4 days (***Note 3***). If the pH of the media changes dramatically (yellowish color), a more often media change might be required;
- Follow with regular media change until the desired maturation is reached to further assays (***Note 4***).

***Note 3:*** *Mature neuronal cultures form a network of interconnected cells that are very prone to detachment. At this stage, perform media change carefully and slowly. If a small part of the neuronal network start to detach, it could be enough to detach the whole culture at once. If a detachment is detected, avoid any disturbance on that part of the plate. Also, the transportation of the plate, as well the opening and closing of the incubator door should all be carefully performed at all stages of the cell culturing. Finally, if detachment is a recurrent issue, it can be ameliorated with half media change, instead of full change: remove half of the old media and slowly and carefully add the new one. This procedure is less disturbing to the differentiated neurons than a full media change (in this case, media change can be required more often).*

***Note 4:*** *Human cortical neurogenesis* in vivo *takes approximately 70 days (Caviness et al. 1995).* In vitro, *the optimal maturation age will depend on some factors. In our hands, synaptic structures are already evident within 4 weeks of differentiation. In some diseased conditions, especially those neurodevelopmental ones, significant differences can be observed compared to controls at early stages of the development, which allows for short time points of differentiation. Other pathologies, as neurodegenerative conditions, may require longer periods of time in culture to reveal differences between affected and control cells. The longer the neuronal culture is kept in culture, the more mature the derived neurons will be. Also, representation of all cortical layers* in vitro *will require longer time in culture, mimicking the development* in vivo *(Schadt et al. 2014). The decision for the most appropriate time point to perform the assays will depend on the best judgment of the researcher, considering his/her knowledge of the disease. Most of the laboratories use between 4 and 8 weeks of differentiation* in vitro.

### *10.2.5 Auxiliary Protocol 1: Coating of Culture Dishes with Poly-L-Ornithine and Laminin*

There are different matrices that allow the growth of neurons. One of the most used nowadays is the double coating with Poly-L-Ornithine (PLO) and Laminin. PLO is a positively charged synthetic amino acid chain widely used to enhance cell attachment over a surface (plastic and glass). Laminin are proteins of the extracellular matrix with high molecular weight. Both substrates have important roles in neurites outgrowth and cell attachment. The coating protocol consists of a bottom layer of PLO and a top layer of Laminin:

1. Prepare enough working solution of PLO to cover all the plates to be used for NPC/neuronal culture. Refers to ***Note 6*** for a convenient volume chart and select the total volume needed by multiplying the number of plates to be coated by the volume per plate;
2. PLO solution: dilute the stock solution (10 mg/ml) (***Note 5***) in sterile $H_2O$ to yield 10 μg/ml (1:1000) for plastic surfaces or 50 μg/ml (1:200) for glass surfaces;
3. Add enough PLO solution to cover the entire surface of the plate, moving the plate in a way that you obtain a homogenous coating surface (refers to the Volume Chart on ***Note 6***);
4. Leave the plates at 37 °C overnight (12–24 h);
5. Next, wash the plates twice with sterile $H_2O$;
6. Thaw the Laminin on ice. To avoid any delay in the preparation, place the Laminin on ice before the step 5. *Do not leave the Laminin at room temperature*;
7. To coat plastic surfaces, dilute Laminin in PBS to yield 5 μg/ml (1:200). For glass surfaces, dilute to 10 μg/ml (1:100). Freeze the leftover of stock Laminin;
8. Add enough Laminin solution to cover the entire dish (refer to Volume Chart on ***Note 6***);
9. Incubate the plates at 37 °C for at least 2 h. After 2 h, the plates are ready to use. You can keep them at humidified 37 °C up to 2 days, but should freeze (−20 °C) the plates for longer periods. Do not remove the Laminin when freezing the plates. To avoid the plates from being accidentally open or to have liquid evaporation, seal the plates with Parafilm before freezing. Plates are good for 6–8 months at −20 °C.

Before using the coated dishes, wash them once with PBS (just wash the dishes that you will use in the experiment).

***Note 5:*** *Preparation of PLO stock solution: Dissolve the PLO powder in sterile $H_2O$ to make a 10 mg/ml stock solution. Filter through a 0.22 μm strainer and aliquot in 1 ml portions. Store aliquots at −20 °C. The PLO in the stock solution is stable to a few cycles of thawing/freezing.*

***Note 6:*** *Volume chart:*

- *Solution of PLO/dish: 7 ml/10 cm; 3 ml/6 cm; 2 ml/well from 6-well plate; 500 μL/well from a 2-well ChamberSlide.*
- *Solution of Laminin/dish: 6 ml/10 cm; 2.5 ml/6 cm; 1.5 ml/well from 6-well plate; 300 μL/well from a 2-well ChamberSlide.*

### 10.2.6 *Auxiliary Protocol 2: Characterization of NPCs and Neurons by Immunostaining*

Immunostaining is a versatile tool widely used in the laboratory to access different aspects of the cellular culture, with applications ranging from characterization of cellular content to identification of phenotypes between affected and control cells. After the generation of target cells of interest (NPCs or neurons), it is crucial to confirm the neuroectoderm identity of these cell lines (Fig. 10.3a–c). Fortunately, canonical markers for these cells are extensively described in the literature and many are accepted as good markers for NPCs and neurons. For a brief reference, please see Table 10.1. Perform the protocol carefully when working with neurons, especially while discarding/aspirating and adding solutions (do it slowly). At this stage, the cells are prone to detachment even after fixation.

1. Discard the media and fix the cells with 4 % paraformaldehyde (PFA) solution for 20 min at room temperature;
2. Wash 3× with PBS. Leave the PBS on the dish for 2 min before removing it between the washes. At this stage, you can stop and continue the next day, but it is recommended to proceed. If stopping at this step, leave the cells with PBS at 4 °C;
3. Permeabilize the cells: Dilute Triton X-100 to a final concentration of 0.1 % in Blocking Solution (5 % donkey serum in PBS) (*Note 7*). Discard the PBS from

**Table 10.1** Antibodies for immunostaining

| Antibodies[a] | Company | Catalog number | Stage[b] | Dilution |
|---|---|---|---|---|
| Sox2 | Cell Signaling | 3579 | NPS | 1:200 |
| Nestin | EM Millipore | ABD69 | NPS | 1:200 |
| Pax6 | Covance | PRB-278P | NPS | 1:200 |
| Map2 | Sigma-Aldrich | M 1406 | NDS | 1:100 |
| Tuj1 | Covance | PRB-435P | NDS | 1:200 |
| Synapsin I | Calbiochem | 574778 | NDS | 1:200 |
| Vglut1 | Synaptic System | 135 302 | NDS | 1:400 |
| PSD95 | Invitrogen | 75-028 | NDS | 1:200 |
| GABA | Sigma-Aldrich | A0310 | NDS | 1:500 |
| GFAP | EM Millipore | AB5541 | NDS | 1:500 |

[a]This list of antibodies is an excellent 'start' to the characterization of the obtained cells. Sox2, Nestin and Pax6 are good markers for NPCs; with Pax6 being a suitable marker for early passage NPCs. Map2 and Tuj1 are excellent markers to access the neuronal morphology/differentiation, with Map2 being the staining of choice in most cases (given its preferential staining of the cell body and dendrites, while Tuj1 preferentially stains the axon). Synapsin I and PSD95 are pre- and post-synaptic markers, respectively. Their co-localization implies mature synapse formation in the system. VGlut1 and GABA are used to identify glutamatergic and GABAergic neurons respectively. VGlut1 is a good marker to quantify the number of glutamatergic synaptic puncta. GFAP is a marker for astrocytes

[b]This column refers to the current developmental stage that the antibody can be useful in discriminating between *NPS* neural progenitor stage, and *NDS* neural differentiated stage

previous step and add enough Permeabilization Solution to cover the cells. Incubate for 15 min at room temperature;
4. Discard the Permeabilization Solution and add fresh Blocking Solution. Incubate for 30 min at room temperature;
5. Discard the Blocking Solution and add primary antibodies (see Table 10.1 for references) and incubate at room temperature for 2 h or at 4 °C overnight under gentle agitation (***Note 8***). The primary antibodies should be diluted in Blocking Solution;
6. Wash 3× with PBS. Leave the PBS on the dish for 2 min before removing it between the washes;
7. Discard the PBS and add fresh Blocking Solution. Incubate for 30 min at room temperature;
8. Discard the Blocking Solution and add the secondary antibodies (diluted in Blocking Solution) (***Note 9***). Incubate for 1 h at room temperature under gentle agitation. Protect from light during incubation;
9. Wash 1× with PBS and incubate with PBS on for 2 min. Protect from light during washes;
10. Proceed to nuclei staining: Dilute DAPI 1:10,000 in PBS and incubate it for 4 min at room temperature. Protect from light during incubation;
11. Wash 3× with PBS. Leave the PBS on the dish for 2 min before removing it between the washes. Protect from light during washes;
12. Mount the slides using any desired mount solution.

***Note 7:*** *Triton X-100 is a very viscous detergent, difficult to pipette accurately. To avoid much inaccuracy, it is advised to prepare a solution of 10 % Triton X-100 in PBS, pipetting it very slowly. After complete dissolution of the detergent, this stock solution of Triton X-100 can be diluted 1:100 in Blocking Solution to reach the final concentration of 0.01 % (working solution).*

***Note 8:*** *Although both incubation methods would work, incubation overnight (16 h) at 4 °C with gentle agitation seems to generate the best ratio signal/background.*

***Note 9:*** *To avoid precipitated fluorophores, briefly centrifuge (spin down) the secondary antibodies before adding to the cells. If accumulated in the slide, precipitated fluorophores can cause a very strong punctual background signal, scattered all over the field of view. Additionally, do not incubate cells with secondary antibodies for more than 2 h, once it also can significantly increase the background signal.*

## 10.3 Potential Use of In Vitro Neurons and Perspectives

The cortical neurons obtained with the protocol described in this chapter can be applied to investigate the nervous system in multiple ways, ranging from understanding fundamental aspects of neurobiology to disease modeling. Although great scientific efforts are allocated in revealing the meanders of the intrinsic biology of the nervous system, many researchers are interested in translating the scientific

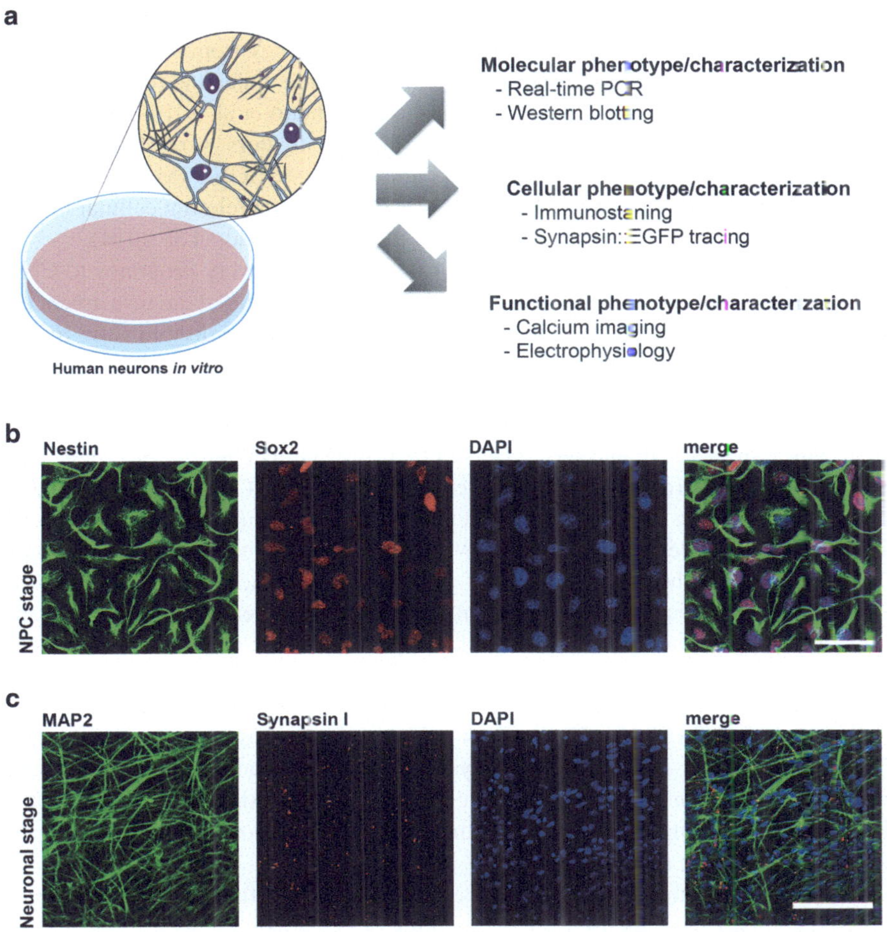


**Fig. 10.3** Characterization of the derived progenitors and neurons. (**a**) A schematic representation of potential assays used to characterize the cells at different levels (molecular, cellular and functional). (**b**) Immunostaining of obtained Neural Progenitor Cells (NPCs) using Nestin (cytosolic) and Sox2 (nuclear). (**c**) Immunostaining of NPC-derived neurons using Map2 (cytosolic) and Synpasin I (located on the synapses). Bar represents 50 μm

findings from the bench to the bedside. With the advent of reprogramming, an unlimited source of disease-specific cells became suddenly available and disease modeling turned into a popular methodology to get insights about the disease's mechanisms (Marchetto et al. 2010a, b). For the first time, human neuronal networks bearing specific mutations can now be used to access phenotypes *in vitro* at different levels: molecular, cellular and functional (Fig. 10.3a).

Monogenic disorders are of particular interest to disease modeling. The genetic defined origin makes the disease suitable to genetic correction in gain- and loss-of-function assays, confirming the contribution of the mutation to the disease phenotype (Brocolli et al. 2014). Rett Syndrome (RTT), for example, is a neurodevelopmental disorder with autism-like phenotype caused by loss-of-function mutations in the *MECP2* gene (Amir et al. 1999). After deriving PSC lines from RTT patients by reprogramming, researchers revealed several molecular, cellular and functional phenotypes of RTT neurons compared to controls (Marchetto et al. 2010a, b; Ananiev et al. 2011; Cheung et al. 2011). Moreover, the investigators were able to rescue some phenotypes using a potential therapeutic candidate, IGF-I (Marchetto et al. 2010a, b), evidencing that iPSC-derived nerve cells may be a relevant cellular model to be used in high-throughput drug screening platforms (Schadt et al. 2014).

In fact, until recently, the low success rate in drug discovery programs to CNS disorders using animal models or irrelevant cell types has limited the investment by pharmacological companies, generating few innovative CNS drugs in the recent decades. Considering the start point of 10,000 possible drug-candidates to treat one disease and the long time (15 years) to have one drug ready to clinical trial, faster, less expensive and more efficient techniques should be applied on drug discover field. Furthermore, it should be considered that some human genetic disorders cannot be properly mirrored in animal models, for instance the Cockayne Syndrome, a rare and devastating disease where patients display both neurodevelopmental deficiencies and neurodegeneration, and in which transgenic mouse models show minor (if any) of these phenotypes. In all of these scenarios, the use of iPSC-derived human neurons emerges as a promising alternative for identification of relevant phenotypes at cellular or molecular level, and guide the development of more effective therapeutic strategies (Khurana et al. 2015). Additionally, patient-derived PSCs also have the potential to be used in cell therapies for neurodegenerative diseases (Marchetto et al. 2010a, b). The goal would be to replace lost cells (neurons/astrocytes) with patient-specific nervous system progenitors, able to repopulate the affected area and integrate in the native neural network. Although there is still a long way before implementation into the clinic, many laboratories are exploring this possibility with animal models.

Finally, recent attempts of creating cellular models that fully recapitulate the *in vivo* counterpart have focused on 3D cultures or organoids (Willyard 2015). The main goal is to drive away from a monolayer cell-specific culture towards a multiple cell types 3D organized structure that more closely resembles the human brain (Lancaster and Knoblich 2014). The "mini-brain" organoid would be more suitable to model complex diseases, where the phenotype is a consequence of a multicellular system interacting in an organized network. Even though in its early ages, this technology has the potential to revolutionize medicine with insights about the brain development. Nonetheless, it will not substitute the standard cellular culture; the synergy between both techniques will be a powerful way to understand the complex biochemical, cellular and network milieu of the nervous system. Meanwhile, with the emergence of the stem cell field, diverse protocols have been continuously developed in a exaustive process of enhancing their previous versions. The protocol described in this chapter has been continuously and successfully used in diverse laboratories around the world. Although it certainly represents a good starting point to any laboratory interested in modeling the CNS *in vitro*, optimization may be required to model the intended (specific) disease, ensuring the complete recapitulation of phenotypes *in vitro* and unleashing the full potential of the stem cells to neuroscience.

## References

Altman J (1962) Are new neurons formed in the brains of adult mammals? Science 135(3509):1127–1128

Amir RE, Van den Veyver IB, Wan M, Tran CQ, Francke U, Zoghbi HY (1999) Rett syndrome is caused by mutations in X-linked MECP2, encoding methyl-CpG-binding protein 2. Nat Genet 23(2):185–188

Ananiev G, Williams EC, Li H, Chang Q (2011) Isogenic pairs of wild type and mutant induced pluripotent stem cell (iPSC) lines from Rett syndrome patients as in vitro disease model. PLoS One 6(9):e25255

Beltrão-Braga PCB, Pignatari GC, Russo FB, Fernandes IR, Muotri AR (2013) In-a-dish: induced pluripotent stem cells as a novel model for human diseases. Cytometry A 83(1):11–17

Brocolli V, Giannelli SG, Mazzara PG (2014) Modeling physiological and pathological human neurogenesis in the dish. Front Neurosci 8:183

Bystron I, Rakic P, Molnar Z, Blakemore C (2006) The first neurons of the human cerebral cortex. Nat Neurosci 9:880–886

Caviness VS Jr, Takahashi T, Nowakowski RS (1995) Numbers, time and neocortical neuronogenesis: a general developmental and evolutionary model. Trends Neurosci 18(9):379–383

Cheung AY, Horvath LM, Grafodatskaya D, Pasceri P, Weksberg R, Hotta A, Carrel L, Ellis J (2011) Isolation of MECP2-null Rett Syndrome patient hiPS cells and isogenic controls through X-chomosome inactivation. Hum Mol Genet 20(11):2103–2115

Doi D, Samata B, Katsukawa M, Kikuchi T, Morizane A, Ono Y, Sekiguchi K, Nakagawa M, Parmar M, Takahashi J (2014) Isolation of human induced pluripotent stem cell-derived dopaminergic progenitors by cell sorting for successful transplantation. Stem Cell Rep 2(3):337–350

Dolmetsch R, Geschwind DH (2011) The human brain in a dish: the promise of iPSC-derived neurons. Cell 145(6):831–834

Dragunow M (2008) The adult human brain in preclinical drug development. Nat Rev Drug Discov 7(8):659–666

Gu H, Yu SP, Gutekunst CA, Gross RE, Wei L (2013) Inhibition of the Rho signaling pathway improves neurite outgrowth and neuronal differentiation of mouse neural stem cells. Int J Physiol Pathophysiol Pharmacol 5(1):11–20

Kandel E, Schwartz J, Jessell T (2000) Principles of neuroscience, 4th edn. McGraw-Hill Companies, New York. ISBN 0838577016, 9780838577011

Khurana V, Tardiff DF, Chung CY, Lindquist S (2015) Toward stem cell-based phenotypic screens for neurodegenerative diseases. Nat Rev Neurol 11(6):339–350

Kirkeby A, Grealish S, Wolf DA, Nelander J, Wood J, Lundblad M, Lindvall O, Parmar M (2012) Generation of regionally specified neural progenitors and functional neurons from human embryonic stem cells under defined conditions. Cell Rep 1(6):703–714

Kriks S, Shim JW, Piao J, Ganat YM, Wakeman DR, Xie Z, Carrillo-Reid L, Auyeung G, Antonacci C, Buch A, Yang L, Beal MF, Surmeier DJ, Kordower JH, Tabar V, Studer L (2011) Dopamine neurons derived from human ES cells efficiently engraft in animal models of Parkinson's disease. Nature 480:547–551

Lancaster MA, Knoblich JA (2014) Generation of cerebral organoids from human pluripotent stem cells. Nat Protoc 9(10):2329–2340

Marchetto MC, Carromeu C, Acab A, Yu D, Yeo GW, Mu Y, Chen G, Gage FH, Muotri AR (2010a) A model for neural development and treatment of Rett syndrome using human induced pluripotent stem cells. Cell 143(4):527–539

Marchetto MC, Winner B, Gage FH (2010b) Pluripotent stem cells in neurodegenerative and neurodevelopmental diseases. Hum Mol Genet 19(R1):R71–R76

Maroof AM, Keros S, Tyson JA, Ying SW, Ganat YM, Merkle FT, Liu B, Goulburn A, Stanley EG, Elefanty EG, Widmer HR, Eggan K, Goldstein PA, Anderson SA, Studer L (2013)

Directed differentiation and functional maturation of cortical interneurons from human embryonic stem cells. Cell Stem Cell 12:559–572

Muratore CR, Srikanth P, Callahan DG, Young-Pearse TL (2014) Comparison and optimization of hiPSC forebrain cortical differentiation protocols. PLoS One 9(8):e105807

Nicholas CR, Chen J, Tang Y, Southwell DG, Chalmers N, Vogt D, Arnold CM, Chen YJ, Stanley EG, Elefanty AG, Sasai Y, Alvarez-Buylla A, Rubenstein JL, Kriegstein A (2013) Functional maturation of hPSC-derived forebrain interneurons requires an extended timeline and mimics human neural development. Cell Stem Cell 12:573–586

O'Rahilly M (1999) The embryonic human brain: an atlas of developmental stages. Wiley-Liss, New York

Schadt EE, Buchanan S, Brennand KJ, Merchant KM (2014) Evolving toward a human-cell based and multiscale approach to drug discovery for CNS disorders. Front Pharmacol 5:252

Takahashi K, Yamanaka S (2006) Induction of pluripotent stem cells from mouse embryonic and adult fibroblast cultures by defined factors. Cell 126(4):663–676

Takahashi T, Nowakowski RS, Caviness VS Jr (1995) Early ontogeny of the secondary proliferative population of the embryonic murine cerebral wall. J Neurosci 15:6058–6068

The Boulder Committee (1970) Embryonic vertebrate central nervous system: revised terminology. Anat Rec 166:257–261

Willyard C (2015) The boom in mini stomachs, brains, breasts, kidneys and more. Nature 523(7562):520–522

# Chapter 11
# Motor Neuron Differentiation from Pluripotent Stem Cells: Development of the Technique, Synopsis of Protocols, and Summary of Current Applications

**Helen Cristina Miranda and Albert R. La Spada**

**Abstract** Motor neurons conduct electrical signals to the musculature via synapses in the central nervous system (CNS). This communication guarantees proper performance of movements and vital functions including: standing, walking, and breathing. Motor neurons can be divided in two subgroups: cortical motor neurons (also known as upper motor neurons) and spinal cord motor neurons (also known as lower motor neurons). Lower motor neurons are responsible for the contraction of more than 500 different skeletal muscles. A defect that affects either upper motor neurons or lower motor neurons (or both) will lead to different types of motor neuron disease. Before the advent of pluripotent stem cells, motor neuron disease could only be studied using transformed cell lines, animal models, or postmortem tissues, which even when utilized together do not fully recapitulate the motor neuron disease phenotypes.

---

H.C. Miranda
Department of Pediatrics, University of California, San Diego, La Jolla, CA 92093, USA

Sanford Consortium for Regenerative Medicine, La Jolla, CA 92093, USA

A.R. La Spada (✉)
Department of Pediatrics, University of California, San Diego, La Jolla, CA 92093, USA

Sanford Consortium for Regenerative Medicine, University of California,
2880 Torrey Pines Scenic Dr. – Room 3003, San Diego, La Jolla, CA 92093, USA

Cellular and Molecular Medicine, University of California, San Diego,
La Jolla, CA 92093, USA

Department of Neurosciences, University of California, San Diego, La Jolla, CA 92093, USA

Division of Biological Sciences, University of California, San Diego,
La Jolla, CA 92093, USA

Institute for Genomic Medicine, University of California, San Diego,
La Jolla, CA 92093, USA
e-mail: alaspada@ucsd.edu


H. Ulrich, P. Davidson Negraes, *Working with Stem Cells*,
DOI 10.1007/978-3-319-30582-0_11

Over the last decade, the motor neuron disease field has made incredible strides by establishing efficient and reliable protocols for generating motor neurons *in vitro* from pluripotent stem cells. In this chapter, we will consider the biology of motor neuron development, describe protocols for generating motor neurons and enriching derived cell populations for motor neurons, and summarize key advances in the motor neuron disease field achieved through the use of stem cell model systems.

**Keywords** Upper motor neurons • Lower motor neuros • Stem cell model • iPSC • Differentiation

## 11.1 Motor Neuron Biology and Development

Motor neurons are efferent neurons that facilitate communication between brain and muscle. Motor neurons carry electrical signals from the brain through the spinal cord, resulting in muscle contraction via the neuromuscular junction. Our understanding of motor neurons has changed considerably throughout the years. What was originally implied to be a uniform cell population is now known to consist of a very heterogeneous group of neurons, which differ in their morphology, the types of muscle fibers that they innervate, and their anatomical localization. The motor system is composed of upper motor neurons and lower motor neurons. Upper motor neurons (UMNs), also known as corticospinal motor neurons, reside in the motor region of the cerebral cortex, whereas lower motor neurons (LMNs), also known as alpha motor neurons, are located in the spinal cord.

### *11.1.1 Upper Motor Neurons*

UMNs form the basis for voluntary movement in the human body; they are large, excitatory glutamatergic projection neurons that extend axons via the internal capsule in the midbrain to the spinal cord, where they synapse with LMNs (Shoemaker and Arlotta 2010). The exact moment in development when UMNs differentiate themselves from other types of neurons in the cortex remains unknown. However, considerable progress has been achieved in defining the molecular mechanisms involved in UMN specification in the developing brain.

Elegant work done about a decade ago identified the molecular determinants of UMNs located in cortical layer V in the brains of mice (Arlotta et al. 2005; Molyneaux et al. 2005). The list of genes identified in these studies to specify UMN fate include: transcription factors (e.g., *Ctip2*, *Bcl6*, *Sox5*, *Fezf2*), cell surface proteins (e.g., *Encephalopsin*, *Itm2a*, *Dafl*), calcium signaling proteins (e.g., *Pcp4*, *S100a10*), cell adhesion proteins (e.g., *Cdh22*, *Cdh13*, *Cntn6*), and axon guidance molecules (e.g., *Neto1*, *Netrin-G1*). These molecules appear to be expressed together in a pattern that uniquely identifies UMNs. Among these genes, the transcription factor *Forebrain embryonic zinc finger-like protein 2* (Fezf2) is absolutely required for the specification of an UMN.

Fezf2 is a highly conserved transcription factor, which contains a six-zinc finger domain (Molyneaux et al. 2005). Fezf2 expression alone is sufficient to generate corticofugal neurons from progenitors fated to become medium spiny neurons (Rouaux and Arlotta 2010). Another protein identified as playing a critical role in UMN development is COUP-TF1 interacting protein 2 (CTIP2). Experiments carried out in *Ctip2* null mutant mice display striking abnormalities of axonal fiber tracts with no apparent widespread death of neuronal populations in the cortex, or neuronal lamination defects (Arlotta et al. 2005). *Ctip2*$^{-/-}$ neurons show highly disorganized, non-fasciculated axonal projection (Arlotta et al. 2005). Interestingly, expression of *Ctip2* appears crucial for the establishment and maintenance of communication between UMNs and LMNs in the spinal cord.

### *11.1.2 Lower Motor Neurons*

LMNs are cholinergic neurons that connect the central nervous system (CNS) to muscle. With axons that can reach well over a meter long, they translate signals coming from UMNs, interneurons (INs), and sensory neurons (SNs) into muscle contraction. LMNs can be classified according to their innervating targets, and are divided into "branchial," "visceral," and "somatic" (reviewed in Stifani 2014). Branchial motor neurons are found in the brainstem and are responsible for the control of skeletal muscles in the neck and face. The visceral motor neurons are part of the autonomic nervous system and designate motor outflow to cardiac muscle, smooth muscle, and various endocrine glands. The visceral motor system can be considered a two-neuron system. The first neuron type is the preganglionic motor neuron, which has its cell body located in the CNS and connects to the second neuron type, known as the postganglionic motor neuron that belongs to the peripheral nervous system. The visceral motor system can be divided in two categories: the sympathetic system and the parasympathetic system.

The sympathetic nervous system is associated with the "fight or flight" response. The motor neurons of the sympathetic system localize in the spinal cord, between thoracic segment 1 (T1) and lumbar segment 2 (L2). Sympathetic motor neurons are responsible for the increase in the level of overall body metabolism in response to stress stimuli (i.e. increase in heart rate, decrease in gastric motility, pupillary dilatation, and increased sweat gland secretion). The parasympathetic nervous system, on the other hand, controls functions generally thought of as "vegetative"; it is involved in the body's recovery from stress, and in functions that occur when the body is in its resting state (i.e. decrease in heart rate, increase in gastric motility, pupillary constriction, and contraction of smooth muscle in the walls of the urinary bladder). Motor neurons from the parasympathetic system are located in the brainstem and in sacral segments 2–4 (S2-S4) (Stifani 2014).

Finally, the somatic motor neurons reside in the brainstem and in the spinal cord, and are the neurons that innervate skeletal muscles responsible for locomotion (Philippidou and Dasen 2013). Somatic motor neurons can be divided in thee different subcategories, according to their targeted muscle fiber (Manuel and Zytnicki 2011; Stifani 2014):

- *Alpha motor neurons (α-MNs)* are responsible for muscle contraction. They possess large cell bodies and well-defined neuromuscular endings. α-MNs innervate extrafusal muscle fibers, the most abundant muscle fiber type.
- *Beta motor neurons (β-MNs)* innervate extrafusal fibers and intrafusal muscle spindles.
- *Gamma motor neurons (γ-MNs)* innervate and control muscle spindle sensitivity independent from the control of the motor units. The specific innervation of intrafusal muscle fibers by gamma motor neurons seems to be the product of more recent evolution, since it appears only in mammals. In lower vertebrates, such as amphibian and reptiles, the intrafusal innervation arises from branches of the β-MNs.

During embryonic development, more specifically during the process of neurulation, the ectoderm folds, forming the neural plates, which will then progress to form the neural tube. The neural tube will ultimately differentiate into the brain and the spinal cord. Several morphogens stimulate the subsequent differentiation process. These signaling molecules include fibroblast growth factors (FGFs), retinoic acid (RA), and members of the wingless-type MMTV integration site family (WNT) (Dasen and Jessell 2009). The dorsal to ventral pattern of expression of bone morphogenic proteins (BMPs) and WNT, in association with the somite expression of RA, determine the ventral to dorsal expression of sonic hedgehog (SHH), which is the major signaling molecule of the notochord and the floor plate (Jessell 2000). SHH then initiates a cascade of events that culminate in the patterning of the motor neurons in the spinal cord. The grades of SHH expression are interpreted by a group of homeodomain proteins (Jessell 2000). The expression of these homeodomain proteins can be repressed or promoted, and according to their response to SHH, they have been divided into two classes: the class I homeodomain proteins include paired box 3/6/7 (PAX3/6/7), developing brain homeobox 1 and 2 (DBX1/2), and Iroquois related homeobox 3 (IRX3), which are ventrally repressed and thus expressed dorsally. Conversely, the class II homeodomain proteins are NK2 homeobox 2 and 9 (NKX2.2/2.9), NK6 homeobox 1 and 2 (NKX6.1/6.2), and oligodendrocyte transcription factor 2 (OLIG2), which are all ventrally expressed (Stifani 2014). These initial patterns are subsequently refined through selective transcriptional cross-repressive regulation between class I and class II homeodomain proteins.

The generation of motor neuron progenitors results from a combinatorial effect of the expression of the basic helix-loop-helix (bHLH) transcription factor OLIG2 and the homeodomain proteins Nkx6.1, Nkx6.2, and PAX6 (Shirasaki and Pfaff 2002). Soon after their generation, spinal motor neurons express a set of homeodomain transcription factors (Hb9, Lhx3, Isl1, and Isl2), which control features common to all spinal motor neurons. Although patterning events mediated by SHH, RA, and FGF signaling define how motor neurons are specified, additional signaling initiated by a regulatory network of nearly two dozen homeobox (HOX) genes are necessary for further diversification into hundreds of distinct motor neuron subtypes (Dasen and Jessell 2009). The mechanisms involved in initiating and maintaining proper communication between CNS and muscles are thus incredibly complex; damage in this communication, either in UMNs or LMNs or both, are the cause of motor neuron disease.

## *11.1.3 Motor Neuron Disease: ALS, SMA and SBMA*

### 11.1.3.1 Amyotrophic Lateral Sclerosis

Amyotrophic Lateral Sclerosis (ALS), also known as Lou Gehrig's disease, was originally described in 1869 by Jean-Martin Charcot and is the most common adult–onset motor neuron disease (MND) (Goodall and Morrison 2006). Currently, there is no cure for this devastating disorder that affects ~30,000 individuals at any one time in the USA (Lepore et al. 2008). ALS is characterized by rapidly progressive motor impairment with muscle atrophy, accompanied by muscular weakness, cramps, and fasciculations. These clinical signs result from UMN degeneration in the cortex, and LMN degeneration in the brainstem and spinal cord. ALS patients succumb to their disease between 2 and 5 years from symptom onset, usually due to respiratory failure. Definitive diagnosis is challenging and is mainly based upon the clinical evaluation in combination with electromyography (EMG), and relies upon the detection of upper and lower motor neuron signs in the limb and/or bulbar territories, together with a history of symptomatic progression (Brooks et al. 2000; Turner et al. 2009). At this time, there is still no reliable biomarker that can be used for diagnosis or for predicting a patient's prognosis. The only approved drug, riluzole, extends the survival about 2–4 months, but shows only a mild neuroprotective effect (Bensimon et al. 1994).

Approximately 95 % of all ALS cases are sporadic (SALS), meaning that it is not possible to identify a unique genetic or environmental cause for disease pathogenesis. The remaining 5 % comprise familial ALS (FALS), in which the disease is transmitted from one generation to the next in typically an autosomal dominant fashion, through autosomal recessive FALS types are occasionally seen (Deidda et al. 2014). While most efforts to understand ALS disease pathogenesis over the last 20 years have focused upon mutations in the ubiquitously expressed cytosolic superoxide dismutase 1 (SOD1) (representing ~2 % of all ALS), the discovery of mutations in more than a dozen genes, including especially TAR DNA-binding protein 43 (TDP-43), Fused in sarcoma (FUS/TLS), and C9orf72, as well as Vamp-associated protein B (VAPB), Optineurin (OPTN), and Senataxin (SETX), has highlighted the importance of RNA processing and protein quality control as two fundamentally important pathways that may converge to yield almost all forms of ALS. Despite recent advances, a parsimonious understanding of how and why ALS occurs specifically in motor neurons remains elusive.

#### 11.1.3.2 Spinal Muscular Atrophy

Spinal Muscular Atrophy (SMA) refers to a spectrum of hereditary degenerative disorders clinically characterized by muscular atrophy and muscular paralysis (Darras 2015). The first SMA case was described in 1891, when the Austrian doctor Werdnig reported two infant brothers with the onset of progressive proximal leg weakness at 10 months of age (Wilkins and Brody 1971). Autopsy studies described severe loss of anterior horn cells in the spinal cord as the main pathological hallmark (Faravelli et al. 2015). The term SMA is most often used to refer a recessive form of motor neuron disease caused by mutations in the gene survival motor neuron 1 (*SMN1*), which is located on chromosome 5q11.2-q13.3.

The frequency of SMA is ~1:10,000, with the carrier frequency for a *SMN1* gene mutation estimated to be as high as 1:40 to 1:60 in the Caucasian European population (Faravelli et al. 2015). SMN is an ubiquitously expressed protein that is involved in pre-mRNA splicing, as evidenced by SMN's ability to stimulate in vitro pre-mRNA splicing (Lorson et al. 1999). Complete loss of SMN is embryonic lethal, but the gene paralog survival motor neuron 2 (*SMN2*), which is located adjacent to the *SMN1* gene on the distal long arm of chromosome 5, is almost identical to *SMN1*, except that *SMN2* has a single nucleotide transition in exon 7, resulting in the skipping of exon 7 in the majority of transcripts, and production of a truncated SMN protein that is rapidly degraded. The absence of SMN1 can be compensated for by increasing copy numbers of hypomorphic *SMN2* alleles (Iascone et al. 2015).

SMA has been divided into four different subtypes, according to clinical course and age of onset (see Darras 2015; Faravelli et al. 2015):

- SMA type 1: This type is also known as Werdnig-Hoffman disease and is the most frequent subtype, accounting for more the 60 % of all SMA cases. The most common genotype in SMA type 1 is homozygous *SMN1* deletion in the presence of two copies of the *SMN2* gene. It is the most severe type and symptom onset occurs usually between 4 and 6 months of age. Children with SMA type 1 in most cases are incapable of achieving motor milestones, never sit unsupported, and progress to respiratory failure and death within 2 years.
- SMA type 2: Otherwise known as Dubowitz disease, SMA2 accounts for ~27 % of all SMA cases. Onset occurs between 6 and 18 months of age. Patients gain the ability to sit independently; however, they do not stand or walk. About 70 % of the patients survive until early adulthood, but lifespan can be highly variable.
- SMA type 3: This subtype is also named Kugelberg-Welander disease, and its clinical features are highly variable. Patients may show disease signs between 6 months and 3 years of age, but they can sit and stand independently and they learn how to walk; however, they lose the ability to walk during puberty and become wheelchair bound. Lifespan is usually normal, and SMA type 3 accounts for ~12 % of all SMA cases.
- SMA type 4: This type of SMA is the only subtype characterized by a late onset (~20–30 years old); it is therefore also known as adult-onset SMA. It accounts

for less than 1 % of all cases of SMA. Although patients experience signs of motor neuron degeneration, the symptoms are mild compared to the other types, and life expectancy is normal.

#### 11.1.3.3 X-Linked Spinal and Bulbar Muscular Atrophy: Kennedy's Disease

X-linked spinal and bulbar muscular atrophy (also known as SBMA or Kennedy's disease) is a neuromuscular disorder with a prevalence of about 1 in 50,000–400,000 males (Beitel et al. 2013). The main neurological manifestations of SBMA are weakness and wasting of bulbar, facial, and limb muscles due to lower motor neuron degeneration (Sopher et al. 2004; La Spada and Taylor 2010). SBMA affects mainly males, whereas females usually display minor symptoms, even if they are homozygous for the mutation. SBMA male patients display signs of androgen insensitivity, including gynecomastia, reduced fertility, and testicular atrophy (Thomas et al. 2006; Cary and La Spada 2008). La Spada and colleagues, in (1991), reported the identification of a trinucleotide CAG repeat expansion, the first mutation of this type discovered in the human, in the first exon of the androgen receptor (*AR*) gene as the mutation responsible for SBMA (La Spada et al. 1991).

AR contains three CAG repeat tracts, of which the first one is polymorphic in length with a range of 4–35 repeats, but with an average length of 16–25 repeats in the Caucasian European population (with repeats <16 CAGs typical in African individuals, and repeats >22 CAGs common in the Asian population); the other two CAG tracts are composed of five and six glutamines respectively (Parodi and Pennuto 2011). Expansion of the first polymorphic CAG tract, to greater than 35 repeats, causes disease. Recently, Grunseich et al. (2014), described the longest expansion to date, 68 CAG repeats, in a 29 years old patient. As CAG encodes the amino acid glutamine (Q), diseases caused by expansion of a CAG repeat tract are also known as polyglutamine (polyQ) repeat diseases. Eight other inherited neurodegenerative disorders were found to be caused by expanded glutamine repeats, hence, in addition to SBMA, the polyQ disease group includes Huntington's disease (HD), dentatorubral-pallidoluysian atrophy (DRPLA), and six forms of spinocerebellar ataxia: SCA1, 2, 3, 6, 7, and 17 (La Spada and Taylor 2010).

SBMA was the first disorder shown to be due to expansion of a polyQ repeat (La Spada et al. 1991). For that reason, SBMA studies contributed to our understanding of polyQ repeat expansion neurotoxicity, impacting research in many areas. After the discovery of the link between AR and SBMA (La Spada et al. 1991), several molecular mechanisms involved in the pathogenesis of SBMA were elucidated, leading to the identification of potential opportunities for therapeutic intervention. Presumed mechanisms of polyQ-AR toxicity include loss of neurotrophic support, mitochondrial dysfunction, protein aggregation, transcriptional dysregulation, and altered autophagy (Rocchi and Pennuto 2013). Different therapies for SBMA are under investigation, but to date, there is no treatment or cure for this disease.

## 11.2 Modeling Motor Neuron Disease Using iPSC Technology

Recent advances in stem cell biology have provided researchers with a powerful tool—induced pluripotent stem cells—for unraveling the disease mechanisms that underpin MND. Induced pluripotent stem cell (iPSC) technology allows the genomes of human subjects afflicted with MND to be studied in pluripotent stem cell lines (Marchetto et al. 2011). Such iPSCs can then be differentiated into motor neurons, skeletal muscles, or glial cells to evaluate whether specific gene defects, present in the patient's genome, affects cell physiology and neuron function. iPSC modeling also offers an avenue for studying disease processes in human cells in a system where disease genes are expressed at endogenous levels under proper regulation. iPSC modeling thus does not rely upon overexpression, as is often the case in cell culture models, and avoids many of the well-known limitations of animal models.

Before the advent of iPSCs, the most commonly used approaches to study neurological diseases included the analysis of human post-mortem tissues, immortalized cell lines, and animal models (Marchetto et al. 2011). Post-mortem tissues are not always well preserved and typically represent the end-stage of disease, while immortalized cell lines seldom fully recapitulate neurodegenerative disease phenotypes. Although animal models have been very useful, the absence of the human genetic background upon which a mutation acts remains a distinct disadvantage. Furthermore, many animal models rely upon placing the disease gene of interest out of genomic context and driving expression at supra-physiological levels in an expression pattern that may not closely resemble the normal expression pattern of the gene being studied. Thus, the creation of motor neurons from pluripotent stem cells, derived from patients with ALS and SMA type 1 in 2008 and 2009 respectively, represents a landmark development in the application of iPSC technology to the study of MND (Dimos et al. 2008; Ebert et al. 2009).

### *11.2.1 A Stem Cell Model for ALS*

Since Dimos et al. (2008) demonstrated that the skin fibroblasts of an 82-year-old ALS patient harboring an SOD1 mutation could be reprogrammed into iPSCs and then successfully differentiated into functional motor neurons, the ALS field has witnessed a number of major breakthroughs predicated upon the use of iPSC technology. Several studies have defined ALS disease-specific phenotypes in motor neuron cell lines derived from patient stem cells, including abnormalities of dendrites, hyperexcitability, protein aggregation, cell death, differential susceptibility to stress, abnormal RNA metabolism, mitochondrial defects, and ER dysfunction (Matus et al. 2014). Recently, one study documented spontaneous hyperexcitability in iPSC-derived motor neurons from patients carrying a SOD1, C9orf72, or FUS mutation, demonstrating that there are shared pathological phenotypes among different ALS disease mutations (Wainger et al. 2014). Kiskinis et al. (2014) detected an increase in oxidative stress, mitochondrial dysfunction, altered subcellular transport, and activation of ER stress pathways that occurred in cell lines derived from patients carrying SOD1 disease mutations. This group also evaluated stem cell-derived motor neurons from ALS patients harboring C9orf72 repeat expansions, and reported that some of these pathological changes are shared in motor neurons derived from all such patients (Kiskinis et al. 2014).

However, despite substantive progress in delineating disease-relevant phenotypes in motor neurons, over the last 10 years, an emerging theme in the motor neuron disease field—the concept of non-cell autonomous degeneration—has challenged the prevailing neurocentric view of neurological disease pathogenesis (Re et al. 2014; Marchetto et al. 2008; Yamanaka et al. 2008; Boille et al. 2006). In familial amyotrophic lateral sclerosis type 1 (ALS1), the inability to recapitulate SOD1 neurotoxicity in transgenic mice upon expression of mutant SOD1 in motor neurons galvanized the field around this issue (Lino et al. 2002; Pramatarova et al. 2001), and led to a line of investigation that has implicated astrocytes and microglia in ALS1 motor neuron degeneration. Based upon such studies done in mice, the role of non-neural cells in MND pathogenesis has been tested in iPSC experimentation. Indeed, the first fully humanized co-culture model composed of human adult primary sporadic ALS (SALS) astrocytes and human embryonic stem-cell-derived MNs found that astrocytes from SALS patients can kill hES-MNs, whereas control astrocytes do not (Re et al. 2014). Therefore, stem cell research has proven to be a useful tool to evaluate cell autonomous as well as non-cell autonomous toxicity in MND.

### *11.2.2 Stem Cell Model for SMA*

Even through Dimos et al. 2008 published the first generation of motor neurons from iPSCs, Ebert et al. (2009) achieved success in accurately modeling a MND through the use of iPSCs. In this study, the authors produced iPSCs from a type 1 SMA patient and his unaffected mother, and showed that these cells retain two defining SMA phenotypes: lack of SMN1 expression and selective motor neuron death (Ebert et al. 2009). Furthermore, stem cell-derived lines also responded to compounds known to increase SMN protein expression. Using a genetic correction approach, another group designed oligonucleotides to SMN2 that induced a permanent modification of a single nucleotide in exon 7 (Corti et al. 2012). When these workers compared motor neurons derived from corrected and untreated SMA-iPSCs, motor neuron disease phenotypes were ameliorated in the former. More recently, Naryshkin et al. (2014) have used iPSC-derived motor neurons in a preclinical trial for drug-screening and identified compounds capable of increasing SMN protein levels, demonstrating the feasibility of using stem cell models for drug screening in SMA.

### *11.2.3 Stem Cell Model for SBMA*

Over the years, SBMA disease pathogenesis has been extensively studied using a variety of different mouse models. For example, in 2004, the La Spada lab created a highly representative YAC transgenic SBMA mouse model. To achieve this, SBMA mice must express the human AR gene with >90 CAG repeats, even through repeat expansion mutations in human SBMA patients range from 37 to 70 CAGs. Furthermore, although SBMA YAC mice (and now more recently SBMA BAC mice) display neurogenic atrophy and motor neuron degeneration, significant motor neuron drop-out is not a prominent feature in the SBMA mice. In contrast, in human SBMA patients, motor neuron cell death is pronounced. Hence, a model system that successfully recapitulates aspects of human SBMA disease not observed in SBMA mice might prove useful. For these reasons, we, and others, have developed iPSC-based stem cell models of SBMA. In one SBMA iPSC model, increased AR aggregate formation in SBMA iPSC-derived motor neurons was reported in comparison to control iPSCs (Nihei et al. 2013). An increase of acetylated α-tubulin and reduced HDAC6 in SBMA iPSC-derived motor neurons with the longest repeats is another disease-specific phenotype uncovered by iPSC modeling (Grunseich et al. 2014). Our approach involved the use of multiple independent clones from neural progenitor cell lines from three different SBMA patients and three different controls to assure that identified phenotypes were truly disease-specific and not just an artifact of the reprogramming process (Cortes et al. 2014). In this work, we documented increased AR protein aggregation and autophagic flux defects as disease-specific phenotypes in SBMA stem cell derivatives. Hence, human iPSC models for SBMA have not only displayed disease-relevant phenotypes, but have also enabled researchers to study different cell types within the same genetic background.

## 11.3 Protocols

In the following section, we describe detailed step-by-step protocols, which enable MN generation from pluripotent stem cells (ESCs or iPSCs) (Figs. 11.1 and 11.2). iPSC generation is previously described in this compendium, and the following MN differentiation protocol can be applied to stem cells cultured in feeder layer or in feeder free conditions. Alternatively, MNs can be directly differentiated from fibroblasts, bypassing the stem cell stage. This alternative MN differentiation method will also be presented.

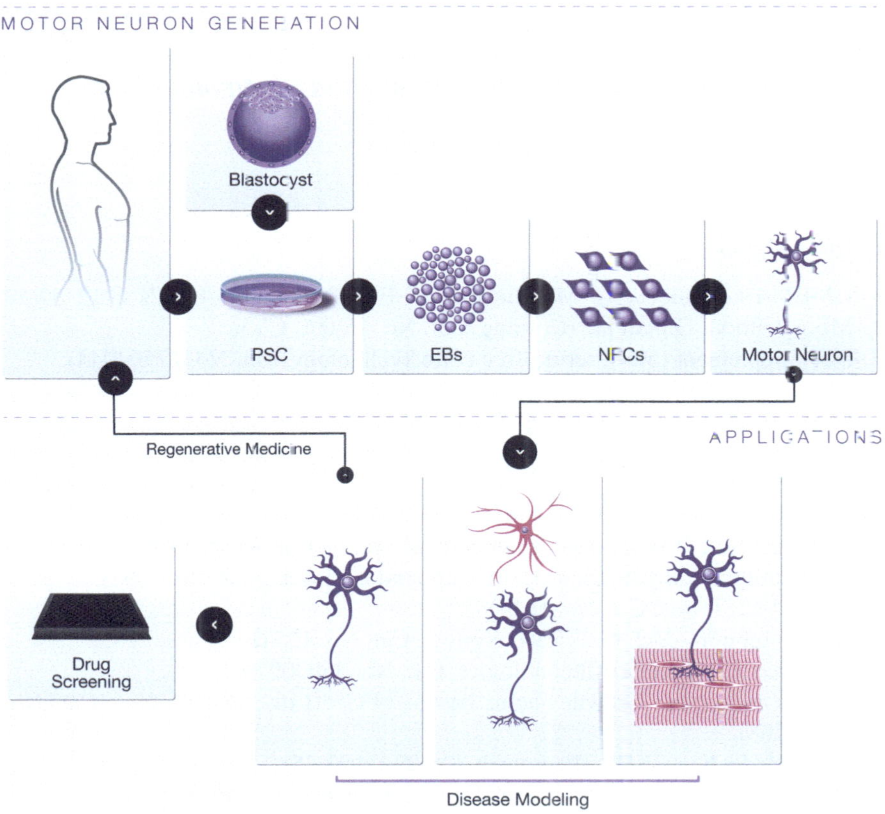


**Fig. 11.1** Motor neuron generation from pluripotent stem cells—overview of techniques and applications. Pluripotent stem cells represent ESCs isolated from the inner cell mass and iPSCs generated from human skin fibroblasts. Colonies form embryoid bodies (EBs) in suspension that further generate neural progenitor cells (NPCs). The NPCs are differentiated into motor neurons. Such derived motor neurons can be used for disease modeling alone or in co-culture with other cell types, such as astrocytes or skeletal muscle. Upon identification of screenable phenotypes, such motor neurons can also be used for drug screening, or possibly transplanted back into patients, as approach referred to as 'Regenerative Medicine'

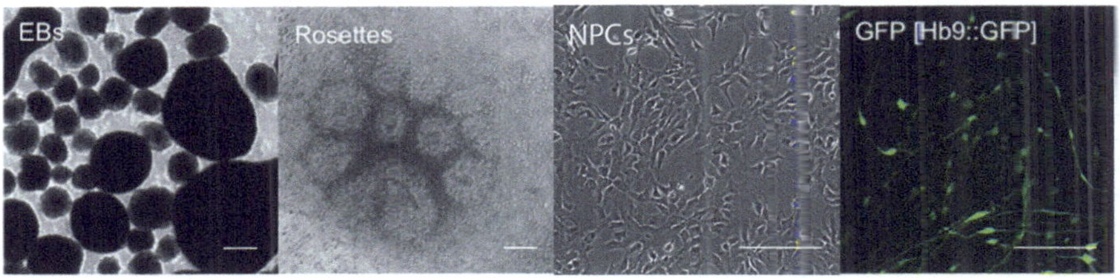


**Fig. 11.2** Motor Neuron differentiation. iPSC colonies form embryoid bodies (EBs) in suspension. Neural rosettes arise 5–8 days afterwards, from attached EBs. Rosettes can be picked and dissociated to form neural progenitor cells (NPCs). After withdraw of FGF and addition of different factors, neurons will mature in culture. Upon lentivirus transduction of the Hb9 promoter driving the GFP reporter gene, motor neurons can be identified in culture. Scale bars = 200 μm

## *11.3.1 Motor Neuron Differentiation from Pluripotent Stem Cells*

### 11.3.1.1 Reagents

**Culture Media**

- DMEM/F12 Dulbecco's Modification of Eagle's Medium/Ham's F-12 50/50 Mix—without Glutamine (Corning, Cat. No. 15-090-CV);
- B27 Supplement (50×), serum-free (Life Technology, Cat. No. 17504044).
- NeuroPlex N2 (Gemini, Cat. No. 400163).
- Matrigel (BD-Biosciences, Cat. No. CB40234).

**Enzyme, Chemicals and Reagents**

- StemPro Accutase (ThermoFisher, Cat. No. A1110501).
- Accumax Cell Dissociation Solution (BioExpress, Cat. No. S-1100-2).
- Dorsomorphin dihydrochloride—Compound C (Tocris, Cat. No. 3093).
- SB431542 (Tocris, Cat. No. 1614).
- Rock inhibitor—Y-27632 (R&D System, Cat. No. 1254).
- Glutamax I, 100× (Life Technologies, Cat. No. 35050079).
- Human Fibroblast Growth Factor basic—bFGF (Life Technologies, Cat. No. PHG0263).
- Sonic Hedgehog—SHH (Peprotech, Cat. No. 100-45).
- Brain-derived neural factor—BDNF (Peprotech, Cat. No. 450-10).
- Growth-derived neural factor—GDNF (Peprotech, Cat. No. 450-10).
- Insulin growth factor-1 (Peprotech, Cat. No. 100-11).

### 11.3.1.2 Experimental Procedure

**STEP 1: Generation of Embryoid Bodies (EBs)**

- **Day 0:** Let undifferentiated cultures of PSCs grow until they reach 70–80 % confluency, change appropriate PSC media (mTeSR, E8, Wi-Cell media) to DMEM/F12 supplemented with Glutamax, N2 (here on refereed to as DMEM + N2) and 5 μM of Rock Inhibitor (Ri);
- **Day 1:** Use a 20G needle to cut the colonies in a chess pattern (this procedure serves the purpose of making the colonies smaller and finally making a more homogeneous set of EBs);
- Remove the media and add 1× PBS at 37 °C for 5 min;
- Remove PBS and add DMEM + N2 media supplemented with 5 μM Ri, 1 μM Dorsomorphin and 10 μM SB431542;
- Use cell lifter to gently detach the colonies from the plate and transfer the PSCs to a low-binding plate or bacterial dish to let the cell grow in suspension in 3D;

- ***Note:*** *Alternatively, the colonies can be transferred to a 6-well plate and be kept under rotation inside the 37 °C incubator; this would be a more efficient way of preventing subsequent attachment;*
- **<u>Day 2–7:</u>** Change the media for DMEM + N2 supplemented with 1 μM Dorsomorphin and 10 μM SB431542 (no Rock Inhibitor) every 2 days;

**STEP 2: Formation of Neural Rosettes**

- **<u>Day 1:</u>** To plate the EBs, first prepare DMEM/F12 media supplemented with N2 B27 (here on referred to as DMEM + NB) and coat 6 cm plates with matrigel following manufacturer's instruction;
- Transfer the EBs from the low-binding plate to a 15 ml conical tube. Let the EBs sink to the bottom of the tube and remove the media;
- Add DMEM + NB supplemented with 20 μg/ml bFGF, then use a 1 ml pipet to mechanically break the EBs until the media turns cloudy;
- Plate the "flattened" EBs onto 6 cm matrigel coated plates and change media every 2 days;

**STEP 3: Picking Neural Rosettes and Generation of Neural Progenitor Cells (NPCs)**

- **<u>Day 6:</u>** To pick neural rosettes, first coat 6 and 10 cm plates with 10 μg/ml poly-ornithine and 3 μg/ml Laminin (pO/Lam);
- Wash the plated EBs with PBS, remove PBS and add DMEM + NB media;
- Use a 20 μl pipetter tip to select and scrape the rosette (see Fig. 11.3 for reference on the morphology of the rosettes that should be picked);
- Transfer all rosettes picked from 1 plate to a 15 ml conical tube, centrifuge the tube at 0.2 RCF for 3 min;
- Remove the media and add 1 ml of Accutase for 10 min at 37 °C to gently dissociate the rosettes;
- Add 9 ml of PBS to dilute the Accutase and centrifuge the rosettes at 0.2 RCF for 3 min;
- Remove PBS and resuspend the cell pellet with 1 ml of DMEM + NB with 20 ng/ml bFGF and gently pipette up and down to break the remaining clamps of rosettes;
- Plate the cells onto the pO/Lam pre-coated 6 cm plates initially, once the cells reach 90 % of confluence they may be expended onto 10 cm plates—these cells are then called neural progenitor cells (NPCs);
- NPCs should be passed when its confluence reaches 90 % using Accutase.

**STEP 4: Motor Neuron Generation**

- **<u>Day 1:</u>** Once the confluence of the NPCs achieve 90 %, change the media for DMEM + NB (without bFGF);
- ***Note:*** *On the first day of differentiation, it is normal to see cell death. The amount of cell death should decrease significantly after the first week*

- **Day 2–7:** Change the media every 2 days for DMEM + NB supplemented with 1 μM of Retinoic Acid (RA);
- **Day 8–14:** Media can now be changed every 3 days with DMEM + NB supplemented with 500 ng/ml Sonic Hedgehog (SHH);
- ***Note:*** *SHH can be substituted for with 1 μM smoothened agonist—SAG (Calbiochem, Billerica, MA)*
- ***Note:*** *From this step on, it is recommended to change only half of the media on the plates, this procedure has two implications: (1) to maintain the factors secreted by the differentiated cells in the media, (2) to prevent the detachment of cells from the plates*
- **Day 15–42:** Change every 3 days with DMEM + NB supplemented with 500 ng/ml Sonic Hedgehog (SHH), 20 ng/ml Brain-derived neural factor (BDNF), 20 ng/ml Growth-derived neural factor (GDNF) and 20 ng/ml Insulin-like growth factor 1 (IGF1);
- ***Note:*** *Between days 7 and 15, the cells can be transduced with a lentivirus harboring the Hb9 promoter driving a GFP or RFP reporter gene (Hb9::GFP or Hb9::RFP) for subsequent FACSorting to enrich the final motor neuron population.*

## *11.3.2 Motor Neuron Enrichment*

The efficiency of this particular motor neuron differentiation protocol is between 30 and 45 %; the variation is due to the intrinsic clonal and line variability experienced in PSC studies. One way to overcome the variable efficiency is by enriching the motor neuron population using fluorescence-activated cell sorting (FACS). The following section describes a FACS protocol that can be coupled with the motor neuron differentiation protocol presented above. To perform the FACSorting, motor neurons should be transduced with the Hb9::GFP or Hb9::RFP lentivirus between day 7 and 15 of the differentiation process. It is also important to have a motor neuron differentiation sample, which has not been transduced, to be used as a negative control.

### 11.3.2.1 Motor Neuron Dissociation and FACS

Material

- Prepare 96-well or 384-well plates coated with poly-ornithine and laminin (pO/Lam).

Reagents

**Solutions**

- FACS Wash Buffer:

PBS 40 ml.

0.5 M EDTA 0.2 ml.

1.0 M HEPES 1.0 ml.

FBS 0.4 ml.

- DNase Wash Buffer:

DMEM 10 ml.

FBS 100 μl.

DNase 20 μl.
**Note:** *DNase I recombinant, RNase-free (Roche, Cat. No. 04716728001).*

- FACS media:

DMEM + NB 10 ml.

0.5 M EDTA 0.05 ml.

FBS 0.1 ml.
**Note:** *Filter media/buffers if components are not sterile.*

Experimental Procedure

**STEP 1: Preparing Cells for Sorting (1–2 h)**

1. Remove growth medium from plates, wash the cells with PBS and incubate at 37 °C for 10 min;
2. Remove PBS, add 1 ml of accutase and 1 ml of accumax (for one 10 cm plate) and incubate at 37 °C for 20–30 min;
3. If cells are still in a sheet or in large clumps, triturate cells gently using 1 ml pipet;

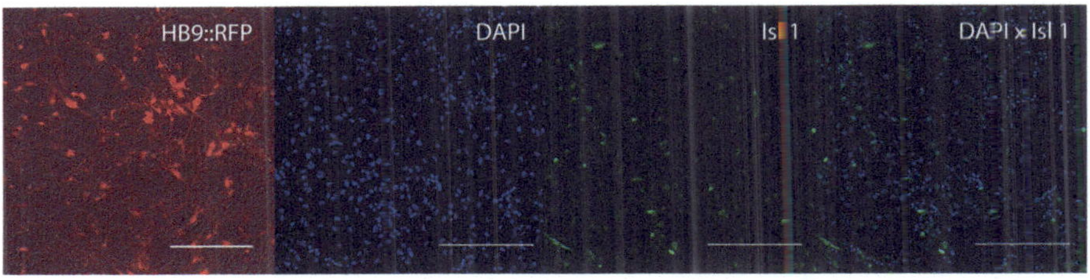


**Fig. 11.3** Sorting and re-plating of motor neurons. Differentiated cells were FACSorted on day 30 and then re-plated onto 96-well plates. On day 42, ~90 % of replated cells were Hb9::RFP positive and found to display neuronal morphology (20× bright field; red fluorescence overlay). Hb9: RFP positive cells also express Islet1, a marker of motor neurons. Scale bars = 200 μm

4. Suspend cells in FACS wash buffer (maximum of 15 ml total volume) and triturate gently using 5 ml serological pipet to further dissolve cell clumps;
   ***Note:*** *In this stage, it is preferable to have some remaining claps as "over-triturating" the cells may cause cell viability to decrease*
5. Pass cell suspension through 70 μm cell strainer into 50 ml conical tube;
   ***Note:*** *It is recommended to pre-wet tubes and cell strainer to improve cell viability;*
6. Transfer cells to a 15 ml conical and spin cells down at 0.2 RCF for 4 min;
7. Remove supernatant and add 3 ml of DNase Wash Media.
8. Carefully pipet up and down a few times to resuspend the pellet and spin cells down at 0.2 RCF for 4 min;
9. Wash cells with 4 ml of FACS Wash Buffer and spin cells down at 0.2 RCF for 4 min;
10. Remove supernatant, re-suspend cell pellet in 750 μl of FACS media and pass through a 40 μm cell strainer into flow cytometry tubes;
11. Cells are ready for sorting, place them on ice;
12. Prepare sorting collection tubes by adding 2.5–4 ml motor neuron media (DMEM + NB supplemented with SHH, BDNF, GDNF and IGF1) into 15 ml conical tubes;
13. Use the non-transduced cells to gate the negative population and sort for $RFP^+$ cells.

***Note:*** *wetting/lubricating the walls of the collection tubes and mixing the collection tubes as cells are collected helps with viability and cell recovery*

**STEP 2: Plating the Sorted Motor Neurons**

1. Spin down the 15 ml collection tubes at 0.2 RCF for 4 min;
   ***Note:*** *The cells are very sensitive to the sheath fluid; therefore, collecting cells as soon as possible will increase cell viability and cell recovery.*
2. Remove the supernatant and re-suspend with warm motor neuron media (DMEM + NB supplemented with SHH, BDNF, GDNF and IGF1) supplemented with 5 μM Ri;
3. Count live cells and plate onto pO/Lam coated plates;
   ***Note:*** *Plate 100,000 cells/well of a 96-well plate or 10,000–20,000 cells/well of 384-well/plate.*
4. Record FACS data and live cell counts for yield and recovery calculations.

### *11.3.3 Alternative Protocol*

Even through the majority of the protocols for motor neuron generation have pluripotent stem cells (ESCs or iPSCs) (reviewed in Nizzardo et al. 2010) as the starting point for the differentiation process, Son et al. (2011) suggested an alternative to such protocols, differentiatig motor neurons directly from fibroblasts. An advantage to this approach is that by bypassing the pluripotent stage, the cells will not reset their development. Therefore, one might argue that directly differentiated cells are more suitable cells to model neurodegenerative diseases; however, this hypothesis is yet to be investigated.

In a similar approach to the one used by Takahashi and Yamanaka in (2006) to generate ESC-like cells, Son et al. (2011) opted for factors acting on cells intrinsically, rather than relying on morphogens that act extrinsically. Therefore, this group initially selected 11 transcription factors to convert fibroblasts into induced motor neurons (iMNs). MEFs were harvested from Hb9::GFP E12.5 embryos and retrovirus transduced with the 11 factors. To determine which of the 11 factors were necessary for generating iMNs, they omitted each factor, one at a time. The final list had seven transcription factors, and these factors were able to induce direct differentiation of mouse and human fibroblasts into motor neurons. The iMNs displayed morphology, gene expression signature, electrophysiology, synaptic functionality in vivo engraftment capacity, and sensitivity to degenerative stimuli that was very similar to embryo-derived motor neurons (Son et al. 2011).

## 11.4 Maintenance and Characterization

The neural progenitor cells generated using this protocol can be expanded for multiple passages; they can also be frozen and thawed when needed. Therefore, to include this extra step with progenitor cell can be advantageous, providing a bank of cells to generate motor neurons without needing to go back to iPSCs for every differentiation. NPCs should be maintained with DMEM+NB with 20 μg/ml bFGF, and media should be changed every 2 days to assure constant levels of bFGF in order to prevent unwanted differentiation. NPCs are characterized by the expression of the proteins SOX2 and NESTIN (Cortes et al. 2014). Further differentiation of NPCs will generate motor neurons that are characterized by the expression of Hb9 and Isl1. This characterization can be performed by gene expression analysis with quantitative RT-PCR, or by immunostaining, analyzing the proteins (Figs. 11.2 and 11.3).

The Hb9::GFP or Hb9::RFP lentivirus construct is a very useful tool, not only for characterization, but also because the fluorescence is driven by the Hb9 promoter rather than the protein; therefore, positive cells will exhibit fluorescence throughout the cell body and processes. The use of Hb9::GFP or Hb9::RFP positive motor neurons became attractive in the context of a co-culture with other cell types, such as astrocytes or skeletal muscles, or in a mixed population, because in both situations, fluorescent motor neurons can be easily identified. Moreover, this permits one to FACSort motor neurons from a mixed population. Identified cells can be re-plated and kept in culture for another 10–15 days. The sorted population can also be assessed for efficiency of enrichment by Isl1 immunostaining (Fig. 11.3). Combining this differentiation protocol with the use of the Hb9::RFP motor neuron-specific reporter for cell sorting can result in a further enrichment to ~90 % (Fig. 11.3).

Another important characteristic of neurons is their capacity to receive electrical signals from other neurons and transmit electrical output signals. For such electrical signaling, neurons must maintain voltage gradients across the cell membrane. Measurement of motor neuron electrical activity is thus an important component of the characterization of these cells, and can be accomplished using patch clamping or a multi-electrode array (MEA) (Kiskinis et al. 2014). The patch clamp technique allows for "whole-cell" stable intracellular recording, whereas the MEA is a system for stimulation and recording of field potentials (extracellular signals) from different areas within a biological sample over time. Importantly, a comprehensive characterization can only be achieved by the combination of the different techniques presented in this section.

## 11.5 Applications

Nine years ago, when Takahashi and Yamanaka demonstrated that mouse fibroblasts can acquire properties similar to embryonic stem cells by introducing the transcription factors Oct3/4, Sox2, Klf4 and c-Myc, the entire biomedical research field was excited and intrigued with the potential use of these cells. Stem cells that can generate virtually every cell and tissue in the human body (with the only exception of extra-embryonic tissues) can now be generated and manipulated *in vitro* from differentiated adult cells. The iPSC technology has thus created extraordinary opportunities (Fig. 11.1), including: (1) regenerative medicine to repair tissues damaged through disease or injury, and (2) studies of disease pathogenesis, for which sample/tissue availability has hindered mechanistic understanding and therapy development. Regenerative medicine with the use of iPSCs is no longer mere speculation, as clinical trials to treat age-related macular degeneration will soon be underway. This strategy will employ retinal pigment epithelium (RPE) cells that were differentiated *in vitro* from iPSCs. Evaluation of iPSC-derived RPE cells, generated as a monolayer of cells without any artificial scaffolds, involved transplantation into monkeys to test if the cells meet clinical use requirements, including quality, quantity, consistency, and safety, and testing for expression of expected RPE markers (Kamao et al. 2014).

In the MND field, scientists have made tremendous progress in the use of iPSCs for drug screening. In a study published in 2012, iPSCs from familial ALS patients, harboring TAR DNA binding protein-43 (TDP-43) mutations, were used to generate motor neurons and compared to unaffected controls (Egawa et al. 2012). The authors identified cytosolic aggregates and shortened neurites, similar to findings observed in postmortem brain tissues and in a zebrafish model. Gene expression analysis indicated that transcription and RNA splicing were perturbed in TDP-43 ALS motor neurons compared to controls. This group then tested four different drugs, which are able to modulate either RNA transcription or splicing. From this focused screening, anacardic acid protected against arsenite-induced death of ALS iPSC-derived motor neurons, presumably by decreasing TDP-43 expression levels and reducing accumulation of its insoluble protein product. Moreover, anarcadic acid rescued the shortened neurite phenotype. This work indicates that stem cell derivatives from human patients can be used not only for disease modeling, but also for the identification of new candidate therapies.

As noted above, another study documented a hyperexcitability phenotype in motor neurons derived from iPSCs from ALS1 patients with SOD1 mutations. This phenotype, which manifested as delayed-rectifier voltage-gated potassium currents, could be rescued upon gene-targeted correction of the mutation, and was detected in ALS patients with mutations in the C9orf72 or FUS genes (Wainger et al. 2014). The identification of such a hyperexcitability phenotype, which had been previously observed in ALS patients (Vucic and Kiernan 2006), indicates that iPSC models can validate clinically relevant phenotypes. Furthermore, recapitulation of the hyperexcitability phenotype permitted testing of the FDA approved anticonvulsant retigabine, an activator of Kv7 potassium channels, and the realization that retigabine could abrogate the hyperexcitability and promote increased ALS motor neuron survival *in vitro* (Wainger et al. 2014). This group subsequently partnered with Glaxo-Smith-Kline and is preparing to launch a placebo-controlled Phase II clinical trial that will involve 192 ALS patients (McNeish et al. 2015; Mullard 2015).

The utility of iPSC modeling for identification of novel therapies for MND has not been restricted to ALS, as another group has employed this strategy for target validation in SMA. After completing a high throughput screen in HEK293H human embryonic kidney cell line harboring an *SMN2* minigene, a set of orally available compounds (SMN-C1,-C2, and -C3) that promote the inclusion of exon 7 in the SMN2 mRNA were discovered (Naryshkin et al. 2014). These small molecules were then validated in motor neurons from SMA patient-derived iPSCs. In this highly relevant human model system, the compounds were found to modulate SMN2 alternative splicing and to increase intracellular levels of SMN protein. Furthermore, the administration of the validated compounds to SMAΔ7 mice markedly improved motor function while promoting the production of full-length SMN protein (Naryshkin et al. 2014). Based upon these results, Roche is currently launching a Phase II trial of RG7800 in 48 SMA patients.

All of these studies clearly demonstrate that iPSC-based disease modeling offers unprecedented potential for *in vitro* investigation of neurological diseases and for the identification of novel therapies. The fact that iPSC models can provide a platform for drug screening and disease pathway discovery is a provocative development, given that this technique is barely a decade old. In many instances, drugs found to alleviate symptoms in animal models have failed in clinical trials. While the reasons for this are not always clear, modeling with disease cell types from affected patients now provides another powerful tool in the drug development armamentarium, and could set the stage for an era of more successful drug discovery. Such progress will only be possible if iPSC disease modeling yields disease phenotypes that are relevant to disease pathogenesis, robust, and reproducible. Although it is impossible to predict if drugs found to abrogate specific disease phenotypes in iPSC-derived motor neurons will ultimately be effective in humans, the application of the iPSC modeling approach for understanding disease biology and pinpointing novel therapies is likely to continue to grow and become more sophisticated in the future.

**Acknowledgements** We would like to thank Leonardo Rabelo for the images and for designing Fig. 11.1. We would also like to thank Dr. Allan Acab for optimizing the dissociation and FACSorting protocol. Our motor neuron disease research is supported by the National Institutes of Health (R01 NS041648 to A.R.L.), and by grants from the Muscular Dystrophy Association and the Robert Packard Center for ALS Research at Johns Hopkins School of Medicine.

## References

Arlotta P, Molyneaux BJ, Chen J, Inoue J, Kominami R, MacKlis JD (2005) Neuronal subtype-specific genes that control corticospinal motor neuron development in vivo. Neuron 45:207–221

Beitel LK, Alvarado C, Mokhtar S, Paliouras M, Trifiro M (2013) Mechanisms mediating spinal and bulbar muscular atrophy: investigations into polyglutamine-expanded androgen receptor function and dysfunction. Front Neurol 4:53

Bensimon G, Lacomblez L, Meininger V (1994) A controlled trial of riluzole in amyotrophic lateral sclerosis. ALS/Riluzole Study Group. N Engl J Med 330(9):585–591

Boille S, Vande Velde C, Cleveland D (2006) ALS: A Disease of Motor Neurons and Their Nonneuronal Neighbors. Neuron 52:39–59

Brooks BR, Miller RG, Swash M, Munsat TL (2000) El Escorial revisited: revised criteria for the diagnosis of amyotrophic lateral sclerosis. In: Amyotrophic lateral sclerosis and other motor neuron disorders: official publication of the World Federation of Neurology, Research Group on Motor Neuron Diseases. pp 293–299

Cary GA, La Spada AR (2008) Androgen receptor function in motor neuron survival and degeneration. Phys Med Rehabil Clin North Am 19:479–494

Cortes C, Miranda H, Frankowski H, Batlevi Y, Young J, Le A, Ivanov N, Sopher B, Carromeu C, Muotri A, Garden G, La Spada A (2014) Polyglutamine-expanded androgen receptor interferes with TFEB to elicit autophagy defects in SBMA. Nat Neurosci 17(9):1180–1189

Corti S, Nizzardo M, Simone C, Falcone M, Nardini M, Ronchi D, Donadoni C, Salani S, Riboldi G, Magri F, Menozzi G, Bonaglia C, Rizzo F, Bresolin N, Comi GP (2012) Genetic correction of human induced pluripotent stem cells from patients with spinal muscular atrophy. Sci Transl Med 4:165ra162

Darras BT (2015) Spinal muscular atrophies. Pediatr Clin North Am 62:743–766

Dasen JS, Jessell TM (2009) Chapter six hox networks and the origins of motor neuron diversity, 1st edn. Elsevier Inc

Deidda I, Galizzi G, Passantino R, Cascio C, Russo D, Colletti T, La Bella V, Guarneri P (2014) Expression of vesicle-associated membrane-protein-associated protein B cleavage products in peripheral blood leukocytes and cerebrospinal fluid of patients with sporadic amyotrophic lateral sclerosis. Eur J Neurol 21:478–485

Dimos JT, Rodolfa KT, Niakan KK, Weisenthal LM, Mitsumoto H, Chung W, Croft GF, Saphier G, Leibel R, Goland R, Wichterle H, Henderson CE, Eggan K (2008) Induced pluripotent stem cells generated from patients with ALS can be differentiated into motor neurons. Science 321:1218–1221

Ebert AD, Yu J, Rose FF, Mattis VB, Lorson CL, Thomson JA, Svendsen CN (2009) Induced pluripotent stem cells from a spinal muscular atrophy patient. Nature 457:277–280

Egawa N, Kitaoka S, Tsukita K, Naitoh M, Takahashi K, Yamamoto T, Adachi F, Kondo T, Okita K, Asaka I, Aoi T, Watanabe A, Yamada Y, Morizane A, Takahashi J, Ayaki T, Ito H, Yoshikawa K, Yamawaki S, Suzuki S, Watanabe D, Hioki H, Kaneko T, Makioka K, Okamoto K, Takuma H, Tamaoka A, Hasegawa K, Nonaka T, Hasegawa M, Kawata A, Yoshida M, Nakahata T, Takahashi R, Marchetto MCN, Gage FH, Yamanaka S, Inoue H (2012) Drug screening for ALS using patient-specific induced pluripotent stem cells. Sci Transl Med 4:145ra104

Faravelli I, Nizzardo M, Comi GP, Corti S (2015) Spinal muscular atrophy—recent therapeutic advances for an old challenge. Nat Rev Neurol 11:351–359

Goodall EF, Morrison KE (2006) Amyotrophic lateral sclerosis (motor neuron disease): proposed mechanisms and pathways to treatment. Expert Rev Mol Med 8:1–22

Grunseich C, Kats IR, Bott LC, Rinaldi C, Kokkinis A, Fox D, Chen KL, Schindler AB, Mankodi AK, Shader JA, Schwartz DP, Lehky TJ, Liu CY, Fischbeck KH (2014) Early onset and novel features in a spinal and bulbar muscular atrophy patient with a 68 CAG repeat. Neuromuscul Disord 24(11):978–981

Iascone DM, Henderson CE, Lee JC (2015) Spinal muscular atrophy: from tissue specificity to therapeutic strategies. F1000Prime Rep 7:1–14

Jessell TM (2000) Neuronal specification in the spinal cord: inductive signals and transcriptional codes. Nat Rev Genet 1:20–29

Kamao H, Mandai M, Okamoto S, Sakai N, Suga A, Sugita S, Kiryu J, Takahashi M (2014) Characterization of human induced pluripotent stem cell-derived retinal pigment epithelium cell sheets aiming for clinical application. Stem Cell Rep 2:205–218

Kiskinis E, Sandoe J, Williams LA, Boulting GL, Moccia R, Wainger BJ, Han S, Peng T, Thams S, Mikkilineni S, Mellin C, Merkle FT, Davis-Dusenbery BN, Ziller M, Oakley D, Ichida J, Di Costanzo S, Atwater N, Maeder ML, Goodwin MJ, Nemesh J, Handsaker RE, Paull D, Noggle S, McCarroll SA, Joung JK, Woolf CJ, Brown RH, Eggan K (2014) Pathways disrupted in human ALS motor neurons identified through genetic correction of mutant SOD1. Cell Stem Cell 14:781–795

La Spada AR, Taylor JP (2010) Repeat expansion disease: progress and puzzles in disease pathogenesis. Nat Rev Genet 11:247–258

La Spada AR, Wilson EM, Lubahn DB, Harding AE, Fischbeck KH (1991) Androgen receptor gene mutations in X-linked spinal and bulbar muscular atrophy. Nature 352:77–79

Lepore AC, Rauck B, Dejea C, Pardo AC, Rao MS, Rothstein JD, Maragakis NJ (2008) Focal transplantation-based astrocyte replacement is neuroprotective in a model of motor neuron disease. Nat Neurosci 11:1294–1301

Lino MM, Schneider C, Caroni P (2002) Accumulation of SOD1 mutants in postnatal motoneurons does not cause motoneuron pathology or motoneuron disease. J Neurosci 22:4825–4832

Lorson CL, Hahnen E, Androphy EJ, Wirth B (1999) A single nucleotide in the SMN gene regulates splicing and is responsible for spinal muscular atrophy. Proc Natl Acad Sci U S A 96:6307–6311

Manuel M, Zytnicki D (2011) Alpha, beta and gamma motoneurons: functional diversity in the motor system's final pathway. J Integr Neurosci 10:243–276

Marchetto MCN, Muotri AR, Mu Y, et al (2008) Non-cell-autonomous effect of human SOD1 G37R astrocytes on motor neurons derived from human embryonic stem cells. Cell Stem Cell 3:649–657

Marchetto MC, Brennand KJ, Boyer LF, Gage FH (2011) Induced pluripotent stem cells (iPSCs) and neurological disease modeling: Progress and promises. Hum Mol Genet 20:109–115

Matus S, Medinas DB, Hetz C (2014) Common ground: stem cell approaches find shared pathways underlying ALS. Cell Stem Cell 14:697–699

McNeish J, Gardner JP, Wainger BJ, Woolf CJ, Eggan K (2015) From dish to bedside: lessons learned while translating findings from a stem cell model of disease to a clinical trial. Cell Stem Cell 17:8–10

Molyneaux BJ, Arlotta P, Hirata T, Hibi M, Macklis JD (2005) Fezl is required for the birth and specification of corticospinal motor neurons. Neuron 47:817–831

Mullard A (2015) Stem-cell discovery platforms yield first clinical candidates. Nat Rev Drug Discov 14:589–591

Naryshkin NA, Weetall M, Dakka A, Narasimhan J, Zhao X, Feng Z, Ling KKY, Karp GM, Qi H, Woll MG, Chen G, Zhang N, Gabbeta V, Vazirani P, Bhattacharyya A, Furia B, Risher N, Sheedy J, Kong R, Ma J, Turpoff A, Lee C, Zhang X, Moon Y, Trifillis P, Welch EM, Colacino JM, Babiak J, Almstead NG, Peltz SW, Eng LA, Chen KS, Mull JL, Lynes MS, Rubin LL, Fontoura P, Santarelli L, Haehnke D, Mccarthy KD, Schmucki R, Ebeling M, Sivaramakrishnan

M, Ko C, Paushkin SV, Ratni H (2014) Motor neuron disease. SMN2 splicing modifiers improve motor function and longevity in mice with spinal muscular atrophy. Science 345:688–693

Nihei Y, Ito D, Okada Y, et al (2013) Enhanced aggregation of androgen receptor in induced pluripotent stem cell-derived neurons from spinal and bulbar muscular atrophy. J Biol Chem 288:8043–52

Nizzardo M, Simone C, Falcone M, Locatelli F, Riboldi G, Comi GP, Corti S (2010) Human motor neuron generation from embryonic stem cells and induced pluripotent stem cells. Cell Mol Life Sci 67:3837–3847

Parodi S, Pennuto M (2011) Neurotoxic effects of androgens in spinal and bulbar muscular atrophy. Front. Neuroendocrinol. 32:416–425

Philippidou P, Dasen JS (2013) Hox genes: choreographers in neural development, architects of circuit organization. Neuron 80:12–34

Pramatarova A, Laganière J, Roussel J, et al (2001) Neuron-specific expression of mutant superoxide dismutase 1 in transgenic mice does not lead to motor impairment. J Neurosci 21:3369–3374

Re DB, Le Verche V, Yu C, et al (2014) Necroptosis drives motor neuron death in models of both sporadic and familial ALS. Neuron 81:1001–8

Rocchi A, Pennuto M (2013) New routes to therapy for spinal and bulbar muscular atrophy. J Mol Neurosci 50:514–523

Rouaux C, Arlotta P (2010) Fezf2 directs the differentiation of corticofugal neurons from striatal progenitors in vivo. Nat Neurosci 13:1345–1347

Shirasaki R, Pfaff SL (2002) Transcriptional codes and the control of neuronal identity. Annu Rev Neurosci 25:251–281

Shoemaker LD, Arlotta P (2010) Untangling the cortex: advances in understanding specification and differentiation of corticospinal motor neurons. BioEssays 32:197–206

Son EY, Ichida JK, Wainger BJ, Toma JS, Rafuse VF, Woolf CJ, Eggan K (2011) Conversion of mouse and human fibroblasts into functional spinal motor neurons. Cell Stem Cell 9:205–218

Sopher BL, Thomas PS, Lafevre-Bernt MA, Holm IE, Wilke SA, Ware CB, Jin LW, Libby RT, Ellerby LM, La Spada AR (2004) Androgen receptor YAC transgenic mice recapitulate SBMA motor neuronopathy and implicate VEGF164 in the motor neuron degeneration. Neuron 41:687–699

Stifani N (2014) Motor neurons and the generation of spinal motor neuron diversity. Front Cell Neurosci 8:1–22

Takahashi K, Yamanaka S (2006) Induction of pluripotent stem cells from mouse embryonic and adult fibroblast cultures by defined factors. Cell 126:663–676

Thomas PS, Fraley GS, Damien V, Woodke LB, Zapata F, Sopher BL, Plymate SR, La Spada AR (2006) Loss of endogenous androgen receptor protein accelerates motor neuron degeneration and accentuates androgen insensitivity in a mouse model of X-linked spinal and bulbar muscular atrophy. Hum Mol Genet 15:2225–2238

Turner MR, Kiernan MC, Leigh PN, Talbot K (2009) Biomarkers in amyotrophic lateral sclerosis. Lancet Neurol 8:94–109

Vucic S, Kiernan MC (2006) Axonal excitability properties in amyotrophic lateral sclerosis. Clin Neurophysiol 117:1458–1466

Wainger BJ, Kiskinis E, Mellin C, Wiskow O, Han SSW, Sandoe J, Perez NP, Williams LA, Lee S, Boulting G, Berry JD, Brown RH, Cudkowicz ME, Bean BP, Eggan K, Woolf CJ (2014) Intrinsic membrane hyperexcitability of amyotrophic lateral sclerosis patient-derived motor neurons. Cell Rep 7:1–11

Wilkins RH, Brody IA (1971) Infantile spinal muscular atrophy. Arch Neuro 25(2):179–186

Yamanaka K, Chun SJ, Boillee S, et al (2008) Astrocytes as determinants of disease progression in inherited amyotrophic lateral sclerosis. Nat Neurosci 11:251–253

# Chapter 12
# Derivation of Dopaminergic Neurons from Human Pluripotent Stem Cells Using a Defined System and/or Small Molecules

**Atossa Shaltouki**

**Abstract** The ability of pluripotent stem cells (PSC), such as embryonic stem cells and induced pluripotent stem cells, to assume the identity of any other cell type makes them a reliable way to derive cells for disease modeling and cell therapy. Reprogramming patient somatic cells into neurons and other therapeutically relevant cell types offers new strategies for the treatment of neurodegenerative diseases, like Parkinson's disease. The lineage-specific differentiation of PSCs to produce a large quantity of healthy and/or genetically engineered neurons is the ultimate goal of stem cell therapy for patients with neurodegenerative disorders. Here, we describe two established protocols for the efficient derivation of dopaminergic neurons from PSCs using defined media and/or small molecules. Dopaminergic neurons can be efficiently generated using either of these protocols on a large scale.

**Keywords** Dopaminergic neurons • Parkinson's disease • Human embryonic stem cells • Induced pluripotent stem cells • Neural stem cells

## 12.1 Introduction

It has been almost a decade since two independent groups showed that post-mitotic adult somatic cells can be rendered pluripotent by exogenous expression of a set of embryonic transcription factors (Takahashi et al. 2007; Thomson et al. 1998). The generation of induced pluripotent stem cells (iPSC) provides an unlimited source of cells from both healthy and diseased individuals with various genetic backgrounds. Utilizing patient-derived iPSCs to reflect the pathological features of a disease allows for *in vitro* genetic disease modeling, drug screening, and cell-based therapy (Wang and Doering 2012). With regard to clinical translational application, replenishing

A. Shaltouki (✉)
Department of Neurosurgery, School of Medicine, Stanford University,
1050 Arastradero Road, Building A, Room A119, Palo Alto, CA 94304, USA
e-mail: ashaltouki@stanford.edu



H. Ulrich, P. Davidson Negraes, *Working with Stem Cells*,
DOI 10.1007/978-3-319-30582-0_12

damaged or degenerated cells with a patient's own reprogrammed cells allows for new neurodegenerative disorder treatments. In Parkinson's disease (PD), iPSC technology allows for the preparation of a desirable population of patient-specific dopaminergic (DA) neurons, the type of cells that are progressively lost in the progression of PD. A large supply of DA neurons from both familial and sporadic PD patients can enhance PD-based disease modeling and accelerate the identification of new therapeutic targets (Byers et al. 2012; Ebert et al. 2012). Efficiently generating the desired type of neural cells has presented a significant challenge to achieving the therapeutic benefit of iPSC-derived neurons; however, the number of existing protocols for ESC/iPSC-derived DA neuron generation is increasing. These include the early differentiation models which relied on the formation of embryoid bodies (EB) (Lee et al. 2000; Zhang et al. 2001), co-culture-based models using PA6 or MS5 stromal cells (Kawasaki et al. 2000; Perrier et al. 2004; Zeng et al. 2004), trans-differentiation of somatic cells (Pfisterer et al. 2011; Efe et al. 2011; Kim et al. 2011), as well as enhanced differentiation of ESC/iPSC-derived neurons by overexpression (Andersson et al. 2006; Chung et al. 2005; Kim et al. 2003).

Over the past few years, the need for more clinically compliant DA neurons has lead to the development of more robust differentiation protocols which use a combination of neural signaling molecules in a more defined system (Kirkeby et al. 2012; Kriks et al. 2011; Liu et al. 2013; Xi et al. 2012). Although these protocols have produced phenotypically reasonable quantities of neurons with DA identity, they vary in their efficiency, timescale of differentiation, maturity of neural population, progeny identity, and *in vivo* survival upon transplantation. Here we describe two published protocols for dopaminergic differentiation of iPSCs (Chambers et al. 2009; Kriks et al. 2011; Swistowski et al. 2009, 2010).

The first protocol (Swistowski et al. 2009) uses a multi-step strategy that mimics *in vivo* DA neural development by directing iPSCs to a neuroepithelial fate using embryoid body (EB) formation and neural rosette isolation. The resulting neural stem cells are then subjected to patterning cues that limit their ability to differentiate, generating a large (30–35 %) proportion of TH-positive neurons. This protocol is xeno-free and allows for expansion or storage of progenitors at several intermediate stages. The second protocol comes from Studer's research group, and is similarly efficient. This protocol is based on dual SMAD inhibition and mimics generation of midbrain A9 DA during embryonic development by using a combination of small molecules such as LDN-193189 and SB431542 to concomitantly target both BMP and lefty/Activin/TGFβ pathways respectively, and to neutralize ESC/iPSC cells (Chambers et al. 2009). These methods allow for the rapid generation of a high percentage (>80 %) of PAX6 positive cells with the potential for a variety of neuronal fates. Patterning to midbrain dopaminergic neurons is achieved through exposure to FGF-8 and modulation of sonic hedgehog (SHH) activity at an early stage of neural induction, resulting in a change of identity from progenitors to floor plate precursors (Fasano et al. 2010; Kriks et al. 2011). Both FGF8 and SHH have been previously implicated in the induction of forkhead box protein A2 (FoxA2) expression in the ventral floor plate, where it displays differentiation-promoting activity (Cooper et al. 2010; Hynes et al. 1995). To adopt a ventral DA neuronal phenotype, expression of LIM homeobox transcription factor 1, alpha (Lmx1a)—another key determinant of ventral midbrain DA neurons—must overlap with that of FoxA2 in floor plate precursors (Chung et al. 2009). To induce Lmx1a expression in floor plate precursors, another group from Studer's lab used CHIR99021, a small molecule that promotes Wnt signaling by inhibiting glycogen synthase kinase β (GSKβ) (Kriks et al. 2011). Differentiation of floor plate precursor towards midbrain DA neurons can be achieved in conventional method by exposing to neurotrophic and survival factor.

Both protocols result in greater than 30 % DA neurons in less than 4 weeks, generate a comparable homogenous population of neurons, and provide evidence of survivability following transplantation in animal models.

## 12.2 Materials and Reagents

- Dulbecco's Phosphate-Buffered Saline (D-PBS) (Cat. No. 14040).
- N-2 Supplement (Cat. No. 17502).
- Antibiotic-Antimycotic Solution (Gibco, Cat. No. 15240).
- Fetal Bovine Serum, ES Cell-Qualified FBS (Cat. No. 15141).
- GlutaMAX-I (Gibco, Cat. No. 35050).
- MEM-Non Essential Amino Acids (Gibco, Cat. No. 11140).
- Geltrex Reduced Growth Factor Basement Membrane Matrix (Life Technologies Cat. No. 12760).
- Poly-L-Ornithine (Sigma, Cat. No. P3655).
- Laminin (Life Technologies, Cat. No. 23017).
- Dibutyryl cAMP (Sigma, Cat. No. D0627).
- Dulbecco's Modified Eagle's Medium (DMEM)/F-12 with GLUTAMAX™ (Gibco, Cat. No. A10565-018).
- Basic Fibroblast Growth Factor, Human Rec (EBioscience, Cat. No. 14-8936).
- Neurobasal Medium (Gibco, Cat. No. 21103).
- B27 supplement, 50× (Gibco, Cat. No. 17504).
- StemPro hESC SFM Growth Supplement, 50× (Gibco, Cat. No. A10006-01).
- Bovine Serum Albumin (BSA) 25 % (Gibco, Cat. No. A10008-01).
- StemPro Accutase (Life Technologies, Cat. No. A11105).
- Dimethylsulfoxide, (Sigma, Cat. No. D8418).
- 2-Mercaptoethanol (Gibco, Cat. No. 219852).
- KnockOut™ Serum Replacement (KSR) (Gibco, Cat. No. 10828-028).

### *12.2.1 Small Molecules and Cytokines*

- Dibutyryl-cAMP (Sigma, Cat. No. D0627).
- CHIR99021 (Stemgent, Cat. No. 04-0004).

- Ascorbic Acid (Sigma, Cat. No. A5960).
- Dorsomorphin (Sigma, Cat. No. P5499).
- SB431542 (Tocris, Cat. No. 1614).
- Recombinant human GDNF (Peprotech, Cat. No. 450-10).
- Recombinant human BDNF (Peprotech, Cat. No. 450-02).
- Recombinant human FGF8a (R&D Syntems, Cat. No. 4745-F8-50).
- Recombinant human SHH(C24II) (R&D Syntems, Cat. No. 1845-SH-100).
- Recombinant human TGFβ3 (R&D Syntems, Cat. No. 243-B3-010).
- DAPT (Tocris, Cat. No. 2634).
- LDN193189 (Stemgent, Cat. No. 040074).
- Purmorphamine (Stemgent, Cat. No. 040009).

### 12.2.2 Antibodies

- Alexa Fluor 488 goat anti-rabbit (Life Technologies, Cat. No. A-11008).
- Alexa Fluor 594 goat anti-mouse (Life Technologies, Cat. No. A-11012).
- Mouse anti-β-Tubulin isotype III clone SDL.3D10, 1:1500 (Sigma, Cat. No. T8660).
- Rabbit anti-TH IgG1 antibody, 1:500 dilution (Pel-Freez Biologicals, Cat. No. P40101).

### 12.2.3 Cell Culture Media

- ***iPSC Medium:*** To make 50 ml of iPSC medium: 37 ml of (DMEM)/F-12 with 10 ml of KSR (f.d.20 %), 0.5 ml GLUTAMAX, 1 ml of 1 % MEM-NEAA, 1 ml of Anti-Anti solution, 0.1 mM of 2-Mercaptoethanol. Sterile-filter the medium.
- ***StemPro Medium:*** To make 50 ml of StemPro medium: add 45.4 ml of DMEM/F-12 with GLUTAMAX™, 3.6 ml of BSA (25 %), 91 μl of 2-Mercaptoethanol. Sterile filter the solution then add 1 ml of StemPro® hESC SFM Growth Supplement (50×) and b-FGF (f.c. 8 ng/ml).
- ***EB Medium:*** StemPro medium without bFGF.
- ***KSR differentiation medium:*** iPSC medium with 15 % KSR
- ***Neural Induction Medium:*** To make 50 ml of Neural induction medium: add 48 ml of DMEM/F-12 with GLUTAMAX™ to 0.5 ml of MEM-NEAA (1 %), and then sterile-filter before adding 500 μl of N-2 Supplement and b-FGF (f.c. 20 ng/ml).
- ***NSC Medium:*** To make 50 ml NSC medium: mix 48 ml of Neurobasal™ Medium, 0.5 ml of MEM-Non Essential Amino Acids, 0.5 ml of GlutaMAX™-I, and 0.5 ml of Antibiotic-Antimycotic solution. Then after sterile-filtering of the media add 1 ml of B27 supplement, 50×, as well as 10 μl of 100 ng/μl bFGF (f.c. 20 ng/ml).

- ***NSC freezing medium:***
  NSC-FREEZE-A: 10 ml of NSC medium.
  NSC-FREEZE-B: 8 ml of NSC medium with 2 ml of DMSO (f.d. 10 %).
- ***Dopaminergic Differentiation Medium (DAD medium):*** DAD medium consists of two separate media: **DAD-A** and **DAD-B**.
  To make a 50 ml of **DAD-A medium,** mix 47.5 ml Neurobasal™ Medium, 0.5 ml of MEM-Non Essential Amino Acids, 0.5 ml of GlutaMAX™-I, and 0.5 ml of Antibiotic-Antimycotic solution. Sterile-filter the solution, then add 1 ml of B27 Supplement (50×), SHH (f.c. 200 ng/ml) and fibroblast growth factor 8 (FGF8) (f.c. 100 ng/ml).
  To make 50 ml of **DAD-B medium:** Follow the recipe for 50 ml of DAD-A medium, substituting sonic hedgehog (SHH) and FGF8 for brain-derived neurotrophic factor (BDNF) (f.c. 20 ng/ml), glial cell line-derived neurotrophic factor (GDNF) (f.c.20 ng/ml), cAMP (f.c. 1 mM), and transforming growth factor type β-3 (TGFβ3) (f.c. 1 μM).
- ***Dopaminergic Freezing Medium.***

  DAD-B A: 10 ml of DAD-B medium.

  DAD-B B: 8 ml of DAD-B medium with 2 ml of DMSO.

## *12.2.4 Substrate-Coating of Cell Culture Cishes*

**Poly-L-Ornithine and Laminin Coating**

1. Following the manufacturer's instructions, prepare a poly-L-ornithine solution using cell culture-grade water (f.c. 20 μg/ml);
2. Add enough Poly-L-ornithine solution to cover the whole surface of the cell culture dishes. Incubate the plates at 37 °C for 2 h. (Alternatively, these plates can be stored at 4 °C overnight in a plastic container);
3. The following day, use cell culture-grade water to perform two rinses on the plates;
4. Prepare a laminin solution of sufficient concentration in cell culture-grade water (f.c. 10 μg/ml) and use this solution to coat the entire surface of the washed plates;
5. Incubate the plates at 37 °C for 2 h;
6. Coated plates that are not immediately used can be stored at 4 °C for up to 2 weeks.

**Geltrex Coating**

1. Thaw Geltrex™ Basement Membrane Matrix at 2–8 °C overnight;
2. Dilute 0.500 ml of Geltrex in 25 ml chilled (2–8 °C) DMEM/F12 medium, pipetting up and down to mix;

3. Add enough Geltrex solution to cover the entire growth surface area. Incubate the coated plates at room temperature for a minimum of 1 h. Afterwards, carefully remove the supernatant above the Geltrex coating and plate cells in pre-equilibrated cell culture medium. The coated dish can be kept for 2 weeks when stored at 2–8 °C and sealed with Parafilm.

**Matrigel Coating**

1. Thaw an aliquot of Matrigel at 2–8 °C overnight;
2. Dilute the appropriate amount of Matrigel (which is lot-dependent) in pre-chilled (2–8 °C) DMEM/F12 medium, pipetting up and down to mix;
3. For coating, follow Geltrex Co (Carpenter et al. 2001).

## 12.3 DA Generation from NSC in Defined Medium (Protocol 1)

### *12.3.1 Generation of ESC/IPSC-Derived NSC Using Defined Media*

The majority of established *in vitro* protocols for derivation of neurons from ESCs utilize a group of cells that resemble neuroepithelial cells in the early stages of neural development (Rathjen et al. 2002). These ESC-derived progenies are self-renewing cells capable of giving rise to variety of neural progenies, astrocytes, and oligodendrocytes (Cai and Grabel 2007; Plachta et al. 2004). The first NSC derivation protocol involved the formation of EBs from mouse ESCs (Zhang et al. 2001; Carpenter et al. 2001; Martin 1981). EBs are formed as the result of aggregated ESC colonies under a low-attachment culture condition in the absence of growth factors. The resulting spherical clusters of cells represent the earliest differentiated progenies of ESCs and have the derivatives of thee germ layers (Keller 2005). The attachment of aggregates in neural-inducing culture conditions promotes their neuroectodermal fate. This is accompanied by morphological changes into columnar-like epithelial cells, which organize themselves into a distinctive rosette (Elkabetz and Studer 2008). Here we describe in detail a previously published protocol for EB formation from both ESCs/iPSCs (Swistowski et al. 2009).

### *12.3.2 EB Formation*

1. Detach ESCs/iPSCs using appropriate detachment enzymes. Detachment enzymes are selected based on the media and the substrate on which PSCs are expanding;
2. Pool the colonies in a 15 ml conical tube and spin at 1000 RPM for 3 min;

3. Remove the detachment enzyme solution (supernatant) and replace with 5 ml of fresh EB medium supplemented with 10 μM ROCK inhibitor;
4. Culture the detached colonies in suspension on a low-attachment culture dish using a 5 ml sterile pipet. Avoid fragmentation of colonies into small pieces;
5. Place the plate at 37 °C with 5 % $CO_2$;
6. Allow colonies to form aggregates in suspension culture, to allow EBs to form over the next 8 days, while changing the medium every other day;
7. Change the medium by collecting the contents of the plate into a 15 ml conical tube. Allow the EBs to settle by gravity and then replace the medium without disturbing pelleted EBs;
8. On day 8, switch the suspension medium from EB to neural induction medium and continue to culture EBs for another 2 days (Fig. 12.1a);
9. On day 10, to allow for the attachment of aggregates, plate the EBs on Matrigel-coated plates (10–15 EBs per 60 mm plate) in neural induction medium;
10. Two to five days after the EBs attach, neural rosettes will appear in culture (Fig. 12.1b);
11. Manually isolate the rosettes, then treat the isolated rosettes with accutase so that they dissociate into single cells. Ten to fifteen rosettes are usually isolated for a dissociation assay;

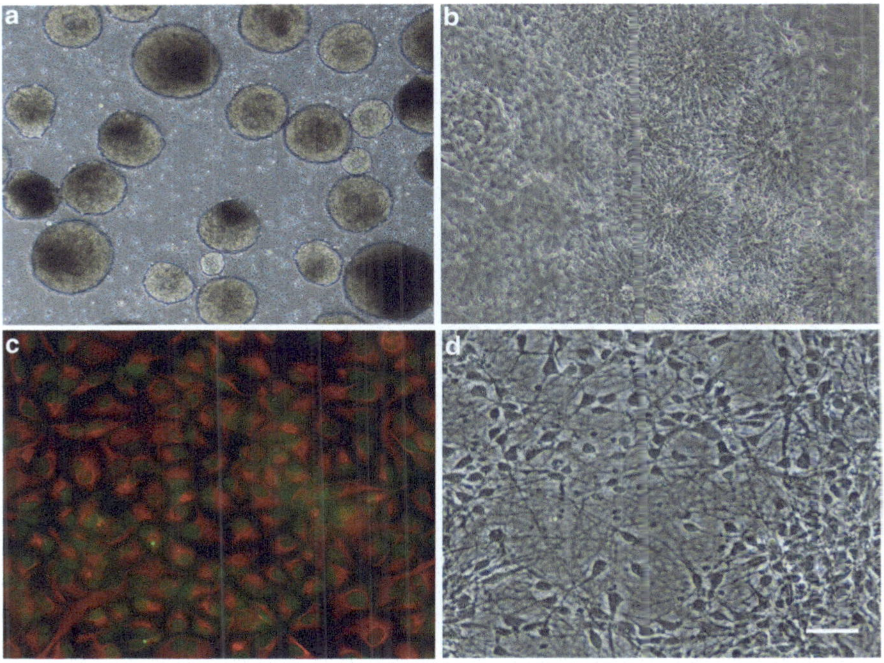

**Fig. 12.1** Differentiation protocol. (**a**) Day 6, embryoid bodies; (**b**) Neuronal rosettes formed 6 days following EB attachments; (**c**) Expanded NSC: Sox1 (*green*) and Nestin (*red*); (**d**) Post mitotic DA neurons (day 25). Scale bar = 30 μm

12. Plate the single neuroepithelial cells onto Geltrex-coated dishes and expand the line in NSC medium.

### *12.3.3 Expansion of ESC/iPSC-Derived NSCs*

NSCs can be maintained and expanded in NSC medium on Geltrex-coated dishes or Matrigel-coated dishes (Fig. 12.1c). For differentiation, it is recommended that NSC be cultured for a minimum of 5 days prior to differentiation. NSC are passaged when they are 80 % confluent.

### *12.3.4 Enzymatic Passage of NSCs*

1. Remove medium from the dish of confluent NSCs;
2. Use appropriate amount of prewarmed accutase to cover the dish surface. Allow 3–5 min for cells to detach in a cell culture incubator;
3. Collect detached cells, and add them into a 15 ml conical tube;
4. Wash the plates with 1–5 ml of prewarmed NE medium and collect media into the same 15 ml conical tube;
5. Centrifuge the detached cells at 0.2 RCF for 3 min;
6. Carefully aspirate the supernatant and re-suspend the cells in 5 ml of NSC medium;
7. Plate the re-suspended cells at $5 \times 10^4/\text{cm}^2$;
8. Evenly distribute NSCs and incubate them at 37 °C and 5 % $CO_2$, being sure to change the medium every other day.

### *12.3.5 Cryopreservation of NSCs*

1. Repeat steps 1–5 from the previous section (*Enzymatic Passage of NSCs*);
2. Remove the supernatant and re-suspend the pellet in NSC-FREEZE-A medium and mix thoroughly;
3. Count the cells;
4. Using the same medium, dilute the cells to obtain desired cell density;
5. Carefully add an equal volume of NSC-FREEZE-B medium containing 20 % DMSO to the media;
6. Pipet 2–3 times to mix gently;
7. Immediately aliquot NSCs into previously labeled cryogenic vials (1 ml/vial);
8. Using a freezing device, freeze the cells at −80 °C. The following day, transfer vials into liquid nitrogen tank.

## 12.3.6 Dopaminergic Differentiation of NSCs

- **Day 0:** Plate NSCs onto either 35 or 60 mm dishes coated with poly-ornithine and laminin ($5\times10^4$/cm$^2$) in 2 or 5 ml of NSC medium and incubate the plates overnight at 37 °C, 5 % $CO_2$.
- **Day 1–10:** Change the medium on the following day to DAD-A medium (2 ml/35 mm dish or 5 ml/60 mm dishes). Change medium every other day.
- **Day 11–35:** Change the medium from DAD-A medium to DAD-B. Continue to replenish the media every other day. Care must be taken so that media is added dropwise in order to avoid disrupting the cells or peeling them off (Fig. 12.1d). Characterization of differentiated cells can be performed throughout the differentiation process to define the stage-specific identity of the cells.

#### 12.3.6.1 Enzymatic Passage of Dopaminergic Neurons

The enzymatic passage of DA neurons is similar to that of NSCs as described here, with the exception of media used for the protocol. Use prewarmed DMEM/F12 medium for collection and washing steps. Be sure to break up the clumps of cells before they attach. Plate the differentiated progenies at 1.5–$2\times10^6$/60 mm.

#### 12.3.6.2 Cryopreservation of Dopaminergic Neurons

The cryopreservation of DA neurons uses a similar protocol to Cryopreservation of NSCs, with the only change being the substitution of DAD-B A and DAD-B B for NSC-FREEZE-A and NSC-FREEZE-B, respectively.

## 12.3.7 DA Differentiation Using Small Molecules

The second DA differentiation protocol outlined here is based on several previous publications from Studer's group. This differentiation approach involves manipulation of several signaling pathways using small molecules and morphogens to achieve midbrain dopaminergic neurons in less than 35 days. The protocol is compatible with both MEF-dependent and feeder-free ESC/iPSC as starting material. This protocol is also sui with defined ES media such as mTeSR (Stem cell technologies) or Essential 8 (Life Technologies). Here, we will describe the protocol for a feeder-free iPSC but similar approaches can be made for ESC.

**iPSC Culture**

It is important to passage the iPSCs several times before differentiating. For passaging, the appropriate detachment enzyme must be used. Detached cells are collected, centrifuged at 1000× RPM and pelleted cells are washed with DMEM/

F12 medium. Cells are then plated at 35–40 k cells/$cm^2$ on Matrigel-coated plates in presence of 10 μM Rock inhibitor in iPSC culture medium (ICM). Note: Here, ICM refers to the medium in which the iPSCs are maintained and could be MEF-CM, mTSER or Essential 8. Allow the cells to become confluent for the next three days (50–80 % confluent). Change the ICM daily and remove differentiated colonies.

- **Day 0:** Treat the cells with 10 μM SB431542 (TGFβ inhibitor) and 100 nM of LDN193189 (BMP inhibitor) in KSR differentiation medium. Change the medium daily.
- **Day 1:** Add 100 ng/ml of SHH C25II, 2 μM of Purmorphamine and 100 ng/ml of FGF8 to Day 0 medium. Change the medium daily.
- **Day 3:** Add 3 μM CHIR99021 (GSKβ inhibitor) to Day 2 medium. Change medium daily.
- **Day 5:** Remove TGFβ inhibitor. Gradually shift the KSR differentiation medium to N2 Medium (DMEM/F12 Medium with 1x of N2 Supplement) by adding 25 % N2 and 75 % KSR differentiation medium supplemented with Day 3 components, except for TGFβ inhibitor. Change medium daily.
- **Day 7:** Add 50 % N2 and 50 % KSR differentiation medium and keep other components, except for SHH, Purmorphamine and FGF8. Change medium daily.
- **Day 9:** Add 75 % N2 and 25 % KSR differentiation medium and keep other components. Change medium daily.
- **Day 11:** Add NSC medium (no bFGF and no NEAA) supplemented with 3 μM CHIR99021 and 20 ng/ml of BDNF, 0.2 mM of ascorbic acid (AA), 20 ng/ml of GDNF, 1 ng/ml TGFβ3, 0.5 mM of dibutyryl cAMP, and 10 nM of DAPT. Change medium daily.
- **Day 13:** Remove CHIR99021 from Day 11 medium. Change medium daily.
- **Day 20:** Remove the media and use accutase to dissociate the cells. Wash the cells by pelleting at 1300× RPM. Remove the supernatant and resuspend the cells in Day 13 medium and plate at 300–400 k cells/$cm^2$ on Poly-L-ornithine/laminin coated plates. Change medium daily. Continue the differentiation up to day 35.

***Note:*** *Differentiation can continue past day 35 depending on experimental design and desired maturity.*

Although this second protocol is not completely free of xeno factors, it is the most widely used approach for *in vitro* DA generation, mainly because of its ease of reproducibility and reasonably high efficiency. Several other laboratories have modified the floor plate protocol, either by changing the type or dose of small molecules, or by combining these methods with other methods of neural induction (Kirkeby et al. 2012; Xi et al. 2012; Sundberg et al. 2013). However, the advantages of NSC-mediated DA generation (first protocol) is that the homogeneity of NSC population can sometimes reach greater than 90 %. This would allow elimination of unwanted non neuronal cells in resulting DA population which is highly desirable in clinical application.

## 12.4 Characterization of Dopaminergic Neurons

Immunohistochemistry analysis can be performed to determine if either protocols narrowed the differentiation potential of iPSC toward the derivation of A9 dopaminergic neurons. The basic phenotypic characterization of generated neurons requires monitoring the sequential expression of a set of stage-specific transcription factors that are implicated in midbrain DA neuron development *in vivo*. These include the NSC markers such as Pax6, Sox2, Sox1 and Nestin, followed by the floor plate progenitor markers such as forkhead family of winged-helix transcription factor 2 (Foxa2) and LIM homeobox transcription factors 1b (Lmx1b) and TuJ1 (Elkabetz et al. 2008; Abeliovich and Hammond 2007). The transition to precursor fate requires additional markers such nuclear receptor related 1 protein (Nurr1), paired-like homeodomain 3 protein (Pitx3), Lmx1a and neurogenin. Finally post mitotic dopaminergic neurons can be identified by increased in expression of tyrosine hydroxylase (TH) and dopamine transporter (DAT) (Abeliovich and Hammond 2007; Cossette et al. 2005).

Here, we describe immunocytochemistry analysis of dopaminergic neurons using TH and Tuj staining protocols;

1. Remove medium from the dish of confluent NSCs. Aspirate differentiation medium from the dish and rinse the plate twice with PBS;
2. Fix the cells by adding adequate amount of 4 % paraformaldehyde (in PBS) to cover the dish and incubate for 20 min;
3. Wash the cells thee times with PBS;
4. Incubate the cells for 1 h in blocking solution composed of 0.3 % Triton-X100 in PBS, 10 % goat serum and 1 % BSA;
5. Dilute the primary antibodies in blocking buffer (mouse anti-β III-tubulin 1:1000, and rabbit anti-TH 1:500). Incubate the cells with primary antibodies solution overnight at 4 °C;
6. Next, aspirate the primary antibody solution and wash the cell thee times with PBS;
7. Dilute the secondary antibodies (Alexa Fluor 488 goat anti-rabbit and Alexa Fluor 594 goat anti-mouse, both at dilution 1:1000) in blocking buffer, and incubate the cells with secondary antibody solution at room temperature for 1 h;
8. Wash the cells thee times with PBS;
9. Apply 15 μl of antifade reagent (prolong gold with Hoechst) on the cell layer and carefully place cover slip on the top while trying to avoid any bubble generation;
10. Visualize the staining using a fluorescence microscope with appropriate filter combinations and image the cells;
11. Cells positive for TH staining should be quantified by expressing TH-stained nuclei as a percentage of total Hoechst-labeled nuclei in a field. For each experiment, at least 700–900 TH positive cells should be counted in randomly chosen fields.

## 12.5 Potential Use of iPSC-Derived DA Neurons

The iPSC technology allows cell replacement therapy to be a realistic approach in the treatment of many neurological diseases, including PD. Engraftment of patients' own target cells overcomes immunological complications and these cells can readily be generated from iPSCs. Improved derivation methods for generating scalable dopaminergic neurons from iPSCs have been instrumental in enhancing the efficacy and survivability of engrafted cells in PD animal models (Zeng and Couture 2013). Despite these advancements, many challenges remain in the clinical application of iPSC-derived neuronal progenies in cell therapy (Fu and Xu 2012). In the absence of standardization protocols for differentiation and transplantation procedures, it is difficult to extrapolate from current available transplantation studies which progenies of iPSCs would be more suited for the long-term functional replacement of lost dopaminergic neurons. For instance, while fetal brain-derived DA neuronal precursors successfully regenerated neurons in clinical trials and animal models of PD, engraftment of similar neuronal precursors in PD animal models displayed tumorigenic properties when derived from iPSCs (Akiyama et al. 2001; Brundin et al. 1986; Nori et al. 2015; Hallett et al. 2015). This variability in transplantation outcomes could be partly explained by differences in neuronal donor sources and their *in vitro* preparation and expansion strategies, but it also obviates the need for more stringent protocols characterizing properties and assessing the purity of clinical-grade iPSC-derived neuronal cells (Clelland et al. 2008).

Patient-derived DA neurons are potentially attractive candidates for screening novel therapeutic targets and identifying drugs tailored to patients' own molecular profiles. Despite the increasing number of iPSC-derived neurons generated from PD patients with various genetic abnormalities, their use in high-throughput drug screening is limited by the absence of PD-related phenotype *in vitro*, and high purity of cell-type population. Although PD progression is associated with aging, full potential of patients-derived neurons in drug discovery is only achieved by improving the already efficient methods of cell type differentiation of iPSCs to better model the disease for use in regenerative medicine (Xu and Zhong 2013).

## References

Abeliovich A, Hammond R (2007) Midbrain dopamine neuron differentiation: factors and fates. Dev Biol 304(2):447–454

Akiyama Y, Honmou O, Kato T, Uede T, Hashi K, Kocsis JD (2001) Transplantation of clonal neural precursor cells derived from adult human brain establishes functional peripheral myelin in the rat spinal cord. Exp Neurol 167(1):27–39

Andersson E, Tryggvason U, Deng Q, Friling S, Alekseenko Z, Robert B, Perlmann T, Ericson J (2006) Identification of intrinsic determinants of midbrain dopamine neurons. Cell 124(2):393–405

Brundin P, Nilsson OG, Strecker RE, Lindvall O, Astedt B, Bjorklund A (1986) Behavioural effects of human fetal dopamine neurons grafted in a rat model of Parkinson's disease. Exp Brain Res 65(1):235–240

Byers B, Lee HL, Reijo Pera R (2012) Modeling Parkinson's disease using induced pluripotent stem cells. Curr Neurol Neurosci Rep 12(3):237–242

Cai C, Grabel L (2007) Directing the differentiation of embryonic stem cells to neural stem cells. Dev Dyn 236(12):3255–3266

Carpenter MK, Inokuma MS, Denham J, Mujtaba T, Chiu CP, Rao MS (2001) Enrichment of neurons and neural precursors from human embryonic stem cells. Exp Neurol 172(2):383–397

Chambers SM, Fasano CA, Papapetrou EP, Tomishima M, Sadelain M, Studer L (2009) Highly efficient neural conversion of human ES and iPS cells by dual inhibition of SMAD signaling. Nat Biotechnol 27(3):275–280

Chung S, Hedlund E, Hwang M, Kim DW, Shin BS, Hwang DY, Kang UJ, Isacson O, Kim KS (2005) The homeodomain transcription factor Pitx3 facilitates differentiation of mouse embryonic stem cells into AHD2-expressing dopaminergic neurons. Mol Cell Neurosci 28(2):241–252

Chung S, Leung A, Han BS, Chang MY, Moon JI, Kim CH, Hong S, Pruszak J, Isacson O, Kim KS (2009) Wnt1-lmx1a forms a novel autoregulatory loop and controls midbrain dopaminergic differentiation synergistically with the SHH-FoxA2 pathway. Cell Stem Cell 5(6):646–658

Clelland CD, Barker RA, Watts C (2008) Cell therapy in Huntington disease. Neurosurg Focus 24(3–4):E9

Cooper O, Hargus G, Deleidi M, Blak A, Osborn T, Marlow E, Lee K, Levy A, Perez-Torres E, Yow A, Isacson O (2010) Differentiation of human ES and Parkinson's disease iPS cells into ventral midbrain dopaminergic neurons requires a high activity form of SHH, FGF8a and specific regionalization by retinoic acid. Mol Cell Neurosci 45(3):258–266

Cossette M, Levesque D, Parent A (2005) Neurochemical characterization of dopaminergic neurons in human striatum. Parkinsonism Relat Disord 11(5):277–286

Ebert AD, Liang P, Wu JC (2012) Induced pluripotent stem cells as a disease modeling and drug screening platform. J Cardiovasc Pharmacol 60(4):408–416

Efe JA, Hilcove S, Kim J, Zhou H, Ouyang K, Wang G, Chen J, Ding S (2011) Conversion of mouse fibroblasts into cardiomyocytes using a direct reprogramming strategy. Nat Cell Biol 13(3):215–222

Elkabetz Y, Studer L (2008) Human ESC-derived neural rosettes and neural stem cell progression. Cold Spring Harb Symp Quant Biol 73:377–387

Elkabetz Y, Panagiotakos G, Al Shamy G, Socci ND, Tabar V, Studer L (2008) Human ES cell-derived neural rosettes reveal a functionally distinct early neural stem cell stage. Genes Dev 22(2):152–165

Fasano CA, Chambers SM, Lee G, Tomishima MJ, Studer L (2010) Efficient derivation of functional floor plate tissue from human embryonic stem cells. Cell Stem Cell 6(4):336–347

Fu X, Xu Y (2012) Challenges to the clinical application of pluripotent stem cells: towards genomic and functional stability. Genome Med 4(6):55

Hallett PJ, Deleidi M, Astradsson A, Smith GA, Cooper O, Osborn TM, Sundberg M, Moore MA, Perez-Torres E, Brownell AL, Schumacher JM, Spealman RD, Isacson O (2015) Successful function of autologous iPSC-derived dopamine neurons following transplantation in a non-human primate model of Parkinson's disease. Cell Stem Cell 16(3):269–274

Hynes M, Porter JA, Chiang C, Chang D, Tessier-Lavigne M, Beachy PA, Rosenthal A (1995) Induction of midbrain dopaminergic neurons by Sonic hedgehog. Neuron 15(1):35–44

Kawasaki H, Mizuseki K, Nishikawa S, Kaneko S, Kuwana Y, Nakanishi S, Nishikawa SI, Sasai Y (2000) Induction of midbrain dopaminergic neurons from ES cells by stromal cell-derived inducing activity. Neuron 28(1):31–40

Keller G (2005) Embryonic stem cell differentiation: emergence of a new era in biology and medicine. Genes Dev 19(10):1129–1155

Kim JY, Koh HC, Lee JY, Chang MY, Kim YC, Chung HY, Son H, Lee YS, Studer L, McKay R, Lee SH (2003) Dopaminergic neuronal differentiation from rat embryonic neural precursors by Nurr1 overexpression. J Neurochem 85(6):1443–1454

Kim J, Efe JA, Zhu S, Talantova M, Yuan X, Wang S, Lipton SA, Zhang K, Ding S (2011) Direct reprogramming of mouse fibroblasts to neural progenitors. Proc Natl Acad Sci U S A 108(19):7838–7843

Kirkeby A, Grealish S, Wolf DA, Nelander J, Wood J, Lundblad M, Lindvall O, Parmar M (2012) Generation of regionally specified neural progenitors and functional neurons from human embryonic stem cells under defined conditions. Cell Rep 1(6):703–714

Kriks S, Shim JW, Piao J, Ganat YM, Wakeman DR, Xie Z, Carrillo-Reid L, Auyeung G, Antonacci C, Buch A, Yang L, Beal MF, Surmeier DJ, Kordower JH, Tabar V, Studer L (2011) Dopamine neurons derived from human ES cells efficiently engraft in animal models of Parkinson's disease. Nature 480(7378):547–551

Lee SH, Lumelsky N, Studer L, Auerbach JM, McKay RD (2000) Efficient generation of midbrain and hindbrain neurons from mouse embryonic stem cells. Nat Biotechnol 18(6):675–679

Liu Q, Pedersen OZ, Peng J, Couture LA, Rao MS, Zeng X (2013) Optimizing dopaminergic differentiation of pluripotent stem cells for the manufacture of dopaminergic neurons for transplantation. Cytotherapy 15(8):999–1010

Martin GR (1981) Isolation of a pluripotent cell line from early mouse embryos cultured in medium conditioned by teratocarcinoma stem cells. Proc Natl Acad Sci U S A 78(12):7634–7638

Nori S, Okada Y, Nishimura S, Sasaki T, Itakura G, Kobayashi Y, Renault-Mihara F, Shimizu A, Koya I, Yoshida R, Kudoh J, Koike M, Uchiyama Y, Ikeda E, Toyama Y, Nakamura M, Okano H (2015) Long-term safety issues of iPSC-based cell therapy in a spinal cord injury model: oncogenic transformation with epithelial-mesenchymal transition. Stem Cell Rep 4(3):360–373

Perrier AL, Tabar V, Barberi T, Rubio ME, Bruses J, Topf N, Harrison NL, Studer L (2004) Derivation of midbrain dopamine neurons from human embryonic stem cells. Proc Natl Acad Sci U S A 101(34):12543–12548

Pfisterer U, Wood J, Nihlberg K, Hallgren O, Bjermer L, Westergren-Thorsson G, Lindvall O, Parmar M (2011) Efficient induction of functional neurons from adult human fibroblasts. Cell Cycle 10(19):3311–3316

Plachta N, Bibel M, Tucker KL, Barde YA (2004) Developmental potential of defined neural progenitors derived from mouse embryonic stem cells. Development 131(21):5449–5456

Rathjen J, Haines BP, Hudson KM, Nesci A, Dunn S, Rathjen PD (2002) Directed differentiation of pluripotent cells to neural lineages: homogeneous formation and differentiation of a neurectoderm population. Development 129(11):2649–2661

Sundberg M, Bogetofte H, Lawson T, Jansson J, Smith G, Astradsson A, Moore M, Osborn T, Cooper O, Spealman R, Hallett P, Isacson O (2013) Improved cell therapy protocols for Parkinson's disease based on differentiation efficiency and safety of hESC-, hiPSC-, and non-human primate iPSC-derived dopaminergic neurons. Stem Cells 31(8):1548–1562

Swistowski A, Peng J, Han Y, Swistowska AM, Rao MS, Zeng X (2009) Xeno-free defined conditions for culture of human embryonic stem cells, neural stem cells and dopaminergic neurons derived from them. PLoS One 4(7):e6233

Swistowski A, Peng J, Liu Q, Mali P, Rao MS, Cheng L, Zeng X (2010) Efficient generation of functional dopaminergic neurons from human induced pluripotent stem cells under defined conditions. Stem Cells 28(10):1893–1904

Takahashi K, Tanabe K, Ohnuki M, Narita M, Ichisaka T, Tomoda K, Yamanaka S (2007) Induction of pluripotent stem cells from adult human fibroblasts by defined factors. Cell 131(5):861–872

Thomson JA, Itskovitz-Eldor J, Shapiro SS, Waknitz MA, Swiergiel JJ, Marshall VS, Jones JM (1998) Embryonic stem cell lines derived from human blastocysts. Science 282(5391):1145–1147

Wang H, Doering LC (2012) Induced pluripotent stem cells to model and treat neurogenetic disorders. Neural Plast 2012:346053

Xi J, Liu Y, Liu H, Chen H, Emborg ME, Zhang SC (2012) Specification of midbrain dopamine neurons from primate pluripotent stem cells. Stem Cells 30(8):1655–1663

Xu XH, Zhong Z (2013) Disease modeling and drug screening for neurological diseases using human induced pluripotent stem cells. Acta Pharmacol Sin 34(6):755–764

Zeng X, Couture LA (2013) Pluripotent stem cells for Parkinson's disease: progress and challenges. Stem Cell Res Ther 4(2):25

Zeng X, Cai J, Chen J, Luo Y, You ZB, Fotter E, Wang Y, Harvey B, Miura T, Backman C, Chen GJ, Rao MS, Freed WJ (2004) Dopaminergic differentiation of human embryonic stem cells. Stem Cells 22(6):925–940

Zhang SC, Wernig M, Duncan ID, Brustle O, Thomson JA (2001) In vitro differentiation of transplantable neural precursors from human embryonic stem cells. Nat Biotechnol 19(12):1129–1133

# Chapter 13
# Differentiation of hiPSC-Derived Cardiomyocytes

**Fabian Zanella and Farah Sheikh**

**Abstract** Recent studies have demonstrated tremendous potential of human induced pluripotent stem cell (hiPSC)-derived cardiomyocytes to model cardiac diseases *in vitro*, as well as to evaluate the cardiac safety of potential new drug candidates. As the field has evolved, it has become more feasible to use hiPSC-derived cardiomyocytes in various molecular and physiological assays, creating a need for differentiation protocols to yield a large number of more mature cardiomyocytes. Currently, the most cost-effective cardiac differentiation protocols have exploited small molecule compounds to manipulate components of the Wnt signaling pathway, which is critical for early cardiac development. In this chapter we explore adaptations made to boost chemically defined cardiac differentiation protocols to further enhance their effectiveness and reproducibility across independent hiPSC clones and lines.

**Keywords** Human induced pluripotent stem cells • Cardiomyocyte • Cardiac differentiation • Cardiac assays • Small molecule • Wnt signaling pathway

## 13.1 Introduction

Human iPSCs (hiPSCs) have been widely used to study behaviors of cardiomyocytes in pathological states associated with cardiac disease (Zanella et al 2014) and in the context of drug screens to evaluate cardiotoxic potential (Peters et al. 2015). Current hiPSC models of cardiac disease include LEOPARD syndrome (Carvajal-Vergara et al. 2010), Long QT syndrome (Itzhaki et al. 2011; Moretti et al. 2010; Lahti et al. 2012; Matsa et al. 2011), Timothy syndrome (Yazawa et al. 2011), catecholaminergic polymorphic ventricular tachycardia (Itzhaki et al. 2012; Fatima et al. 2011; Jung et al. 2012; Kujala et al.

F. Zanella • F. Sheikh (✉)
Cardiovascular Medicine Division, Department of Medicine, University of California-San Diego, 9500 Gilman Drive, Mail code: 0613C, La Jolla, CA 92093, USA
e-mail: fasheikh@ucsd.edu



H. Ulrich, P. Davidson Negraes, *Working with Stem Cells*,
DOI 10.1007/978-3-319-30582-0_13

2012), familial dilated cardiomyopathy (Sun et al. 2012), hypertrophic cardiomyopathy (Lan et al. 2013; Han et al. 2014), and arrythmogenic right ventricular cardiomyopathy (Ma et al. 2012; Kim et al. 2013; Caspi et al. 2013), as well as an overlapping cardiac syndrome that encompasses the cardiac $Na^+$ channel disease (Davis et al. 2012). Simultaneously, hiPSC-derived cardiomyocytes have been used to better understand the mechanisms impacting drug cardiotoxicity, serving as guides for more ambitious drug development strategies (Abassi et al. 2012; Liang et al. 2013; Navarrete et al. 2013; Peters et al. 2015; Lamore et al. 2013; Sallam et al. 2015).

It is well established that cardiotoxicity is a frequent and primary cause of novel drugs failing in later stages of clinical trials as well as removal of existing drugs from the market (Sharma et al. 2013). Given their integral role in translating findings to the clinic, the demands for larger yields of functional and mature cardiomyocytes keep evolving. In order to meet those needs, robust small molecule-based protocols for the cardiac differentiation of embryonic and hiPSCs have been developed (Burridge et al. 2014; Lian et al. 2012, 2013; Zanella and Sheikh 2016). Manipulation of the Wnt signaling pathway, which triggers mesoderm specification and commitment during early cardiac development has been at the corner stone of successful cardiac differentiation protocols (Lian et al. 2013; Burridge et al. 2014).

In this chapter, we explore the addition of insulin in order to enhance cell survival in the early steps of differentiation to a previously developed small molecule-based protocol (Burridge et al. 2014). We have observed that this subtle adaptation enables higher cardiomyocyte throughputs and better reproducibility between several independent clones of several hiPSC lines. hiPSC-derived cardiomyocyte maturity is often pointed to as a limitation for their use in physiological assays. Nonetheless, studies have suggested that long-term culture promotes a higher degree of maturation (Lundy et al. 2013). Since cell death and detachment can be problematic in longer-term hiPSC derived cardiomyocyte cultures, we have adapted media formulations capable of supporting longer-term healthy hiPSC-derived cardiomyocyte cultures, enabling further maturation.

## 13.2 Materials Needed

### *13.2.1 Commercially Available Reagents*

See Table 13.1.

**Table 13.1** Commercially available reagents

| Description | Manufacturer | Catalog number |
|---|---|---|
| **Extracelullar matrices** | | |
| Growth factor-reduced matrigel | Corning | 354230 |
| Laminin | Life Technologies | 23017-015 |
| **Cell culture media** | | |
| DMEM/F12 with glutamine and HEPES | Life Technologies | 11330-032 |
| RPMI 1640 with L-glutamine | Life Technologies | 11875093 |
| RPMI 1640 with L-glutamine and without glucose | Life Technologies | 11879020 |
| M199 with L-glutamine | Corning | 10-060-CV |
| Cryostor CS-10 cell cryopreservation media | Sigma | C2874-100ML |
| **Media supplements** | | |
| L-Ascorbic acid 2-phosphate (AA) | Sigma | A8960 |
| Recombinant human serum albumin (HSA) | Sigma | A0237 |
| 60 % Sodium DL-lactate | Sigma | L4263 |
| 100× Penicillin/streptomycin (P/S) | Corning | 30-002-CI |
| **Buffers** | | |
| Dulbecco's phosphate-buffered saline (PBS) without $Ca^{2+}$ and $Mg^{2+}$ | Corning | 21-031-CV |
| Dulbecco's phosphate-buffered saline (PBS) with $Ca^{2+}$ and $Mg^{2+}$ | Corning | 21-030-CV |
| Hanks' balanced salt solution (HBSS) without $Ca^{2+}$ and $Mg^{2+}$ | Corning | 21-022-CV |
| 0.5 M EDTA | Life Technologies | 15575-020 |
| BD perm/wash buffer | BD Biosciences | 554723 |
| **Cell dissociation reagents** | | |
| Collagenase type II | Worthington | LS004174 |
| Taurine | Sigma | T8691-100G |
| EGTA | Sigma | E-4378-25G |
| **Small molecules** | | |
| Y27632/Rock inhibitor (Rock$_i$) | Selleckchem | S1049 |
| CHIR99021 (CH) | Selleckchem | S1263 |
| Wnt-C59 (C59) | Selleckchem | S7037 |

### 13.2.2 Small Molecules

- ***Rock inhibitor/Y27632 (Rocki):*** For a 10 mM (2000×) solution, re-suspend 10 mg vial in 3.122 ml of DMSO. Mix well, vortex and make 25–50 μl aliquots, to be stored at −20 °C. Aliquots may be re-used up to three times.
- ***CHIR99021 (CH):*** For a 12 mM (2000×) solution, re-suspend 25 mg vial in 4.152 ml of DMSO. Mix well, vortex and make 25–50 μl aliquots, to be stored at −20 °C. Aliquots may be re-used up to three times.

**Table 13.2** Volumes of ECM coating to be used in cell culture plates

| Culture format | ECM coating volume |
|---|---|
| 3.5 $cm^2$ plate | 1.5 ml |
| 6-Well plate | 1.5 ml/well |
| 12-Well plate | 0.5 ml/well |

- ***Wnt-C59 (C59)*:** For a 2 μM (1000×) solution, re-suspend 5 mg vial in 6.588 ml of DMSO. Mix well, vortex and make 25–50 μl aliquots, to be stored at −20 °C. Aliquots may be re-used up to three times.

## *13.2.3 Extracellular Matrices (ECMs)*

Add the corresponding volume of diluted ECMs to coat different tissue culture plates as specified in Table 13.2.

### 13.2.3.1 Growth Factor-Reduced (GFR) Matrigel Stock Solution

GFR matrigel stock solution is prepared as a 1:2 dilution of the commercially available reagent. In order to prepare stock solution aliquots, a 10 ml vial of GFR-matrigel can be thawed overnight at 4 °C, or for 3 h on ice. Chill a 10 ml disposable serological pipette by immersing it in cold DMEM/F-12 for 10–15 s and aspirating the medium 3–4 times to chill the pipette. Subsequently, 10 ml of cold DMEM/F12 medium are added quickly to the commercial matrigel solution and mixed thoroughly. Dispense 500 μl aliquots into 1.5 ml microcentrifuge tubes and store at −20 °C.

#### GFR Matrigel Working Solution

Thaw a 500 μl aliquot GFR-matrigel stock solution overnight at 4 °C or on ice for 1 h. Dilute into 50 ml of cold DMEM/F12 in a conical tube and quickly mix thoroughly by inverting the tube several times. Plates coated with GFR matrigel working solution may be stored at 37 °C for up to 5 days prior to cell seeding.

### 13.2.3.2 Laminin Stock Solution

Prepare aliquots of laminin stock solution directly from the commercial stock reagent by thawing a 1 ml vial on ice and dividing its contents into 50 μl aliquots. Store at −80 °C.

Laminin Working Solution

Thaw a 50 μl laminin aliquot on ice and dilute it into 5 ml of PBS with $Ca^{2+}$ and $Mg^{2+}$ (1:100 dilution). Mix thoroughly and add to culture plates as required, using Table 13.2 as a reference. Incubate 4–24 h at 37 °C. Coated plates can be stored at 4 °C for up to 7 days.

### *13.2.4 Cell Culture Media*

#### 13.2.4.1 hiPSC Culture Media

We have observed that Essential 8 (E8) medium (Chen et al. 2011) favors pluripotency at baseline while priming hiPSCs for cardiac differentiation (Zanella and Sheikh 2016). This medium can be obtained commercially or prepared following instructions, as previously reported (Chen et al. 2011; Zanella and Sheikh 2016). Importantly, in order to conserve its ability to sustain hiPSC pluripotency and viability, E8 medium must not be warmed to a temperature higher than room temperature (25 °C maximum).

#### 13.2.4.2 Cardiac Differentiation Medium (CDM): RPMI Medium with AA and HSA

Prepare stock solutions and make aliquots as indicated in Table 13.3.

i. Take a 500 ml bottle of RPMI and add one 2 ml aliquot of AA and one 2 ml aliquot of HAS.
ii. Add one 5 ml aliquot of P/S.
iii. Mix well. If it is predicted that the medium will be used within 10 days, keep the bottle at 4 °C and pour 50 ml aliquots to be warmed when needed. Otherwise, prepare 50 ml aliquots, to be stored at −20 °C.

**Table 13.3** Stock solutions of media supplements required for cardiac differentiation

| Stock solution | Stock concentration (mg/ml) | Aliquot volume (ml) | Quantity needed (g) | Re-suspension volume (ml) | Final concentration (μg/ml) |
|---|---|---|---|---|---|
| AA | 53.2385 | 2 | 5 | 93.917 | 213 |
| HSA | 125 | 2 | 10 | 80 | 500 |

**Table 13.4** Plate formats and corresponding volumes of MIM medium and components required, per unit for each plate

| Format | Well volume (ml) | CDM volume (ml) | Insulin (μl) | CH (μl) |
|---|---|---|---|---|
| 3.5 cm plate | 3 | 3 | 7.5 | 1.5 |
| 6-Well plate | 3 | 18 | 45 | 9 |
| 12-Well plate | 2 | 24 | 60 | 12 |

Mix all components well before adding to cells

**Table 13.5** Plate formats and corresponding volumes of CSM medium and components required, per unit for each plate

| Format | Well volume (ml) | CDM volume (ml) | C59 volume (μl) |
|---|---|---|---|
| 3.5 cm plate | 3 | 3 | 3 |
| 6-Well plate | 3 | 18 | 18 |
| 12-Well plate | 2 | 24 | 24 |

Mix all components well before adding to cells

#### 13.2.4.3 Mesoderm Initiation Medium (MIM): CDM with Insulin and CH

This medium must be prepared immediately before use. Take an appropriate size aliquot of CDM and add insulin to a final concentration of 10 μg/ml and CH to a final concentration of 6 μM, as indicated in Table 13.4.

#### 13.2.4.4 Cardiac Specification Medium (CSM): CDM with C59

This medium must be prepared immediately before use. Take an appropriate size aliquot of CDM and add C59 to a final concentration of 2 μM, as indicated in Table 13.5.

#### 13.2.4.5 Cardiomyocyte Selection Medium (SEL): RPMI Without Glucose But with HSA, AA and Lactate

i. Take a 500 ml bottle of RPMI without glucose and add one 2 ml aliquot of HSA and one 2 ml aliquot of AA.
ii. Add 467 μl of 60 % lactate stock solution (final lactate concentration is 5 mM).
iii. Add one 5 ml aliquot of P/S.
iv. Mix well. If the medium will be used within 10 days, keep at 4 °C and pour 50 ml aliquots to be warmed as needed. Otherwise, prepare 50 ml aliquots, to be stored at −20 °C.

#### 13.2.4.6 Cardiomyocyte Maturation Medium (CMM): 75 % DMEM/25 % M199 Medium with AA, HSA and Insulin

i. Take a fresh bottle of DMEM, remove 135.25 ml and add 125 ml of M199.
ii. Add one 2 ml aliquot of HSA and one 2 ml aliquot of AA.
iii. Add 1.25 ml of Insulin (final concentration is 10 μg/ml).
iv. Add one 5 ml aliquot of P/S.
v. Mix well. If the medium will be used within 10 days, keep the bottle at 4 °C and pour 50 ml aliquots to be warmed when needed. Otherwise, prepare 50 ml aliquots, to be stored at −20 °C.

### *13.2.5 Cell Dissociation Reagents*

#### 13.2.5.1 hiPSC Dissociation Solution: 0.5 mM EDTA

Add 500 μl of 0.5 M EDTA to a 500 ml bottle of PBS without $Ca^{2+}$ and $Mg^{2+}$.

#### 13.2.5.2 Collagenase Type II Solution

Cardiomyocytes are dissociated in a solution containing 200 units/ml of HBSS without $Ca^{2+}$ and $Mg^{2+}$. Refer to Table 13.9 to estimate the required volume of collagenase II solution. For convenience this volume can be doubled, as the Collagenase II solution is the main component required to prepare the Cardiomyocyte Dissociation Solution (described in Sect. 13.2.5.3). Calculate and weigh the mass as necessary to reach 200 units/ml. Dissolve into HBSS without $Ca^{2+}$ and $Mg^{2+}$ and filter sterilize. The solution can be kept at 4 °C for up to 1 week.

#### 13.2.5.3 Cardiomyocyte Dissociation Solution

Prepare Cardiomyocyte Dissociation solution immediately before use. Refer to Table 13.9 to estimate the required volumes of Cardiomyocyte Dissociation Solution, and prepare the stock solutions as described in Table 13.6, using tissue culture-grade water.

**Table 13.6** Components and their respective volumes required to prepare cardiomyocyte dissociation solution

| Component | Stock concentration | Take from stock | Final concentration |
|---|---|---|---|
| Collagenase II solution | 200 units/ml | 854 μl | 200 units/ml |
| Taurine | 200 mM | 100 μl | 2 mM |
| EGTA | 20 mM | 10 μl | 0.2 mM |
| HSA | 125 mg/ml | 16 μl | 2 mg/ml |

## 13.3 Methods

### *13.3.1 hiPSC Culture and Seeding for Cardiac Differentiation*

hiPSCs are grown as high density monolayers on 6 $cm^2$ plates that have been pre-coated with GFR-matrigel. Passaging is performed at a ratio of 1:4–1:12 every 3–5 days and achieved as outlined below. hiPSCs should achieve confluence within the same number of days observed for the last three passages at the same ratios. Have the desired number of target plates pre-coated with GFR-matrigel as described in Table 13.2, and proceed as outlined below:

i. Determine the target numbers of plates/wells and prepare an appropriate volume of E8 medium containing 5 μM $Rock_i$.
ii. Warm appropriate size aliquots of PBS without $Ca^{2+}$ and $Mg^{2+}$ and hiPSC dissociation solution to 37 °C, according to Table 13.7.
iii. Aspirate E8 medium and wash cells once with warm PBS without $Ca^{2+}$ and $Mg^{2+}$.
iv. Aspirate PBS without $Ca^{2+}$ and $Mg^{2+}$ and add hiPSC dissociation solution. Incubate for 5 min at 37 °C. Monitor cell detachment during incubation time and if large clusters of detaching cells are observed, remove hiPSC dissociation immediately.
v. Aspirate hiPSC dissociation solution and re-suspend cells by pipetting up and down 3–6 times in E8 medium containing 5 μM $Rock_i$ as prepared in (i). If cells don't detach completely use a cell lifter to gently scrape them and break larger clusters by pipetting an additional 3–6 times.
vi. Split cell suspension and plate hiPSCs as required. Move plates in perpendicular directions 4–8 times to ensure even distribution of cells.

### *13.3.2 Cardiac Differentiation*

The protocol described here relies on media changes and pathway manipulations at specific time points and sequential stages (Fig. 13.1).

- <u>**Day 0:**</u> *Mesoderm Initiation with MIM.*
- Cardiac differentiation should be initiated when cells reach 95–98 % confluence (within 3–5 days of plating). Proceed as outlined below:

**Table 13.7** Volumes of solutions required for hiPSC passaging

| Format | Well volume | PBS without $Ca^{2+}$ and $Mg^{2+}$ | hiPSC dissociation solution |
|---|---|---|---|
| 3.5 cm plate | 2 ml | 2 ml | 1 ml |
| 6-Well plate | 2 ml | 2 ml per well | 1 ml per well |
| 12-Well plate | 1 ml | 2 ml per well | 1 ml per well |

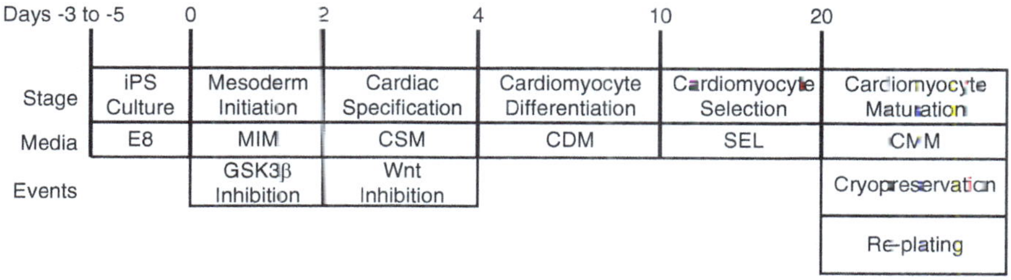


**Fig. 13.1** Schematic representation of the cardiac differentiation protocol, highlighting its key time points, media used and events. *E8* essential 8 medium, *MIM* mesoderm induction medium, *CSM* cardiac specification medium, *CDM* cardiac differentiation medium, *SEL* selection medium, *CMM* cardiac maturation medium, *GSK3β* glycogen synthase kinase 3 beta, *Wnt* wingless-type MMTV integration site family member

i. Prepare the required volume of MIM as outlined in Table 13.4.
ii. Add MIM into each well. Record the time at which CH is added.

- **Day 2:** *Wash and switch to CSM.*
- Mild to considerable cell death is expected to occur following incubation with MIM. Proceed as outlined below:

  i. Prepare the required volume of CSM as outlined in Table 13.5.
  ii. Warm an appropriate volume of RPMI and CSM media.
  iii. 48 h after MIM was added:

     (a) Wash cells once with warm RPMI medium.
     (b) Add appropriate volume of CSM to each plate/well.

- **Day 4:** *Medium change with CDM.*
  After C59 treatment considerable cell death is expected. Proceed with the protocol as outlined below:

  i. Warm an appropriate volume of CDM medium.
  ii. 96 h after MIM was added, aspirate CSM and add the appropriate volume of CDM to each plate/well.

- **Day 6:** *Medium change with CDM*

  i. Warm an aliquot of CDM medium.
  ii. 144 h after MIM was added, aspirate CDM and add the appropriate volume of fresh CDM to each plate/well.

- **Day 8:** *Medium change with CDM*

  i. Warm an aliquot of CDM medium.
  ii. 192 h after MIM was added, aspirate CDM and add the appropriate volume of fresh CDM to each plate/well.

- **Day 10:** *Start Cardiomyocyte selection. Medium switch to SEL medium (Optional)*

**Notes**

(a) *At this time point, sheets and/or clusters of beating cells should be detected. Beating can first be observed between days 7 and 11 for most tested hiPSC lines.*
(b) *If a considerable number of non-beating cells are observed, metabolic selection with lactate may be used to eliminate those undesirable cells. It is noteworthy that this selection does have an impact on cardiomyocyte health, thus its use must be carefully evaluated. This could be a more important issue for disease modeling studies. If selection will not be performed, continue medium changes between days 12–20 switching CDM every 2–3 days as previously described.*

i. Warm an aliquot of RPMI and one aliquot of SEL medium.
ii. 240 h after MIM was added:

(a) Remove CDM and wash cells once with RPMI medium.
(b) Add the appropriate volume of SEL to each plate/well.

- <u>**Days 12–20:**</u> *Cardiomyocyte selection with SEL medium (Optional).*
- Replace SEL medium every 2–3 days until day 20.

i. Refer to Table 13.8 to determine the appropriate SEL medium volume required.
ii. Warm the required volume of SEL medium.
iii. Aspirate SEL and add the appropriate volume of fresh SEL to each plate/well according to Table 13.8.

- <u>**Day 20–45+:**</u> *Cardiomyocyte Maturation. Switch medium to CMM.*

- Change CMM medium every 2–3 days, according to Table 13.8, until desired time points are reached.

i. Warm an aliquot of Maturation medium.
ii. Aspirate spent medium and add the appropriate volume of CMM to each plate/well.
iii. Change medium every 2–3 days with CMM until cardiomyocytes have achieved desired time points, according to Table 13.8.

**Table 13.8** Medium volume required for different changing frequencies

| Experiment format | 2-Day medium change (Mon & Wed) | 3-Day medium change (Fri) |
|---|---|---|
| 3.5 $cm^2$ plate | 3 ml | 4 ml |
| 6-Well plate | 3 ml/well; 18 ml/plate | 4 ml/well; 24 ml/plate |
| 12-Well plate | 2 ml/well; 24 ml/plate | 3 ml/well; 36 ml/plate |

**Table 13.9** Volumes of solutions required for cardiomyocyte dissociation

| Format | Well volume | HBSS without $Ca^{2+}$ and $Mg^{2+}$ | Collagenase II solution | Cardiomyocyte dissociation solution |
|---|---|---|---|---|
| 3.5 cm plate | 2 ml | 3 ml | 1 ml | 1 ml |
| 6-Well plate | 3 ml | 3 ml per well | 1 ml per well | 1 ml per well |
| 12-Well plate | 2 ml | 2 ml per well | 500 µl per well | 500 µl per well |

### 13.3.3 Cryo-preservation of hiPSC-Derived Cardiomyocytes

hiPSC-derived cardiomyocytes can be successfully cryopreserved from days 10–20 of differentiation. Considerable cell death is expected, thus one entire 6-well or 12-well plate can be used to generate 1 vial of cryopreserved cardiomyocytes

i. Add 10 µM $Rock_i$ to cardiomyocyte cultures and incubate for 3 h at 37 °C.
ii. Refer to Table 13.9 to estimate the required volumes of HBSS without $Ca^{2+}$ and $Mg^{2+}$, Collagenase II solution and Cardiomyocyte Dissociation Solution.
iii. Wash cells once with HBSS without $Ca^{2+}$ and $Mg^{2+}$ previously warmed to 37 °C.
iv. Add appropriate volume of Collagenase II solution. Add 10 µM $Rock_i$ and incubate for 30 min at 37 °C, gently swirling the plate every 10 min.
v. Add appropriate volume of dissociation solution dropwise. Mix gently.
vi. Re-suspend cells by pipetting up and down 3–6 times. If cells don't detach completely use a cell lifter to gently scrape them.
vii. Re-suspend cells by pipetting repeatedly every 5–10 times to break larger clusters and transfer cells to a 15 ml conical tube.
viii. Pellet the cells by centrifuging at $150\times g$ for 5 min at room temperature.
ix. Aspirate supernatant and gently re-suspend cells in 1 ml of cold CryoStor CS10 solution.
x. Transfer cells to a cryo-tube and place them in an appropriate cell freezing apparatus.

### 13.3.4 Cardiomyocyte Dissociation

Beating clusters or sheets of hiPSC-derived cardiomyocytes can be dissociated and cryo-preserved, or used for specific downstream applications, which require re-placing.

i. Follow steps in Sect 13.3.3 (i–vi).
ii. Triturate cells through a 20G syringe needle slowly, 3–6 times until larger clumps can no longer be observed by eye. If a single cell suspension is required pass cells through a 100 µm cell strainer.
iii. Spin cells at $150\times g$ for 5 min.

### 13.3.5 Cardiomyocyte Re-plating After Thawing or Dissociation

***Note:*** *Coat culturing surfaces with GFR matrigel (plastic) or laminin (glass) and incubate for the appropriate time before beginning thawing or dissociation.*

If thawing cardiomyocytes, begin at step (i). If starting from dissociated cardiomyocytes, begin from step (viii).

i. Warm a water bath to 37 °C.
ii. For each vial to be thawed, prepare a 15 ml aliquot of Cardiomyocyte Wash Medium (CWM): 75 % DMEM/25 % M-199 by mixing 11.250 ml of DMEM with 3.75 ml of M-199. Allow CWM to warm to room temperature.
iii. Add 1 ml of CWM to an empty 15 ml conical tube;
iv. Bring cryopreserved cardiomyocytes from dry ice to water bath. Immerse 2/3 of the cryotube in the water and swirl gently for 1.5–2 min, until cell solution is visibly thawed and a few ice crystals remain;
v. Thoroughly wipe cryovial to remove bath water and thoroughly spray the vial with 70 % ethanol before bringing it to the tissue culture hood;
vi. Gently re-suspend cells in 1 ml of CWM. Transfer cell suspension to 15 ml conical tube and slowly add them to 1 ml of CWM added in step (iii). Carefully and slowly add 12 ml of CWM dropwise to suspension in the conical tube. Mix gently.
vii. Pellet cells by centrifugation at 150 × *g* for 5 min;
viii. Re-suspend cells in 1 ml of CMM. Add 10 μM $Rock_i$.
ix. Count live cells and re-plate at an appropriate density. Favorable re-plating densities range from 1 to $6 \times 10^5$ cells/$cm^2$.

***Note:*** *Re-plated cardiomyocytes may take up to 3–5 days to start beating again. Carefully change media after 24 h and thereafter every 2–3 days.*

### 13.3.6 Characterization of hiPSC-Derived Cardiomyocytes

#### 13.3.6.1 Flow Cytometry and Immunofluorescence Microscopy

hiPSC-derived cardiomyocytes can be easily and accurately identified by antibody co-staining with contractile proteins, such as sarcomeric alpha-actinin (Sarc αAct) and the transcription factor NK2 homeobox 5 (NKX2.5). This combination of markers works robustly in flow cytometry and immunofluorescence analysis (Zanella and Sheikh 2016).

Preparation of hiPSC-derived cardiomyocytes for immunofluorescence analysis can be achieved by 4 % paraformaldehyde fixation, followed by permeabilization with 0.2 % Triton-X in PBS with calcium and magnesium for 10 min at room temperature. Blocking is performed with 5 % donkey serum in a buffer containing

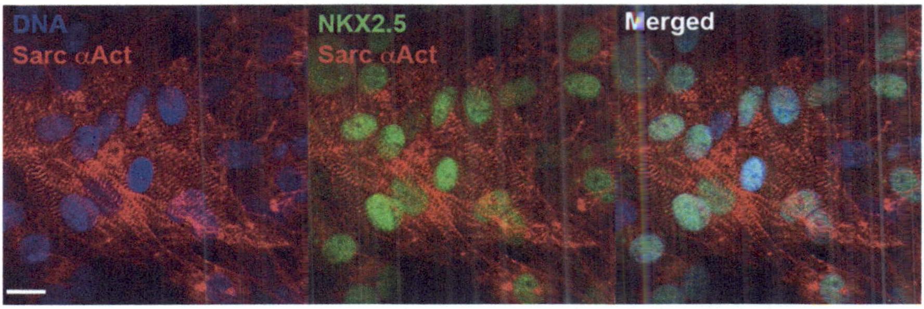


**Fig. 13.2** Immunofluorescence analysis of hiPSC-derived Cardiomyocytes. *DNA* deoxyribonucleic acid stained with Hoeschst 33342, *Sarc αAct* sarcomeric alpha actinin, *NKX2.5* NK2 homeobox 5

**Table 13.10** Reagents used in immunofluorescence characterization of hiPSC-derived cardiomyocytes

| Reagent | Manufacturer | Catalog number |
|---|---|---|
| Donkey serum | Lampire Biological Lab | 7332100 |
| Mouse anti-Sarc αAct antibody | Sigma | A7811 |
| Rabbit anti-NKX2.5 antibody | Santa Cruz | sc-14033 |
| Hoeschst 33342 | Life Technologies | H1399 |
| Alexa Fluor-488 Donkey anti-mouse antibody | Life Technologies | A21202 |
| Alexa Fluor-568 Donkey anti-rabbit antibody | Life Technologies | A10042 |
| Fluorescence mounting medium | Dako | S3023802 |

20 mM Tris–HCl pH=7.4, 155 mM NaCl, 2 mM EGTA and 2 mM $MgCl_2$ for 30 min at room temperature. Primary antibodies against sacomeric alpha actinin and NKX2.5 are diluted 1:100 in blocking buffer and incubated overnight at 4 °C. Subsequently, cells are washed three times with PBS containing calcium and magnesium and incubated with secondary antibodies at 1:100 dilutions and 5 μg/ml Hoeschst 33342 for 2 h at room temperature, in the dark. After washing three times with PBS containing calcium and magnesium cells are mounted for analysis. Figure 13.2 depicts an immunofluorescence analysis of hiPSC-derived cardiomyocytes following the above protocol in an Olympus Fluoview FV-1000 confocal microscope. All reagents used to perform this analysis are described in Table 13.10.

#### 13.3.6.2 Electrophysiological Assays

Several platforms have been developed to analyze the electrical activity, calcium release and contractile behavior of hiPSC-derived cardiomyocytes. Table 13.11 provides examples of available instruments successfully employed to perform these analyses while overriding the need for advanced electrophysiology training.

**Table 13.11** Instruments used to analyze electrophysiological behavior of hiPSC-derived cardiomyocytes and their respective throughputs

| Type of analysis | Detectable phenotypes | Manufacturer | Instrument name | Pacing capable | Throughput |
|---|---|---|---|---|---|
| Field potentials (multi-electrode arrays) | Frequency and duration of spontaneous field potentials<br>Prolongation of depolarization-repolarization interval | Axion Byosystems | Muse | Yes | 1, 12, 48 and 96 wells |
| | | | Maestro | No | 1, 12, 48 and 96 wells |
| | | Panasonic | MED64 | Yes | 1–12 Samples per run |
| Calcium transients | Frequency and amplitude of calcium release | Molecular Devices | FLIPR Tetra | No | 384 Wells |
| Calcium transients and contractility | Frequency and amplitude of calcium releases<br>Frequency and amplitude of contraction | Ionoptix | Myocyte calcium and contractility system | Yes | 1 Sample per run |
| Field potentials and contractility | Frequency and duration of spontaneous field potentials<br>Prolongation of depolarization-repolarization interval<br>Frequency and amplitude of contraction | Nanion | CardioExcyte 96 | Yes | 96 Wells, field potential duration and contractility measured sequentially |
| | | Acea Biosciences | xCELLigence RTCA CardioECR | Yes | 48 Wells, field potential duration and contractility measured simultaneously |
| Contractility | Frequency and amplitude of contraction | Acea Biosciences | xCELLigence RTCA Cardio | No | 96 Wells |

### *13.3.7 Applications of hiPSC-Derived Cardiomyocytes*

Amongst the most attractive applications for hiPSC-derived cardiomyocytes are probing cardiotoxicity of small molecule compounds and modeling of diseases affecting the heart. As an example of how hiPSC-derived cardiomyocytes can reveal the cardiotoxic effect of small molecules, a set of drugs with known cardio-toxic mechanism of action can be used to challenge hiPSC-derived cardiomyocytes (Fig. 13.3). Measurements of spontaneous field potentials provide valuable insights on the global patterns of electrical activity of hiPSC-derived cardiomyocytes while impedance-based measurements provide a good indication on the contractile behavior of cardiomyocytes. We have used the Acea Biosciences xCELLigence

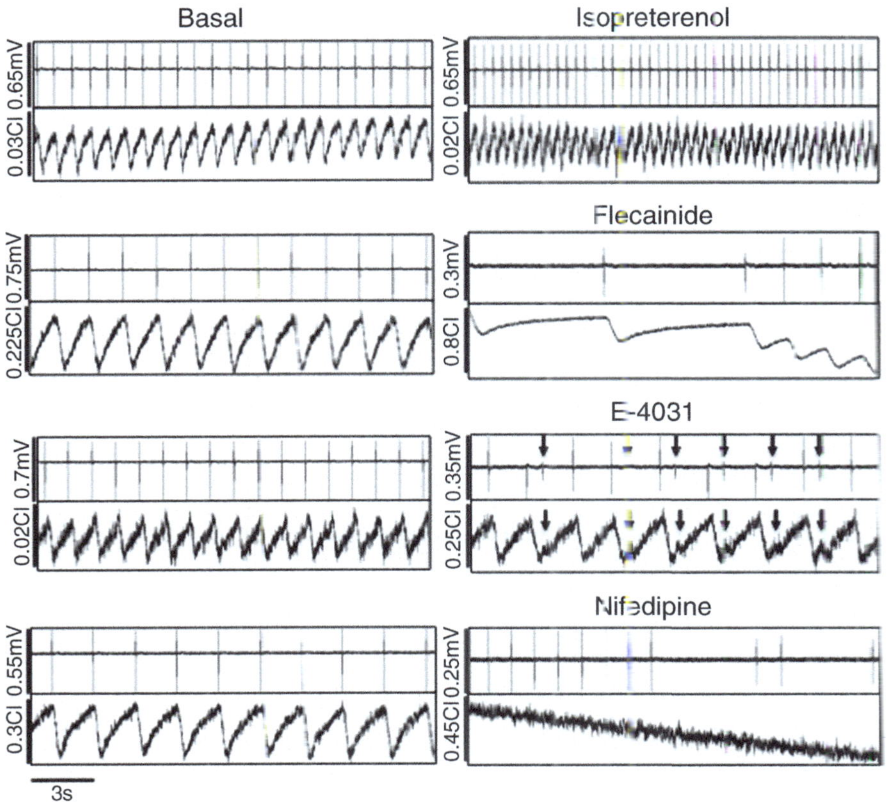


**Fig. 13.3** A panel of small molecules with known mechanism of action upon ion channels is tested on hiPSC-derived cardiomyocytes. *Arrows* highlight abnormal action potentials and respective contractions. *CI* cellular index, a measurement of Impedance

RTCA Cardio Platform to perform dual simultaneous analysis of field potentials and contractile behavior to evaluate the effects of the tested compounds in hiPSC-derived cardiomyocytes. For this assay, cardiomyocytes at day 20 of differentiation were dissociated and plated onto RTCA cardio plates coated with 50 μl GFR matrigel per well at a density of $6 \times 10^4$ viable cells per well in a final volume of 200 μl. Cells were allowed to recover for 5 days before assays were performed.

As expected, the sodium channel inhibitor, flecainide caused a marked reduction in the amplitude of spontaneous field potentials as well as abnormalities in rhythm frequency, also observed in the contraction profiles. The potassium channel blocker E-4031 induced marked inversions in field potential profiles while also causing arrhythmic behavior. Finally, the calcium channel blocker, nifedipine caused drastic changes in the occurrence of field potentials, while completely abrogating the occurrence of contractions. Altogether, these results reinforce the application of hiPSC-derived cardiomyocytes in interrogating key cardiac ion channel pathways, which makes them a formidable platform for the investigation of cardiotoxicity of new drugs.

**Acknowledgments** F.Z. was a recipient of the 2013–2014 American Heart Association Postdoctoral Fellowship Award and the 2011 San Diego Foundation Blasker-Miah Grant for Science and Technology. F.S. is supported by grants from the American Heart Association (GRNT22940045) and NIH/NHLBI (HL095780).

## References

Abassi YA, Xi B, Li N, Ouyang W, Seiler A, Watzele M, Kettenhofen R, Bohlen H, Ehlich A, Kolossov E, Wang X, Xu X (2012) Dynamic monitoring of beating periodicity of stem cell-derived cardiomyocytes as a predictive tool for preclinical safety assessment. Br J Pharmacol 165:1424–1441

Burridge PW, Matsa E, Shukla P, Lin ZC, Churko JM, Ebert AD, Lan F, Diecke S, Huber B, Mordwinkin NM, Plews JR, Abilez OJ, Cui B, Gold JD, Wu JC (2014) Chemically defined generation of human cardiomyocytes. Nat Methods 11:855–860

Carvajal-Vergara X, Sevilla A, D'Souza SL, Ang YS, Schaniel C, Lee DF, Yang L, Kaplan AD, Adler ED, Rozov R, Ge Y, Cohen N, Edelmann LJ, Chang B, Waghay A, Su J, Pardo S, Lichtenbelt KD, Tartaglia M, Gelb BD, Lemischka IR (2010) Patient-specific induced pluripotent stem-cell-derived models of LEOPARD syndrome. Nature 465:808–812

Caspi O, Huber I, Gepstein A, Arbel G, Maizels L, Boulos M, Gepstein L (2013) Modeling of arrhythmogenic right ventricular cardiomyopathy with human induced pluripotent stem cells. Circ Cardiovasc Genet 6:557–568

Chen G, Gulbranson DR, Hou Z, Bolin JM, Ruotti V, Probasco MD, Smuga-Otto K, Howden SE, Diol NR, Propson NE, Wagner R, Lee GO, Antosiewicz-Bourget J, Teng JM, Thomson JA (2011) Chemically defined conditions for human iPSC derivation and culture. Nat Methods 8:424–429

Davis RP, Casini S, van den Berg CW, Hoekstra M, Remme CA, Dambrot C, Salvatori D, Oostwaard DW, Wilde AA, Bezzina CR, Verkerk AO, Freund C, Mummery CL (2012) Cardiomyocytes derived from pluripotent stem cells recapitulate electrophysiological characteristics of an overlap syndrome of cardiac sodium channel disease. Circulation 125:3079–3091

Fatima A, Xu G, Shao K, Papadopoulos S, Lehmann M, Arnaiz-Cot JJ, Rosa AO, Nguemo F, Matzkies M, Dittmann S, Stone SL, Linke M, Zechner U, Beyer V, Hennies HC, Rosenkranz S, Klauke B, Parwani AS, Haverkamp W, Pfitzer G, Farr M, Cleemann L, Morad M, Milting H, Hescheler J, Saric T (2011) In vitro modeling of ryanodine receptor 2 dysfunction using human induced pluripotent stem cells. Cell Physiol Biochem 28:579–592

Han L, Li Y, Tchao J, Kaplan AD, Lin B, Li Y, Mich-Basso J, Lis A, Hassan N, London B, Bett GC, Tobita K, Rasmusson RL, Yang L (2014) Study familial hypertrophic cardiomyopathy using patient-specific induced pluripotent stem cells. Cardiovasc Res 104:258–269

Itzhaki I, Maizels L, Huber I, Zwi-Dantsis L, Caspi O, Winterstern A, Feldman O, Gepstein A, Arbel G, Hammerman H, Boulos M, Gepstein L (2011) Modelling the long QT syndrome with induced pluripotent stem cells. Nature 471:225–229

Itzhaki I, Maizels L, Huber I, Gepstein A, Arbel G, Caspi O, Miller L, Belhassen B, Nof E, Glikson M, Gepstein L (2012) Modeling of catecholaminergic polymorphic ventricular tachycardia with patient-specific human-induced pluripotent stem cells. J Am Coll Cardiol 60:990–1000

Jung CB, Moretti A, Mederos y Schnitzler M, Iop L, Storch U, Bellin M, Dorn T, Ruppenthal S, Pfeiffer S, Goedel A, Dirschinger RJ, Seyfarth M, Lam JT, Sinnecker D, Gudermann T, Lipp P, Laugwitz KL (2012) Dantrolene rescues arrhythmogenic RYR2 defect in a patient-specific stem cell model of catecholaminergic polymorphic ventricular tachycardia. EMBO Mol Med 4:180–191

Kim C, Wong J, Wen J, Wang S, Wang C, Spiering S, Kan NG, Forcales S, Puri PL, Leone TC, Marine JE, Calkins H, Kelly DP, Judge DP, Chen HS (2013) Studying arrhythmogenic right ventricular dysplasia with patient-specific iPSCs. Nature 494:105–110

Kujala K, Paavola J, Lahti A, Larsson K, Pekkanen-Mattila M, Viitasalo M, Lahtinen AM, Toivonen L, Kontula K, Swan H, Laine M, Silvennoinen O, Aalto-Setala K (2012) Cell model of catecholaminergic polymorphic ventricular tachycardia reveals early and delayed afterdepolarizations. PLoS One 7:e44660

Lahti AL, Kujala VJ, Chapman H, Koivisto AP, Pekkanen-Mattila M, Kerkela E, Hyttinen J, Kontula K, Swan H, Conklin BR, Yamanaka S, Silvennoinen O, Aalto-Setala K (2012) Model for long QT syndrome type 2 using human iPS cells demonstrates arrhythmogenic characteristics in cell culture. Dis Model Mech 5:220–230

Lamore SD, Kamendi HW, Scott CW, Dragan YP, Peters MF (2013) Cellular impedance assays for predictive preclinical drug screening of kinase inhibitor cardiovascular toxicity. Toxicol Sci 135:402–413

Lan F, Lee AS, Liang P, Sanchez-Freire V, Nguyen PK, Wang L, Han L, Yen M, Wang Y, Sun N, Abilez OJ, Hu S, Ebert AD, Navarrete EG, Simmons CS, Wheeler M, Pruitt B, Lewis R, Yamaguchi Y, Ashley EA, Bers DM, Robbins RC, Longaker MT, Wu JC (2013) Abnormal calcium handling properties underlie familial hypertrophic cardiomyopathy pathology in patient-specific induced pluripotent stem cells. Cell Stem Cell 12:101–113

Lian X, Hsiao C, Wilson G, Zhu K, Hazeltine LB, Azarin SM, Raval KK, Zhang J, Kamp TJ, Palecek SP (2012) Robust cardiomyocyte differentiation from human pluripotent stem cells via temporal modulation of canonical Wnt signaling. Proc Natl Acad Sci U S A 109:E1848–E1857

Lian X, Zhang J, Azarin SM, Zhu K, Hazeltine LB, Bao X, Hsiao C, Kamp TJ, Palecek SP (2013) Directed cardiomyocyte differentiation from human pluripotent stem cells by modulating Wnt/beta-catenin signaling under fully defined conditions. Nat Protoc 8:162–175

Liang P, Lan F, Lee AS, Gong T, Sanchez-Freire V, Wang Y, Diecke S, Sallam K, Knowles JW, Wang PJ, Nguyen PK, Bers DM, Robbins RC, Wu JC (2013) Drug screening using a library of human induced pluripotent stem cell-derived cardiomyocytes reveals disease-specific patterns of cardiotoxicity. Circulation 127:1677–1691

Lundy SD, Zhu WZ, Regnier M, Laflamme MA (2013) Structural and functional maturation of cardiomyocytes derived from human pluripotent stem cells. Stem Cells Dev 22:1991–2002

Ma D, Wei H, Lu J, Ho S, Zhang G, Sun X, Oh Y, Tan SH, Ng ML, Shim W, Wong P, Liew R (2012) Generation of patient-specific induced pluripotent stem cell-derived cardiomyocytes as a cellular model of arrhythmogenic right ventricular cardiomyopathy. Eur Heart J 34(15):1122–1133

Matsa E, Rajamohan D, Dick E, Young L, Mellor I, Staniforth A, Denning C (2011) Drug evaluation in cardiomyocytes derived from human induced pluripotent stem cells carrying a long QT syndrome type 2 mutation. Eur Heart J 32:952–962

Moretti A, Bellin M, Welling A, Jung CB, Lam JT, Bott-Flugel L, Dorn T, Goedel A, Hohnke C, Hofmann F, Seyfarth M, Sinnecker D, Schomig A, Laugwitz KL (2010) Patient-specific induced pluripotent stem-cell models for long-QT syndrome. N Engl J Med 363:1397–1409

Navarrete EG, Liang P, Lan F, Sanchez-Freire V, Simmons C, Gong T, Sharma A, Burridge PW, Patlolla B, Lee AS, Wu H, Beygui RE, Wu SM, Robbins RC, Bers DM, Wu JC (2013) Screening drug-induced arrhythmia [corrected] using human induced pluripotent stem cell-derived cardiomyocytes and low-impedance microelectrode arrays. Circulation 128:S3–S13

Peters MF, Lamore SD, Guo L, Scott CW, Kolaja KL (2015) Human stem cell-derived cardiomyocytes in cellular impedance assays: bringing cardiotoxicity screening to the front line. Cardiovasc Toxicol 15:127–139

Sallam K, Li Y, Sager PT, Houser SR, Wu JC (2015) Finding the rhythm of sudden cardiac death: new opportunities using induced pluripotent stem cell-derived cardiomyocytes. Circ Res 116:1989–2004

Sharma A, Wu JC, Wu SM (2013) Induced pluripotent stem cell-derived cardiomyocytes for cardiovascular disease modeling and drug screening. Stem Cell Res Ther 4:150

Sun N, Yazawa M, Liu J, Han L, Sanchez-Freire V, Abilez OJ, Navarrete EG, Hu S, Wang L, Lee A, Pavlovic A, Lin S, Chen R, Hajjar RJ, Snyder MP, Dolmetsch RE, Butte MJ, Ashley EA, Longaker MT, Robbins RC, Wu JC (2012) Patient-specific induced pluripotent stem cells as a model for familial dilated cardiomyopathy. Sci Transl Med 4:130ra147

Yazawa M, Hsueh B, Jia X, Pasca AM, Bernstein JA, Hallmayer J, Dolmetsch RE (2011) Using induced pluripotent stem cells to investigate cardiac phenotypes in Timothy syndrome. Nature 471:230–234

Zanella F, Sheikh F (2016) Generation and characterization of cardiac cells. In: Nagy A, Turksen K (eds) Patient-specific induced pluripotent stem cell models, vol 1353, Methods in molecular biology. Springer, New York, pp 147–162

Zanella F, Lyon RC, Sheikh F (2014) Modeling heart disease in a dish: from somatic cells to disease-relevant cardiomyocytes. Trends Cardiovasc Med 24:32–44

# Chapter 14
# Endoderm Differentiation from Human Pluripotent Stem Cells

**Nathan Kumar, David Brafman, and Karl Willert**

**Abstract** Human pluripotent stem cells (hPSCs) provide a virtually unlimited raw material to derive and engineer mature cell types with therapeutic value, including cell transplantation, disease modeling and drug screening. The first step to differentiate hPSCs into such cell types involves specification towards one of the three main embryonic cell populations, ecto-, endo- and mesoderm. Efficient induction into the correct lineage is critical to the success of subsequent differentiation steps and to the final yield of desired cells. Here we describe methods to generate definitive endoderm (DE), the progenitor cell population for such tissues as the thymus, liver, pancreas, stomach and intestine. In addition, we will provide methods to characterize and monitor the efficiency of DE differentiation, including expression of DE markers at the gene and protein level. Flow cytometry based methods described in this chapter can also be extended to isolate and purify cells with DE properties. Such enrichment strategies are useful to eliminate undesired cell populations, especially undifferentiated hPSCs, which harbor the potential risk for seeding tumors upon transplantation. Several of the methods for the manipulation of hPSCs and for their analysis outlined here are of general utility and are applicable to other hPSCs derivative cell populations.

**Keywords** Human pluripotent stem cell (hPSC) • Human embryonic stem cell (hESC) • Induced pluripotent stem cell (iPSC) • Definitive endoderm (DE) • SOX17 • CXCR4 • Wnt signaling

---

N. Kumar
Department of Bioengineering, UCSD, La Jolla, CA 92093, USA

D. Brafman
School of Biological and Health Systems Engineering, Arizona State University, Tempe, AZ 85287, USA

K. Willert (✉)
Department of Cellular and Molecular Medicine, UCSD, La Jolla, CA 92093, USA

Sanford Consortium for Regenerative Medicine,
2880 Torrey Pines Scenic Dr. – Room 3009, La Jolla, CA 92093, USA
e-mail: kwillert@ucsd.edu


H. Ulrich, P. Davidson Negraes, *Working with Stem Cells*,
DOI 10.1007/978-3-319-30582-0_14

## 14.1 Introduction

In mammalian development, definitive endoderm (DE) gives rise to the epithelial lining of the respiratory and digestive tract as well as several major organs including the liver, pancreas, thymus and thyroid. The generation of these tissue types makes DE distinct from primitive endoderm, an extraembryonic tissue that gives rise to visceral endoderm and parietal endoderm. During gastrulation, undifferentiated cells from specific regions of the epiblast ingress to the posterior structure known as the primitive streak where they undergo an epithelial-to-mesenchymal transition, ultimately bringing about the mesodermal and endodermal tissues of the organism. This close relationship between mesoderm and endoderm development suggests the existence of a common bipotential precursor, known as the mesendoderm, however, a single embryonic cell with bipotential properties has yet to be substantiated in mammals. Complexity of this early embryonic system is in large part attributed to the numerous growth factors produced by the primitive streak that act in various combinations to induce either endoderm or mesoderm (Beddington and Smith 1993; Conlon et al. 1994; Faust and Magnuson 1993; Tam and Behinger 1997).

The study of human embryonic germ layer and organ development, as well as the generation of mature and functional cell populations, is made possible by the availability of human pluripotent stem cells (hPSCs), which include human embryonic and induced pluripotent stem cells (hESC and iPSC). These cells can be readily differentiated into the thee main germ layers, ecto-, endo- and mesoderm. Protocols for hPSC differentiation into these distinct lineages have been developed, however, most of these protocols are highly variable in efficiency, are poorly scalable and require undefined and animal-derived reagents, rendering many derivative cell populations unsuitable for eventual therapeutic applications. Here, we present an *in vitro* differentiation protocol based on the modulation of two critical embryonic signaling pathways, TGF-β and WNT signaling, that allows for the consistent and highly efficient generation of DE cells from hPSCs. The resulting DE cell population serves as an important starting point for the subsequent differentiation of more mature cell types and tissues, such as cells of the lung, liver and pancreas.

The earliest stages of DE commitment *in vivo* require the signaling molecule Nodal, a member of the TGFβ superfamily (Vincent et al. 2003). Since active Nodal protein is currently not readily available, another TGFβ family member, Activin A, which is readily sourced and binds to the same receptors as Nodal, can be substituted to mimic Nodal activity for DE commitment from hPSCs (D'Amour et al. 2005). Another embryonic signaling pathway required for the formation of the primitive streak and subsequent DE specification is the WNT signaling pathway. Genetic analyses in mice have shown that disruption of the WNT pathway prevents formation of the primitive streak and, subsequently, DE (Haegel et al. 1995; Kelly et al. 2004; Liu et al. 1999). The human genome contains 19 *WNT* genes, which encode secreted lipid-modified glycoproteins (Willert and Nusse 2012) that bind to the Frizzled (FZD) family of receptors to activate intracellular signaling cascades, most notably the WNT/β-catenin pathway (commonly referred to as the canonical WNT signaling pathway). Of the many WNT proteins, Wnt3a has been purified to homogeneity (Willert et al. 2003; Willert 2008) and applied *in vitro* to promote differentiation of hPSCs to DE (Brafman et al. 2013b; D'Amour et al. 2006). An alternative method to stimulate the WNT/β-catenin pathway in cell culture is with glycogen synthetase kinase 3β (GSK3β) inhibitors (GSK3i) (Sato et al. 2004), many of which are readily available. Triggering these two signaling pathways, TGF-β and WNT, with Activin A and Wnt3a (or GSK3i) respectively, at the proper time intervals efficiently promotes differentiation of hPSCs cells into DE (Fig. 14.1).

Here, we present differentiation protocols based on the modulation of TGF-β and WNT signaling that allow for the consistent and highly efficient generation of DE cells from hPSCs. These differentiation methods generate cell populations that serve as building blocks for future application in tissue engineering, regenerative medicine therapies, disease modeling, and drug discovery. A large number of publications have described the generation of this DE population as an intermediate step to more specialized tissues. For example, several groups have used similar DE differentiation methods en route to generating pancreatic cell types, including pancreatic progenitors (D'Amour et al. 2005, 2006; Kroon et al. 2008) and β-cells (Cheng et al. 2012; Pagliuca et al. 2014). In addition, this DE population can be further instructed to acquire properties of lung, liver, thymus thyroid and intestine, as described in the literature (for examples, see Ameri et al. 2010; Brafman et al. 2013a; Cai et al. 2007; Kadzik and Morrisey 2012; Longmire et al. 2012; Parent et al. 2013; Spence et al. 2011; Su et al. 2015; Touboul et al. 2010).

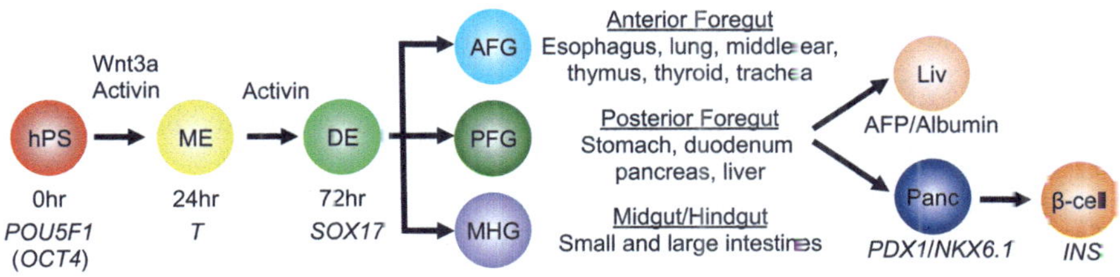


**Fig. 14.1** Schematic of hPSC differentiation into definitive endoderm and derivative cell types. Timed exposure of hPSCs (hPS) to Wnt3a and Activin-A (Activin) promotes definitive endoderm (DE) formation via an intermediate mesendodermal (ME) cell population. DE is capable of differentiating into tissues of the anterior foregut (AFG), posterior foregut (PFG) and midgut/hindgut (MHG). PFG subsequently can give rise to more mature cell types, such as liver (Liv) and pancreas (Panc) cells. Multiple protocols have been published describing the generation of mature β-cells capable of glucose stimulated insulin secretion

## 14.2 Materials

### 14.2.1 *Equipment and Supplies*

- Biological safety cabinet.
- $CO_2$ incubator with humidity and gas controls to maintain a stable environment of 37 °C, >95 % humidity, and 5 % $CO_2$.
- 37 °C water bath.
- Benchtop cell culture centrifuge.
- Pipet controller.
- Serological pipettes (1, 5, 10, and 25 ml), sterile.
- 10-, 20-, 200-, and 1000 μl micropipettes.
- 10-, 20-, 200-, and 1000 μl micropipette tips, sterile.
- Tissue culture treated polystyrene dishes: 6-well, 12-well, and 24-well and 100 mm, sterile.
- 1.5 ml microcentrifuge tubes, sterile.
- Polystyrene conical tubes: 15- and 50-ml, sterile.
- Falcon No. 2052 and No. 2054 tubes or equivalent, sterile.
- Hemocytometer.
- Inverted light microscope (EVOS Cell Imaging System or equivalent).
- Real-Time PCR Detection System (BioRad CFX96 Touch or equivalent).
- UV spectrophotometer (ThermoFisher Scientific NanoDrop or equivalent).
- Flow cytometer (e.g. BD FACSCanto or equivalent).
- Inverted fluorescence microscope (e.g. Olympus FluoView FV-1000).

### 14.2.2 *Stock Solutions and Reagents*

- Human pluripotent stem cells (hPSCs), such as HUES1 (NIH Registration Number 0014), HUES9 (0022), H9/WA09 (0062) or any induced pluripotent stem cell (iPSC) cell line.
- E8 medium (Chen et al. 2011) for maintenance of undifferentiated hPSCs.
- RPMI 1640 (Mediatech, Catalog number: 15-040-CV).
- Knockout DMEM (ThermoFisher Scientific, Cat. No. 10829-018).
- Defined Fetal Bovine Serum (FBS) (HyClone, Cat. No. SH30070.01).
- MEM non-essential amino acids solution (100×) (NEAA) (ThermoFisher Scientific, Cat. No. 11140-050). Make aliquots of 5 ml and store at 4 °C.
- Penicillin-Sterptomycin, 5000 U/ml (P/S) (ThermoFisher Scientific, Cat. No. 15070-063). Make aliquots of 5 ml and store at −20 °C.
- GlutaMAX supplement (ThermoFisher Scientific, Cat. No. 35050-061). Make aliquots of 5 ml and store at −20 °C.
- Phosphate-Buffered Saline (PBS), pH 7.4 (ThermoFisher Scientific, Cat. No. 10010023).

- Bovine Serum Albumin (BSA) (Sigma, Cat. No. 2058).
- StemPro Accutase cell dissociation reagent (ThermoFisher Scientific, Cat. No. A1110501). Make aliquots of 10 ml and store at −20 °C.
- Matrigel hESC qualified matrix (BD Biosciences, Cat. No. 354277). Make aliquots of 250 μl and store at −20 °C.
- Trypan Blue Solution, 0.4 % (ThermoFisher Scientific, Cat. No. 15250-061).
- Rho-associated protein kinase inhibitor (ROCKi) Y-27632 (EMD Millipore, Cat. No. SCM075). Dissolve in DMSO at a concentration of 5 mM. Make aliquots of 50 μl in 1.5 ml microcentrifuge tubes and store at −20 °C. Stability after reconstitution: ~1 year at −20 °C, protected from light.
- Activin A, 100 μg/ml, store at −20 °C (R&D Systems, Cat. No. 338-AC-010).
- CHIR99021 (CHIR), 25 mM, store at −20 °C (Stemgent, Cat. No. 04-0004).
- Wnt3a, 100 μg/ml, store at −20 °C (R&D Systems, Cat. No. 1324-WN-002).
- Fibronectin (FN) from human plasma, 1 mg/ml solution (Sigma, Cat. No. F0895). Make aliquots of 100 μl and store at −20 °C.
- Vitronectin (VN) from human plasma, lyophilized (Sigma, Cat. No. V8379). Reconstitute with 1 ml tissue culture grade water and sterilize by filtration. Make aliquots of 100 μl and store at −20 °C.
- qPCR primers (see Table 14.1).
- 2× SYBR Green PCR Master Mix (ThermoFisher Scientific, Cat. No. 4309155)
- TRIzol® Reagent (ThermoFisher Scientific, Cat. No. 15596-026).
- 48-, 96-, or 384-well optical plate (ThermoFisher Scientific, Cat. No. 4375816, N8010560, 4343370).
- Optical adhesive cover (ThermoFisher Scientific, Cat. No. 4360954).
- High-Capacity RNA-to-cDNA™ Kit (ThermoFisher Scientific, Cat. No. 4387406).
- Fixation Buffer: 4 % Paraformaldehyde/PBS (Paraformaldehyde; Sigma, Cat. No. P6148).
- Blocking Solution: PBS, 0.1 % Triton X-100, 2 % BSA (Triton X-100; Sigma, Cat. No. X100).
- APC anti-human CXCR4 Antibody (BD Biosciences, Cat. No. 555976).
- Isotype control antibody, e.g. BV421 Mouse IgG2b, k (BD Biosciences, Cat. No. 562748).
- Polyclonal Goat IgG anti-human SOX17 antibody (R&D Systems, Cat. No. AF1924).
- Polyclonal Goat IgG anti-human FOXA2 antibody (R&D Systems, Cat. No. AF2400).
- Polyclonal Rabbit IgG anti-human POU5F1 antibody (Abcam, Cat. No. ab19857).
- Polyclonal Mouse IgG anti-human Ki-67 antibody (Cell Signaling Technology, Cat. No. 9449S).
- Donkey anti-Mouse IgG (H+L) Secondary Antibody, Alexa Fluor® 647 conjugate (ThermoFisher Scientific, Cat. No. A31571).
- Donkey anti-Goat IgG (H+L) Secondary Antibody, Alexa Fluor® 488 conjugate (ThermoFisher Scientific, Cat. No. A11055).

**Table 14.1** Primers needed for definitive endoderm characterization

| Gene | Primer sequence | Amount used |
|---|---|---|
| *RNA18S5 (18S)* | Fwd: GTAACCCGTTGAACCCCATT | 1 μl/sample |
| | Rev: CCATCCAATCGGTAGTAGCG | |
| *POU5F1* | Fwd: CTTGAATCCCGAATGGAAAGGG | 1 μl/sample |
| | Rev: GTGTATATCCCAGGGTGATCCTC | |
| *NANOG* | Fwd: TTTGTGGGCCTGAAGAAAACT | 1 μl/sample |
| | Rev: AGGGCTGTCCTGAATAAGCAG | |
| *SOX2* | Fwd: GGGGAAAGTAGTTTGCTGCC | 1 μl/sample |
| | Rev: CGCCGCCGATGATTGTTATT | |
| *FOXA2* | Fwd: GGAGCAGCTACTATGCAGAGC | 1 μl/sample |
| | Rev: CGTGTTCATGCCGTTCATCC | |
| *SOX1* | Fwd: GGCTTTTGTACAGACGTTCCC | 1 μl/sample |
| | Rev: AACCCAAGTCTGGTGTCAGC | |
| *MESP1* | Fwd: CTCGGGCTCGGCATAAAGC | 1 μl/sample |
| | Rev: CGCAGAGAGCATCCAGGACT | |
| *MIXL1* | Fwd: AGTTGCTGGAGCTCGTCTTC | 1 μl/sample |
| | Rev: AGGGCAATGGAGGAAAACTC | |
| *SOX17* | Fwd: GTGGACCGCACGGAATTTG | 1 μl/sample |
| | Rev: GGAGATTCACACCGGAGTCA | |
| *NESTIN* | Fwd: GTCTCAGGACAGTGCTGAGCCTTC | 1 μl/sample |
| | Rev: TCCCCTGAGGACCAGGAGTCTC | |
| *T* | Fwd: CTATTCTGACAACTCACCTGCAT | 1 μl/sample |
| | Rev: ACAGGCTGGGGTACTGACT | |

Concentration of primer stocks are 10 μM

- Goat anti-Rabbit IgG (H + L) Secondary Antibody, Cy3 conjugate (ThermoFisher Scientific, Cat. No. A10520).
- DAPI solution (ThermoFisher Scientific, Cat. No. 62248).

## 14.3 Derivation of Definitive Endoderm

DE derived from hPSCs has the potential to further differentiate to cells of any endodermal derivatives, which include liver, lungs, intestines, pancreas, thymus and thyroid. Efficient differentiation of hPSCs into DE is the first critical step to generating these more mature endodermal tissues. Several protocols to derive DE have been described. For example, Kubo and colleagues successfully differentiated mouse embryonic stem (mES) cells in embryoid bodies (EBs) into DE with high doses of Activin A (Kubo et al. 2004). These EBs consisted of more than 50 % cells expressing the endodermal marker forkhead box A2 (Foxa2), which is also expressed in extra-embryonic visceral endoderm and axial mesoderm, and therefore is not specific to DE (Ang et al. 1993; Sasaki and Hogan 1993). Another study employed a mES cell reporter line in monolayer culture treated with high Activin A doses to generate Goosecoid (Gsc) and SRY-box 17 (Sox17) expressing DE (Yasunaga et al. 2005). The use of this reporter line allowed for the identification of DE as Gsc-positive and Sox17-positive ($Gsc^+$ $Sox17^+$) and visceral endoderm as Gsc-negative and Sox17-positive ($Gsc^-Sox17^+$). This distinction enabled the development of culture conditions that permit for selective differentiation to either DE or visceral endoderm. This study also led to the identification of CXCR4 as a cell surface molecule marking the $Gsc^+$ population. To extend these studies to human cells, D'Amour and colleagues used Wnt3a and Activin A to efficiently differentiated hESCs in monolayer into DE (D'Amour et al. 2005, 2006).

In this protocol, initial mesendoderm differentiation of hESCs was induced using low concentrations of fetal bovine serum (FBS) supplemented with Wnt3a and Activin A, resulting in greater than 80 % of the cell population expressing the endodermal markers SOX17 and FOXA2. Following is a detailed protocol that is based on these prior studies. This protocol will effectively promote DE formation of any hPSC line, however, it should be noted that efficiencies will vary between individual cell lines.

## 14.3.1 *Passaging of hPSCs*

**Coating Plates with Matrigel Matrix**

1. Thaw undiluted Matrigel on ice;
2. Make a 1:100 dilution of Matrigel in cold Knockout DMEM. The working solution of Matrigel should be maintained on ice and can be stored at 2–8 °C for up to 2 weeks;

   ***Note:*** *The optimal dilution of Matrigel will depend on the lot number: refer to manufacturer's instructions.*
3. Coat the desired number of plates with Matrigel (approximately 7 ml per 10 cm plate and 0.5 ml per well of a 24-well plate) and incubate for 2 h at room temperature, or for 30 min at 37 °C;
4. Aspirate the Matrigel solution and wash once with PBS immediately prior to plating the cells;

**Cell Dissociation**

5. Warm the E8 medium and Accutase solution in a 37 °C water bath;
6. Aspirate medium from the growing hPSC culture and gently wash with sterile PBS. Add Accutase cell detachment solution to each plate or well (5 ml per 10 cm culture dish). Incubate at 37 °C for 3–5 min or until cells begin to detach from the plate, as observed under a standard inverted light microscope. Gently tap the sides of the plate against a solid surface to ensure complete cell dissociation. If using cells from multiple plates, work in small batches (1–2 plates at a time) to minimize the amount of time cells are incubated with Accutase
7. Add 5 ml of E8 medium to the plate. Using a 10 ml serological pipette, gently pipette over the plate until all cells have detached;

8. Gently triturate the cell suspension until all noticeable cell clumps are broken up;
9. Transfer the cell suspension to a 15 ml centrifuge tube;
10. Use 9 μl of the cell suspension and 1 μl of Trypan Blue on a hemocytometer to determine the concentration of viable cells;
11. Centrifuge the cell suspension at 200 × *g* for 4 min;
12. Carefully aspirate the medium from the cell pellet;

**Cell Plating**

13. Resuspend the cell pellet in the appropriate amount of E8 medium to produce a final cell concentration of $1.65 \times 10^6$ cells/ml;
14. Add ROCKi to the cell suspension to a final concentration of 10 μM;

    ***Note:*** *ROCKi is only added during passaging to aid in hPSC survival.*

15. Plate the cell suspension (step 13) 1:10 to obtain a density of $3.0 \times 10^4$ cells/cm$^2$. Add 9 ml E8 medium to the prepared Matrigel-coated 10-cm plates. To this, add 1 ml of the cell suspension (step 13). Alternatively, for a 6-well plate, add 300 μl of the cell suspension to 2.7 ml E8 medium per well. For a 24-well plate, add 50 μl of the cell suspension to 450 μl E8 medium per well;
16. Place the plate in the $CO_2$ incubator. Gently move the plate right-to-left and back-and-forth (do not swirl) to disperse the cells evenly across the culture surface;
17. Incubate overnight at 37 °C and 5 % $CO_2$;
18. The next day, each plate should contain tightly packed colonies of cells. Change the medium daily by aspirating the old medium and adding 10 ml E8 medium;
19. After 72 h of growth, the cells should be 70 % confluent (Fig. 14.2) and ready for endoderm induction. Monitor the cell density using an inverted light microscope.

**Note:** *As an alternative to plating cells and initiating differentiation 72 h later as described above, cells can be plated at a higher density ($2.1 \times 10^5$ cells/cm$^2$) with differentiation initiated the following day. However, this method generally leads to lower efficiency of DE differentiation.*

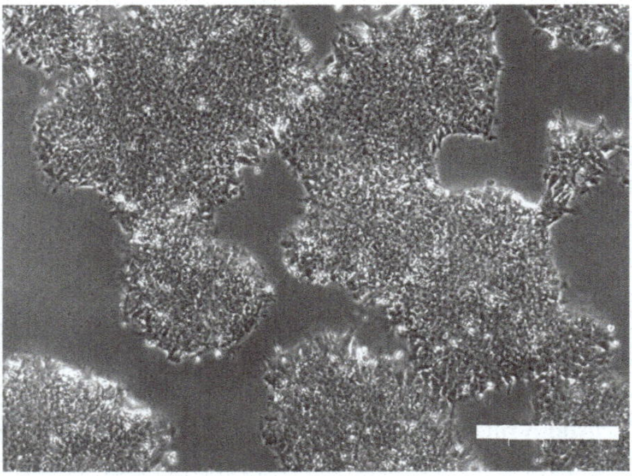

**Fig. 14.2** Human pluripotent stem cells. A phase contrast image depicting the morphology and density of hPSCs (H9/WA09) prior to the induction towards definitive endoderm. Scale bar = 200 μm

### 14.3.2 Differentiation to Definitive Endoderm Using Growth Factors

***Note:*** *This protocol is for the differentiation of a 10 cm dish of hPSCs at a density of 70 % confluence. Adjust media volumes accordingly if smaller or larger culture dishes are used.*

**Day 1 of Differentiation**

1. Prepare 15 ml Day 1 Differentiation Medium: 15 ml RPMI, 150 μl GlutaMAX (100×), and 15 μl Activin-A (100 μg/ml);
2. Warm *Day 1* Differentiation Medium in a 37 °C water bath;
3. Remove *Day 1* Differentiation Medium from water bath;
4. Aspirate the E8 medium from the cells (from *Sect. 14.3.1, step 19*);
5. Gently wash cells once with 10 ml PBS;
6. Add the prewarmed 15 ml *Day 1* Differentiation Medium (from *step 3*) to the cells;
7. Add 5 μl Wnt3a (100 μg/ml stock) and swirl the plate gently to mix. Note that Wnt3a protein is highly labile when diluted into medium, and it is therefore recommended to add Wnt3a directly to the cells;
8. Place the cells in the $CO_2$ incubator and incubate overnight at 37 °C and 5 % $CO_2$. Note that some cell death will be apparent after the first 24 h;

**Day 2 and 3 Differentiation**

9. The next day, prepare 10 ml *Day 2* Differentiation Medium: 10 ml RPMI, 100 μl GlutaMAX, 200 μl FBS and 10 μl 100 ng/ml Activin-A;
10. Warm *Day 2* Differentiation Medium in a 37 °C water bath;
11. Aspirate *Day 1* Differentiation Medium from cells and add *Day 2* Differentiation Medium;
12. Repeat steps 9–11 the next day. Note that the same Differentiation Medium is used on *Days 2* and *3*, however, the medium should be prepared fresh the day of feeding.
13. On the following day (72 h of differentiation) cells are ready for analysis (see *Sect. 14.4* below) or for further differentiation to downstream cell types (please refer to other publications).

### 14.3.3 Differentiation to Definitive Endoderm in Defined Conditions

The protocol described in *Sect. 14.3.2* requires Matrigel, an undefined cell matrix composed of multiple extracellular matrix proteins (ECMPs) with high degrees of batch-to-batch variability. This variability is one factor that can significantly affect the efficiency of DE differentiation. In a previous study, we employed a cellular microarray screening platform to develop a fully defined matrix composed of two

ECMPs, Fibronectin (FN) and Vitronectin (VTN), that supports the efficient and uniform differentiation of multiple hESC lines, H9, HUES1, and HUES9 to DE {Brafman, 2013 #5}. Similar to Matrigel, this defined and optimized matrix in the presence of medium supplemented with Wnt3a and Activin A promotes highly efficient DE differentiation. Here, we describe the protocol using this defined matrix.

***Note:*** *This protocol is for the differentiation of a 10 cm dish of hPSCs at a density of 70 % confluence. Adjust media volumes accordingly if smaller or larger culture dishes are used.*

**Coating Plates with FN-VTN Matrix**

1. Keeping all solutions on ice, prepare the FN-VTN Matrix solution: 7 ml pre-chilled PBS, 70 μl FN (1 mg/ml stock), 70 μl VTN (1 mg/ml stock);
2. Immediately coat a 10-cm plate with the FN-VTN Matrix solution and incubate for 2 h at 37 °C. Alternatively, plates can be coated overnight at 4 °C. Coated plates can be stored for up to 2 weeks: seal them with parafilm to avoid evaporation and keep at 4 °C;
3. Remove the FN-VTN matrix solution by aspiration immediately prior to plating the cells;
4. Dissociate and plate cells as described in *Sect. 14.3.1, steps 5–15*;
5. Seed $1.65 \times 10^6$ cells in 10 ml E8 medium onto the 10-cm FN-VTN-coated dish to obtain a density of $3.0 \times 10^4$ cells/cm$^2$;
6. Two to three days after seeding, cells should be at 70 % confluence. See Fig. 14.2 for representative images of cells at the appropriate cell density;
7. Perform DE differentiation as described in *Sect. 14.3.2*.

### *14.3.4 Differentiation to Definitive Endoderm Using Small Molecules*

Small molecules have been extensively studied for their ability to induce endoderm formation from hPSCs (Borowiak et al. 2009; Chen et al. 2009; Hoveizi et al. 2014; Takeuchi et al. 2014; Zhu et al. 2009). The use of small molecules offers a robust and scalable method to efficiently and reproducibly direct hPSCs toward a desired fate. In the case of DE differentiation, the activity of Wnt3a protein can be mimicked with GSK-3β inhibitors (GSK3i) (Bone et al. 2011; Naujok et al. 2014). These GSK3i, such as CHIR98014 and CHIR99021, can be more carefully dosed than Wnt3a protein, which is highly labile in serum-free media at dilute concentrations. Low GSK3i concentrations promoted endodermal and mesodermal differentiation, while high GSK3i concentrations restricted the differentiation to mesodermal tissues (Bone et al. 2011). Likewise, we observed that the GSK3i CHIR99021 (CHIR) in conjunction with Activin A promoted DE formation in a dose dependent manner as monitored by CXCR4 expression (Fig. 14.3a) with an intermediate dose of 50–100 nM yielding the highest efficiencies of DE differentiation. At this GSK3i concentration, DE differentiation was more effective than with the optimal dose of

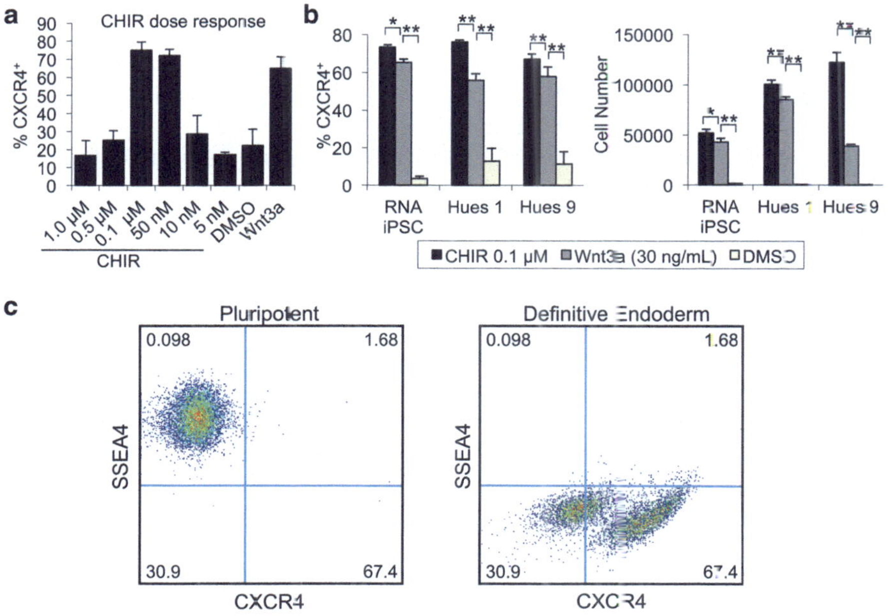


**Fig. 14.3** Differentiation to definitive endoderm using small molecules. (**a**) A dose response of CHIR99021 (CHIR) indicates that intermediate CHIR doses induce the highest percentage of cells expressing CXCR4 expression, as monitored by flow cytometry. This analysis was performed on HUES9 cells at 72 h after endoderm induction. (**b**) The optimal dose of CHIR, 0.1 μM, was tested across three different cell lines and assayed for CXCR4 expression and cell number by flow cytometry. Cell numbers indicated are for one well of a 24-well plate. Wnt3a at 30 ng/ml was used as a positive control and DMSO was used as a negative control. (**c**) Scatter plots showing flow cytometric analysis of undifferentiated HUES9 (*left panel*) and of HUES9 cells differentiated to DE (*right panel*). Cells were stained with SSEA4 and CXCR4. $^{*}p<0.05$, $^{**}p<0.005$

Wnt3a protein (Fig. 14.3b). Notably, CXCR4 expression does not overlap with markers of pluripotency, such as SSEA4 (Fig. 14.3c). These studies demonstrate the utility of small molecules for *in vitro* differentiation protocols, such as DE derivation. Following we describe a modified protocol that utilizes the GSK3i CHIR99021 rather than Wnt3a for the induction of DE from undifferentiated hPS cells. Other GSK3i can likely be substituted for CHIR99021 (e.g. CHIR98014, 6-Bromoindirubin-3′-oxime [BIO], LiCl), however the optimal concentration will have to be determined empirically.

***Note:*** *This protocol is for the differentiation of a 10 cm dish of hPSCs at a density of 70 % confluence. Adjust media volumes accordingly if smaller or larger culture dishes are used.*

- Perform *Passaging of hPSCs,* Cell Dissociation and Cell Plating as described above, *Sect. 14.3.1.*

**Day 1 of Differentiation**

1. Prepare 15 ml *Day 1* Differentiation Medium: 15 ml RPMI, 150 µl GlutaMAX, 0.6 µl CHIR (stock 250 µM), and 15 µl Activin A (stock 100 µg/ml);
2. Warm Day 1 Differentiation Medium in a 37 °C water bath;
3. Aspirate the E8 medium from the 10 cm plate previously seeded with hPSCs (as described in *Sects. 14.3.1* and *14.3.3*);
4. Gently wash the cells once with 10 ml PBS;
5. Add the prewarmed 15 ml *Day 1* Differentiation Medium (from *step 2*) to the cells;
6. Place the cells in the $CO_2$ incubator and incubate overnight at 37 °C and 5 % $CO_2$. ***Note:*** *Some cell death will occur in the first 24 h.*
7. Continue differentiation as described in *Sect. 14.3.2, steps 9–13*.

## 14.4 Characterization of Definitive Endoderm

The DE population generated as described above expresses several defining genes, which can be monitored by reverse transcription quantitative PCR (qPCR). In addition, DE will express marker proteins that can be detected with antibodies by immunofluorescence (IF) or flow cytometry (FC). Below, we describe methods to assess the efficiency of DE differentiation, both at the population level (qPCR) and at the single cell level (IF and FC).

### *14.4.1 Population-Level Characterization*

The transcriptome of hPSCs changes dramatically during DE differentiation (Wang et al. 2011, 2012). As hPSCs exit the pluripotent state, markers of pluripotency, such as *NANOG*, *SOX2* and *POU5F1* (*OCT4*), are potently downregulated (Fig. 14.4a, b). *T* (*T brachyury transcription factor*) is potently upregulated at 24 h and is subsequently downregulated in nascent DE (Fig. 14.4c). *SOX17*, *FOXA2*, *CXCR4* and *GSC* are among the genes that are potently activated and their expression is commonly monitored to confirm DE differentiation (Fig. 14.4d). Determining expression of DE and pluripotency genes by qPCR serves as a reliable readout to confirm the presence of cells with properties of DE. Since differentiated cell populations are often not uniform, it is advisable to monitor expression of genes associated with the other germ layers: *T* (Fig. 14.4c), *MESP1* and *MEOX1* for mesoderm and *SOX1*, *PAX6* and *NES* for ectoderm (not shown). Lack of expression of these genes serves as an indicator of the relative purity of DE. The following protocol outlines steps to monitor gene expression by qPCR. Table 14.1 provides a list of primers to detect genes of interest.

1. Isolate total RNA from undifferentiated hPS cells and DE (cell culture at *step 13* of *Sect. 14.3.2*). Several kits for total RNA extraction are available, for example TRIzol Reagent (ThermoFisher Scientific);

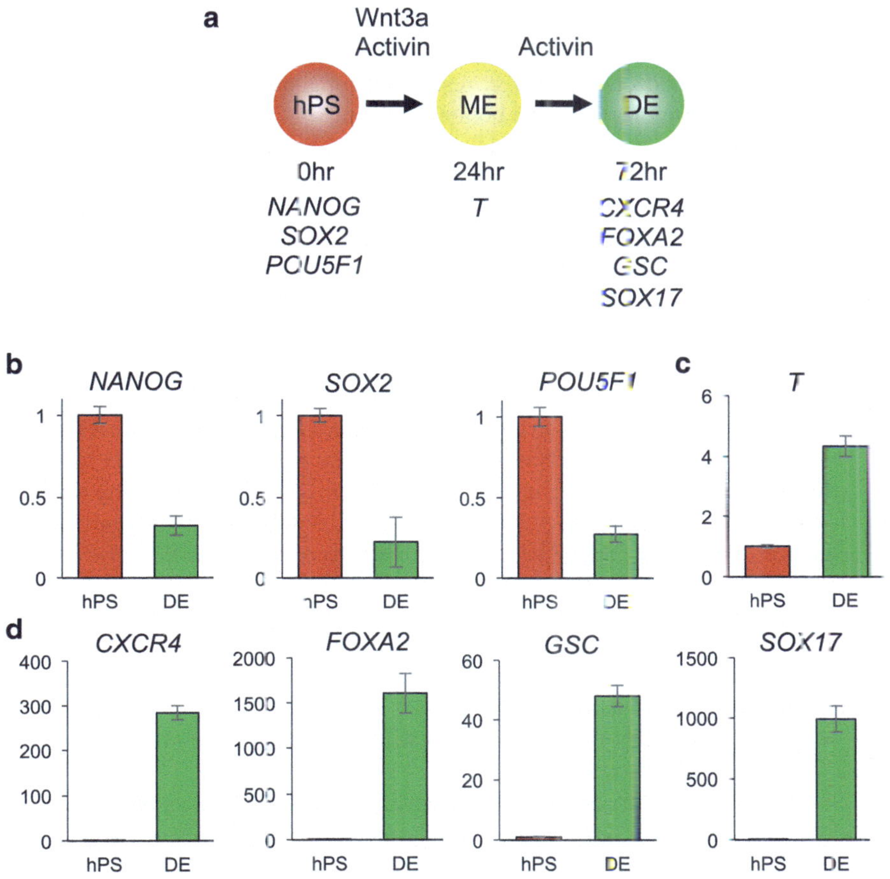


**Fig. 14.4** Characterization of hPSCs and DE by qPCR. (**a**) A schematic depicting the differentiation scheme used to generate DE cells from hPSCs. (**b**) qPCR analysis of pluripotency markers *NANOG*, *SOX2* and *POU5F1* (*OCT4*). The expression of these genes is potently downregulated in DE relative to undifferentiated hPSCs. (**c**) qPCR analysis of *T* (Brachyury). This marker is highly expressed in the intermediate mesendodermal cell population and subsequently maintains some, albeit lower, expression in DE. (**d**) qPCR analysis of DE markers *CXCR4, FOXA2, GSC* and *SOX17*. The expression of these genes is highly elevated in DE relative to undifferentiated hPSCs

2. Perform reverse transcription to obtain cDNA from RNA samples. Several reverse transcription kits are available, for example High-Capacity RNA-to-cDNA Kit (ThermoFisher Scientific);
3. Evaluate the cDNA concentration and quality using a UV spectrophotometer (ThermoFisher Scientific NanoDrop or equivalent). For high quality cDNA, the $A_{260}/A_{280}$ should be between 1.7 and 1.9. Use 10 ng cDNA per 20-μl amplification reaction. Use the same amount of cDNA for each reaction;

*Note:* *cDNA can be stored at −20 °C. Make smaller aliquots to avoid repeated freeze-thaw cycles.*

4. Thaw primers and cDNA on ice, resuspend by gentle vortexing, and briefly centrifuge to bring liquid to the bottom of the tube;
5. Calculate the number of reactions. It is recommended that you perform four technical replicates per sample/gene. For normalization purposes, an endogenous control gene with minimal expression changes across the differentiation is included. We commonly use 18S (Table 14.1). For further information on accurate normalization of qPCR data the reader should consult the literature (e.g. Vandesompele et al. 2002). For example, for an analysis of two samples (DE and pluripotent) and five genes (2 DE genes, 2 pluripotency genes, 1 normalization gene) with four technical replicates, a total of 40 reactions need to be performed;
6. Each reaction will contain 10 μl 2× SYBR Green Master Mix (ThermoFisher Scientific, Cat. No. A25741), 4 μl cDNA template (10 ng), 1 μl Forward Primer, 1 μl Reverse Primer, 4 μl RNase-free water. Make a primer mastermix for each primer set instead of setting up each reaction individually by pipetting the following into a nuclease-free 1.5 ml microcentrifuge tube: 1.1 × (total number of reactions per primer set) × (10 μl 2× SYBR Green Master Mix, 1 μl Forward Primer, 1 μl Reverse Primer, and 4 μl RNase-free water);
7. Cap the tubes and mix by gentle inversion. Centrifuge the tubes briefly to spin down the contents and eliminate any air bubbles;
8. Load the plate. Transfer 16 μl of appropriate primer mastermix into each well of a 48-, 96-, or 384-well optical plate;
9. Add the sample cDNA template. Pipet 4 μl cDNA template (10 ng) into the appropriate PCR reaction;
10. Seal the plate with an optical adhesive cover. Centrifuge the plate briefly to spin down the contents. Load the plate into the Real-Time PCR Detection System;
11. Run the instrument according to manufacturer's recommendations;
12. Analyze gene expression using software provided with the Real-Time PCR Detection System (VanGuilder et al. 2008).

### 14.4.2 Single-Cell-Level Characterization: Flow Cytometry (FC)

Although qPCR analysis serves as a reliable indication of DE formation, this method does not provide information on the percentage of cells that have acquired the desired properties. To monitor for the presence of DE markers at the single-cell level and quantify the percentage of DE cells within a given mixed population, individual cells can be labeled with antibodies and detected using flow cytometry (FC) or immunofluorescence (IF). Useful reagents for this type of analysis include antibodies to CXCXR4, FOXA2 and SOX17. *CXCR4* encodes a alpha-chemokine receptor specific for stromal-derived-factor-1 (SDF-1) (Yasunaga et al. 2005), and its cell surface expression, which is specific to DE and not primitive/visceral endoderm (D'Amour et al. 2005), can be detected on live cells. This is especially useful for enrichment of CXCR4-positive DE cells using flow cytometry based cell sorting. In contrast, the detection of *FOXA2* and *SOX17*, both of which encode transcription factors localized to the nucleus, requires fixation and permeabilization prior to antibody labeling. The following protocol is for detection of CXCR4 on live cells.

1. Dissociate DE cells from step 13 of Sect. 14.3.2 into single cells as described in *Sect. 14.3.1 steps 5–9*;
2. Centrifuge the cell suspension at 200 × *g* for 4 min.
3. Carefully aspirate the medium from the cell pellet;
4. Gently resuspend cell pellet with 2 ml cold PBS, then centrifuge as above;
5. Gently resuspend the cells in 0.5 ml flow cytometry buffer (PBS, 1 % bovine serum albumin, 0.1 % $NaN_3$);
6. Determine cell concentration using a hemocytometer.
7. Dispense $1 \times 10^5$ to $1 \times 10^7$ cells into Falcon tubes on ice. Adjust volume to 100 μl with flow cytometry buffer;
8. Add 20 μl of APC-conjugated CXCR4 antibody or isotype control antibody;
9. Incubate the mixture for 30 min on ice protected from light;
10. Centrifuge the cell suspension at 200 × *g* for 4 min, 4 °C;
11. Carefully aspirate the medium from the cell pellet;
12. Resuspend the cells in 100–500 μl cold flow cytometry buffer.
13. Analyze labeled cells using a BD FACSCanto or equivalent.

***Note:*** *This same protocol can be used to isolate and purify CXCR4-positive cells on a suitable instrument (e.g. BD FACSAria).*

### 14.4.3 Single-Cell-Level Characterization: Immunofluorescence (IF)

An alternative method to FC for single cell resolution is IF. An advantage of IF is that it does not require dissociation, which may damage epitopes used for detection. In addition, this method also provides information on the sub-cellular localization of the protein. Generally, cells are fixed prior to labeling with primary antibodies Although fixation is not necessary if analysis is performed immediately and epitopes are on the cell surface. Direct immunofluorescence uses a primary antibody directly conjugated to a fluorescent dye while indirect immunofluorescence uses a fluorochrome-tagged secondary antibody to detect the primary antibody.

Immunolocalization of transcription factors FOXA2 and SOX17 are good measurements for DE formation. It should be noted that *FOXA2* is also expressed in axial mesoderm, and the co-expression of the two transcription factors FOXA2 and SOX17 is most indicative of the DE phenotype. Co-staining for other proteins, the expression of which is down-regulated in DE, such as *POU5F1* (*OCT4*), is useful to corroborate the efficiency of differentiation (Fig. 14.5). Ki-67, a marker of

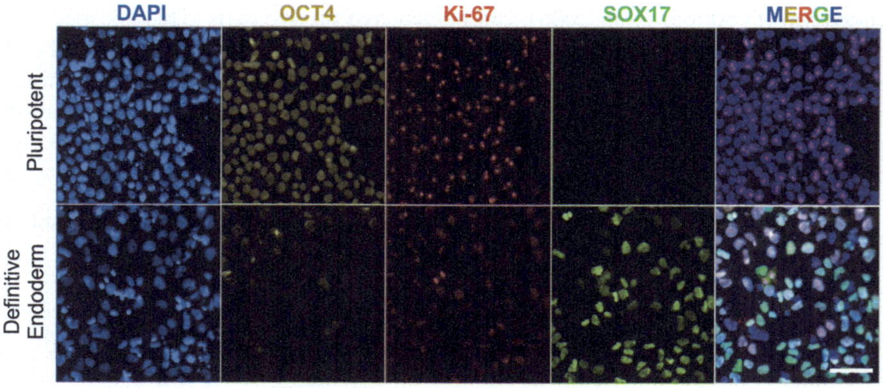


**Fig. 14.5** Immunofluorescence of pluripotent stem cells and DE cells. Fixed and permeabilized HUES9 cells (top row = undifferentiated hPS cells; bottom row = definitive endoderm) were stained for DAPI, OCT4, Ki-67, and SOX17. These images illustrate that the proliferative marker Ki-67 is significantly reduced in DE relative to undifferentiated cells. In addition, residual OCT4 staining observed in DE does not overlap significantly with high SOX17 labeling. Scale bar = 25 μm

proliferating cells, identifies cells that are actively cycling among the differentiated population; this stain should co-localize with OCT4-positive cells and to a lesser extent with DE cells (SOX17- or FOXA2-positive).

The following protocol is for the detection of nuclei (DAPI), SOX17, OCT4 and Ki-67, as shown in Fig. 14.5. Note that for optimal IF conditions, cells are generally plated on glass slides. For the purposes of this protocol, it is sufficient to fix and stain cells directly in the cell culture dish, rather than replating them onto glass slides. One important consideration is that the microscope used for IF can be configured to hold the cell culture dish.

***Note:*** *This protocol is for DE differentiation of hPSCs in a 24-well plate. Adjust volumes accordingly if smaller or larger culture surfaces are used.*

1. Aspirate the medium from the differentiated cells (*Sect. 14.3.2, step 13*);
2. Gently wash cells three times with 0.5 ml PBS;
3. Fix the cells by adding 0.5 ml 4 % paraformaldehyde/PBS and incubate for 10 min;
4. Aspirate the fixative and then wash each well three times with 0.5 ml PBS, 5 min per wash;
5. Add 0.5 ml Blocking Solution (PBS, 2 % BSA) for 1 h at room temperature;
6. Aspirate the Blocking Solution and add the primary antibody(ies) diluted 1:200 in 200 μl Blocking Solution. As a negative control, to one of the wells add only Blocking Solution without a primary antibody;
7. Incubate the cells with the primary antibody for 1 h at room temperature or overnight at 4 °C. Be sure to include a negative control sample incubating in Blocking Solution without primary antibody;
8. Remove the antibody solution and wash each well with PBS three times, 5 min per wash;

**Table 14.2** Primary and secondary antibody combinations for immunofluorescence

| Primary antibody/suggested vendor | Secondary antibody/suggested vendor |
|---|---|
| Polyclonal Rabbit POU5F1 (Abcam®) | Goat anti-Rabbit IgG (H+L) Secondary Antibody, Cy3 conjugate (ThermoFisher Scientific) |
| Polyclonal Mouse Ki-67 (Cell Signaling Technology®) | Donkey anti-Mouse IgG (H+L) Secondary Antibody, Alexa Fluor® 647 conjugate (ThermoFisher Scientific) |
| Polyclonal Goat SOX17 (R&D Systems) | Donkey anti-Goat IgG (H+L) Secondary Antibody, Alexa Fluor® 488 conjugate (ThermoFisher Scientific) |

9. Add the appropriate secondary antibodies (see Table 14.2 for reference) diluted 1:200 in 200 μl Blocking Solution and incubate at room temperature protected from light for 45 min;
10. Prepare working solution of DAPI by diluting 1 mg/ml stock 1:500 in PBS. Add 200 μl DAPI working concentration to the secondary antibody solution in each well to stain nuclei and incubate at room temperature protected from light for an additional 10 min;
11. Remove the secondary antibody/DAPI solution and wash each well three times with PBS, 5 min per wash. Be sure to protect samples from light;
12. Add 0.5 ml PBS per well;
13. Visualize stained cells using an inverted fluorescence microscope.

## References


Ameri J, Stahlberg A, Pedersen J, Johansson JK, Johannesson MM, Artner I, Semb H (2010) FGF2 specifies hESC-derived definitive endoderm into foregut/midgut cell lineages in a concentration-dependent manner. Stem Cells 28:45–56

Ang SL, Wierda A, Wong D, Stevens KA, Cascio S, Rossant J, Zaret KS (1993) The formation and maintenance of the definitive endoderm lineage in the mouse: involvement of HNF3/forkhead proteins. Development 119:1301–1315

Beddington RS, Smith JC (1993) Control of vertebrate gastrulation: inducing signals and responding genes. Curr Opin Genet Dev 3:655–661

Bone HK, Nelson AS, Goldring CE, Tosh D, Welham MJ (2011) A novel chemically directed route for the generation of definitive endoderm from human embryonic stem cells based on inhibition of GSK-3. J Cell Sci 124:1992–2000

Borowiak M, Maehr R, Chen S, Chen AE, Tang W, Fox JL, Scheiber SL, Melton DA (2009) Small molecules efficiently direct endodermal differentiation of mouse and human embryonic stem cells. Cell Stem Cell 4:348–358

Brafman DA, Moya N, Allen-Soltero S, Fellner T, Robinson M, McMillen ZL, Gaasterland T, Willert K (2013a) Analysis of SOX2-expressing cell populations derived from human pluripotent stem cells. Stem Cell Rep 1:464–478

Brafman DA, Phung C, Kumar N, Willert K (2013b) Regulation of endodermal differentiation of human embryonic stem cells through integrin-ECM interactions. Cell Death Differ 20:369–381

Cai J, Zhao Y, Liu Y, Ye F, Song Z, Qin H, Meng S, Chen Y, Zhou R, Song X, Guo Y, Ding M, Deng H (2007) Directed differentiation of human embryonic stem cells into functional hepatic cells. Hepatology 45:1229–1239

Chen S, Borowiak M, Fox JL, Maeh R, Osafune K, Davidow L, Lam K, Peng LF, Scheiber SL, Rubin LL, Melton D (2009) A small molecule that directs differentiation of human ESCs into the pancreatic lineage. Nat Chem Biol 5:258–265

Chen G, Gulbranson DR, Hou Z, Bolin JM, Ruotti V, Probasco MD, Smuga-Otto K, Howden SE, Diol NR, Propson NE, Wagner R, Lee GO, Antosiewicz-Bourget J, Teng JM, Thomson JA (2011) Chemically defined conditions for human iPSC derivation and culture. Nat Methods 8:424–429

Cheng X, Ying L, Lu L, Galvao AM, Mills JA, Lin HC, Kotton DN, Shen SS, Nostro MC, Choi JK, Weiss MJ, French DL, Gadue P (2012) Self-renewing endodermal progenitor lines generated from human pluripotent stem cells. Cell Stem Cell 10:371–384

Conlon FL, Lyons KM, Takaesu N, Barth KS, Kispert A, Herrmann B, Robertson EJ (1994) A primary requirement for nodal in the formation and maintenance of the primitive streak in the mouse. Development 120:1919–1928

D'Amour KA, Agulnick AD, Eliazer S, Kelly OG, Kroon E, Baetge EE (2005) Efficient differentiation of human embryonic stem cells to definitive endoderm. Nat Biotechnol 23:1534–1541

D'Amour KA, Bang AG, Eliazer S, Kelly OG, Agulnick AD, Smart NG, Moorman MA, Kroon E, Carpenter MK, Baetge EE (2006) Production of pancreatic hormone-expressing endocrine cells from human embryonic stem cells. Nat Biotechnol 24:1392–1401

Faust C, Magnuson T (1993) Genetic control of gastrulation in the mouse. Curr Opin Genet Dev 3:491–498

Haegel H, Larue L, Ohsugi M, Fedorov L, Herrenknecht K, Kemler R (1995) Lack of beta-catenin affects mouse development at gastrulation. Development 121:3529–3537

Hoveizi E, Khodadadi S, Tavakol S, Karima O, Nasiri-Khalili MA (2014) Small molecules differentiate definitive endoderm from human induced pluripotent stem cells on PCL scaffold. Appl Biochem Biotechnol 173:1727–1736

Kadzik RS, Morrisey EE (2012) Directing lung endoderm differentiation in pluripotent stem cells. Cell Stem Cell 10:355–361

Kelly OG, Pinson KI, Skarnes WC (2004) The Wnt co-receptors Lrp5 and Lrp6 are essential for gastrulation in mice. Development 131:2803–2815

Kroon E, Martinson LA, Kadoya K, Bang AG, Kelly OG, Eliazer S, Young H, Richardson M, Smart NG, Cunningham J, Agulnick AD, D'Amour KA, Carpenter MK, Baetge EE (2008) Pancreatic endoderm derived from human embryonic stem cells generates glucose-responsive insulin-secreting cells in vivo. Nat Biotechnol 26:443–452

Kubo A, Shinozaki K, Shannon JM, Kouskoff V, Kennedy M, Woo S, Fehling HJ, Keller G (2004) Development of definitive endoderm from embryonic stem cells in culture. Development 131:1651–1662

Liu P, Wakamiya M, Shea MJ, Albrecht U, Behinger RR, Bradley A (1999) Requirement for Wnt3 in vertebrate axis formation. Nat Genet 22:361–365

Longmire TA, Ikonomou L, Hawkins F, Chistodoulou C, Cao Y, Jean JC, Kwok LW, Mou H, Rajagopal J, Shen SS, Dowton AA, Serra M, Weiss DJ, Green MD, Snoeck HW, Ramirez MI, Kotton DN (2012) Efficient derivation of purified lung and thyroid progenitors from embryonic stem cells. Cell Stem Cell 10:398–411

Naujok O, Diekmann U, Lenzen S (2014) The generation of definitive endoderm from human embryonic stem cells is initially independent from activin A but requires canonical Wnt-signaling. Stem Cell Rev 10:480–493

Pagliuca FW, Millman JR, Gurtler M, Segel M, Van Dervort A, Ryu JH, Peterson QP, Greiner D, Melton DA (2014) Generation of functional human pancreatic beta cells in vitro. Cell 159:428–439

Parent AV, Russ HA, Khan IS, LaFlam TN, Metzger TC, Anderson MS, Hebrok M (2013) Generation of functional thymic epithelium from human embryonic stem cells that supports host T cell development. Cell Stem Cell 13:219–229

Sasaki H, Hogan BL (1993) Differential expression of multiple fork head related genes during gastrulation and axial pattern formation in the mouse embryo. Development 118:47–59
Sato N, Meijer L, Skaltsounis L, Greengard P, Brivanlou AH (2004) Maintenance of pluripotency in human and mouse embryonic stem cells through activation of Wnt signaling by a pharmacological GSK-3-specific inhibitor. Nat Med 10:55–63
Spence JR, Mayhew CN, Rankin SA, Kuhar MF, Vallance JE, Tolle K, Hoskins EE, Kalinichenko VV, Wells SI, Zorn AM, Shoyer NF, Wells JM (2011) Directed differentiation of human pluripotent stem cells into intestinal tissue in vitro. Nature 470:105–109
Su M, Hu R, Jin J, Yan Y, Song Y, Sullivan R, Lai L (2015) Efficient in vitro generation of functional thymic epithelial progenitors from human embryonic stem cells. Sci Rep 5:9882
Takeuchi H, Nakatsuji N, Suemori H (2014) Endodermal differentiation of human pluripotent stem cells to insulin-producing cells in 3D culture. Sci Rep 4:4488
Tam PP, Behinger RR (1997) Mouse gastrulation: the formation of a mammalian body plan. Mech Dev 68:3–25
Touboul T, Hannan NR, Corbineau S, Martinez A, Martinet C, Branchereau S, Mainot S, Strick-Marchand H, Pedersen R, Di Santo J, Weber A, Vallier L (2010) Generation of functional hepatocytes from human embryonic stem cells under chemically defined conditions that recapitulate liver development. Hepatology 51:1754–1765
Vandesompele J, De Preter K, Pattyn F, Poppe B, Van Roy N, De Paepe A, Speleman F (2002) Accurate normalization of real-time quantitative RT-PCR data by geometric averaging of multiple internal control genes. Genome Biol 3:RESEARCH0034
VanGuilder HD, Vrana KE, Freeman WM (2008) Twenty-five years of quantitative PCR for gene expression analysis. Biotechniques 44:619–626
Vincent SD, Dunn NR, Hayashi S, Norris DP, Robertson EJ (2003) Cell fate decisions within the mouse organizer are governed by graded Nodal signals. Genes Dev 17:1646–1662
Wang P, Rodriguez RT, Wang J, Ghodasara A, Kim SK (2011) Targeting SOX17 in human embryonic stem cells creates unique strategies for isolating and analyzing developing endoderm. Cell Stem Cell 8:335–346
Wang P, McKnight KD, Wong DJ, Rodriguez RT, Sugiyama T, Gu X, Ghodasara A, Qu K, Chang HY, Kim SK (2012) A molecular signature for purified definitive endoderm guides differentiation and isolation of endoderm from mouse and human embryonic stem cells. Stem Cells Dev 21:2273–2287
Willert KH (2008) Isolation and application of bioactive Wnt proteins. Methods Mol Biol 468:17–29
Willert K, Nusse R (2012) Wnt proteins. Cold Spring Harb Perspect Biol 4:a007864
Willert K, Brown JD, Danenberg E, Duncan AW, Weissman IL, Reya T, Yates JR 3rd, Nusse R (2003) Wnt proteins are lipid-modified and can act as stem cell growth factors. Nature 423:448–452
Yasunaga M, Tada S, Torikai-Nishikawa S, Nakano Y, Okada M, Jakt LM, Nishikawa S, Chiba T, Era T, Nishikawa S (2005) Induction and monitoring of definitive and visceral endoderm differentiation of mouse ES cells. Nat Biotechnol 23:1542–1550
Zhu S, Wurdak H, Wang J, Lyssiotis CA, Peters EC, Cho CY, Wu X, Schultz PG (2009) A small molecule primes embryonic stem cells for differentiation. Cell Stem Cell 4:416–425

# Chapter 15
# Pancreatic Differentiation from Human Pluripotent Stem Cells

**Nicholas Vinckier, Jinzhao Wang, and Maike Sander**

**Abstract** The ability to produce human pancreatic cells *in vitro* would open new possibilities for developing improved therapies through cell transplantation, disease modeling, and drug screening. Of particular medical importance are the insulin-producing beta cells of the pancreas, which are lost or dysfunctional in diabetes. Furthermore, an *in vitro* model of human exocrine cells could help devise new therapies for pancreatic exocrine disease, most notably pancreatic cancer. In the past decade much progress has been made in developing protocols to generate multipotent pancreatic progenitor cells from human pluripotent stem cells (hPSCs) that are capable of differentiating into both endocrine and exocrine cells. The sole approach that has proven successful is to reproduce essential steps of *in vivo* development *in vitro* through directed step-wise differentiation of hPSCs. The directed differentiation entails sequential exposure of hPSCs to different signaling factors, thereby moving cells through several developmental intermediates towards the pancreatic fate. Upon implantation into mice, hPSC-derived pancreatic progenitor cells spontaneously differentiate into endocrine and exocrine cells. Here, we describe a detailed protocol for the generation of pancreatic progenitor cells from hPSCs. We provide methods for the directed differentiation as well as the characterization of pancreatic progenitor cells and lineage intermediates by immunofluorescence staining and flow cytometry. With recently developed protocols, these pancreatic progenitor cells can be further differentiated *in vitro* into beta-like cells that functionally resemble immature human beta cells.

**Keywords** Human pluripotent stem cell (hPSC) • Human embryonic stem cell (hESC) • Definitive endoderm (DE) • Gut tube (GT) • Posterior foregut (FG) • Pancreatic endoderm (PE) • SOX17 • HNF4A • PDX1 • SOX9

---

N. Vinckier • J. Wang
Departments of Pediatrics and Cellular and Molecular Medicine, Pediatric Diabetes Research Center, University of California San Diego, La Jolla, CA 92093, USA

M. Sander (✉)
Departments of Pediatrics and Cellular and Molecular Medicine, Pediatric Diabetes Research Center, University of California San Diego, La Jolla, CA 92093, USA

Sanford Consortium for Regenerative Medicine,
2880 Torrey Pines Scenic Dr. – Room 3006, La Jolla, CA 92093, USA
e-mail: masander@ucsd.edu



H. Ulrich, P. Davidson Negraes, *Working with Stem Cells*,
DOI 10.1007/978-3-319-30582-0_15

## 15.1 Introduction

The development of protocols for deriving mature cell types from hPSCs has opened opportunities for modeling human disease *ex vivo* as well as for discovery of novel drugs (Merkle and Eggan 2013). In addition, functional cell types derived from hPSCs could provide an unlimited source to replace tissue cells lost in disease through cell transplantation. hPSC-based cell models are particularly promising in the context of diseases where a single cell type is affected. One such example is diabetes, which is characterized by loss or dysfunction of the insulin-producing beta cells in the pancreas (Nathan 2015). There are two major forms of diabetes, type 1 and type 2 diabetes. While the causes of both types of diabetes are distinct, the manifestation is similar: insufficient production of insulin results in inadequate glucose uptake by peripheral tissues, which in turn leads to elevated blood glucose levels. Type 1 diabetes is caused by autoimmune-mediated beta cell destruction and comprises 5 % of diabetes cases (Cnop et al. 2005). The remaining 95 % of diabetes cases are classified as type 2, and arise due to a combination of insufficient insulin production by the beta cells and insulin resistance of insulin target tissues (Johnson and Luciani 2010; Kahn et al. 2006; Stumvoll et al. 2005). Because both type 1 and type 2 diabetes eventually result in beta cell loss, transplantation of beta cells produced *in vitro* from hPSCs could help normalize blood glucose levels in patients with diabetes. In the pancreas, beta cells reside in so-called islets of Langerhans, where beta cells form functional units with other hormone-producing cells involved in blood glucose regulation, most notably the glucagon-producing alpha cells and somatostatin-producing delta cells. Since there is paracrine signaling between the endocrine cell types within the islet (Caicedo 2013), transplantation of functional islets might be clinically more beneficial than transplanting beta cells alone. There is already precedence that transplantation of cadaveric islets can be curative for diabetes, at least in the short term (Meloche 2007; Robertson et al. 2000). hPSC-derived beta cells or islets could obviate both the limited supply of cadaveric islets and the need for immunosuppression owing to the allogeneic origin of cadaveric islets. In fact, the first clinical trial to determine the safety and efficacy of a hPSC-based islet cell replacement therapy is currently ongoing (ViaCyte 2014). The current clinical trial involves encapsulation of human embryonic stem cell (hESC)-derived multipotent pancreatic progenitor cells in small devices that are implanted subcutaneously. Animal studies have provided proof-of principle that these pancreatic progenitor cells will further differentiate into islet-like structures containing all endocrine cell types when implanted under the skin (Kelly et al. 2011; Kroon et al. 2008; Nostro et al. 2015; Rezania et al. 2012; Schulz et al. 2012; Xie et al. 2013).

While beta cell loss is the hallmark of type 1 diabetes, type 2 diabetes, at least in its early stages, is characterized by beta cell dysfunction rather than destruction. Type 2 diabetes has a strong genetic component; yet how genetic risk factors cause beta cell dysfunction remains largely unknown. While mouse models have provided important insight into the regulation of insulin production and secretion, comparisons between mouse models of diabetes and the human disease have also revealed species differences. One condition that exemplifies such differences is a monogenetic form of type 2 diabetes, called maturity onset diabetes of the young (MODY). In humans, inheritance of one mutant allele is sufficient to cause MODY, whereas mice carrying the same heterozygous mutation do not acquire the disease (Haumaitre et al. 2006; Lee et al. 1998; Mayer et al. 2008; Yamagata et al. 1996). *In vitro*-generated beta cells and their precursors would provide a powerful model to study how MODY-associated gene mutations cause diabetes. With recent breakthroughs in genome editing technologies, known mutations, like those seen in MODY, could be introduced into hPSCs and beta cell differentiation and function could be analyzed *in vitro*. Alternatively, induced hPSCs could be derived from MODY patients to create patient-specific disease models. Similar strategies could be implemented to determine how genetic variants, identified via genome-wide association studies, affect beta cell development and/or function and cause disease.

A scalable differentiation system for human beta cells or islets would also provide an ideal drug-screening platform. In addition to enabling new drug discovery, such a system could help customize diabetes treatment in a patient-specific manner. It is known that not all patients with type 2 diabetes respond similarly to existing anti-diabetic drugs (Standl and Fuchtenbusch 2003). Responsiveness to different drugs could be tested by deriving induced hPSCs from patients and producing patient-specific beta cells or islets *in vitro*. In the long term, the collective data from such experiments might allow us to predict drug responsiveness based on the absence or presence of specific genetic markers, as already used to predict responsiveness to cancer therapy.

The past decade has seen remarkable progress in the development of protocols to generate pancreatic cells from hPSCs (Mfopou et al. 2010; Pagliuca and Melton 2013). Multiple differentiation protocols aimed at deriving pancreatic progenitors and beta cells have been published with mounting levels of success. The underlying principle of all successful approaches is to closely mimic embryonic development *in vitro*. Efficacious protocols have taken a step-wise approach in which cells are guided through defined developmental stages by sequential exposure to the growth factors that are also relevant during *in vivo* development. Earlier protocols yielded insulin-expressing cells *in vitro* that were not functional, expressed multiple hormones, and failed to restore blood glucose levels when implanted into diabetic mice (D'Amour et al. 2006; Kelly et al. 2011; Kroon et al. 2008; Nostro et al. 2011; Rezania et al. 2012). In contrast, when pancreatic progenitor cells were implanted into mice, the implants rescued diabetes within ~16 weeks after implantation (Kroon et al. 2008; Nostro et al. 2015; Rezania et al. 2012; Schulz et al. 2012). During this time, the progenitors differentiate into islet-like structures, containing mature functional beta cells as well as other endocrine cell types. A small fraction of pancreatic progenitors also differentiates into exocrine acinar and ductal cells after implantation (Kroon et al. 2008; Xie et al. 2013).

Here, we present a step-wise protocol for the generation of pancreatic progenitor cells by directed differentiation of hESCs. The protocol was adapted from several previously published protocols (Schulz et al. 2012; Rezania et al. 2013, 2014). The protocol is optimized for the H1 hESC line from WiCell (WiCell Research Institute, WA01) and employs a planar culture system in which the cells are adhered to Matrigel-coated tissue culture plates throughout differentiation. Recently, protocols have been published that report strategies to further differentiate hPSC-derived pancreatic progenitor cells into functional beta-like cells entirely *in vitro* (Pagliuca et al. 2014; Rezania et al. 2014; Russ et al. 2015). After implantation into diabetic mice, these beta-like cells lead to normalization of blood glucose levels within 2 weeks.

## 15.2 Materials

### *15.2.1 Cell Culture Supplies and Equipment*

- Biological safety cabinet.
- $CO_2$ incubator capable of maintaining 5 % $CO_2$ and >95 % humidity at 37 °C.
- 37 °C water bath.
- Sterile plastic tissue culture treated cell culture dishes:
  - 6-Well plate; 9.5 $cm^2$ cell growth area (Corning, Cat. No. 3516).
  - 12-Well plate; 3.8 $cm^2$ cell growth area (Corning, Cat. No. 3513).
  - 10 cm plate; 56.5 $cm^2$ cell growth area (Corning, Cat. No. 351029).
- Sterile ultra-low attachment 6-well plates (Corning, Cat. No. 3471), for alternative suspension culture (see *Sect. 15.3.5* and ***Note 5***).
- Cell lifter/scraper (VWR, Cat. No. 3008).
- Hemocytometer or other cell counting tool.
- Sterile 15 ml conical tubes.
- Benchtop centrifuge capable of spinning 15 ml conical tubes.
- Sterile 1.5 ml microcentrifuge tubes.
- Benchtop microcentrifuge capable of spinning 1.5 ml microcentrifuge tubes.
- Serological pipette controller and 10 ml sterile pipettes.
- 2, 20, 200, and 1000 μl pipettes and accompanying sterile pipette tips.
- Stericup-GP, 0.22 μm, polyethersulfone, 500 ml, radio-sterilized vacuum filters (Millipore, SCGPU05RE).
- Falcon 5 ml round bottom polystyrene test tube with cell strainer snap cap (Corning, Cat. No. 352235).
- Inverted light microscope (Zeiss Axio Vert or equivalent).
- Inverted fluorescence microscope (Zeiss Axio Observer or equivalent).
- Flow cytometer (BD FACSCanto™ or equivalent).

### *15.2.2 Reagents*

- Matrigel Growth Factor Reduced Basement Membrane Matrix (Corning, Cat. No. 356231).
- RHO/ROCK Pathway Inhibitor Y-27632 (StemCell, Cat. No. 72307).
- TrypLE™ Express Enzyme (Life Technologies, Cat. No. 12604-021).

- Accutase (eBioscience, Cat. No. 00-4555-56).
- Sterile DPBS (Dulbecco's Phosphate Buffered Saline) without $Ca^{2+}$ and $Mg^{2+}$ (VWR, Cat. No. 45000-434).
- DMEM/F12 50:50 mix without glutamine (VWR, Cat. No. 45000-346).
- Bovine Serum Albumin (BSA) Fraction V (7.5 % solution) (Life Technologies, Cat. No. 15260-037), for alternative suspension culture (see *Sect. 15.3.5* and ***Note 6***).
- Reagent Grade BSA—pH 7.0, >98 % purity (Lampire Biological Laboratories, Cat. No. 7500804), for use in differentiation media.
- H1 hESC line (WiCell Research Institute, WA01). Other hPSC lines may be used, but may require further optimization.
- Essential 8 (E8) pluripotency maintenance medium (Life Technologies, Cat. No. A1517001). The E8 culture medium is a xeno-free pluripotency maintenance medium (Chen et al. 2011) used here for feeder-free growth and expansion of hESCs.
- MCDB 131 cell culture medium (Life Technologies, Cat. No. 10372-019). The MCDB 131 medium is used here as the base medium throughout differentiation and is supplemented with different components and factors depending on the stage of differentiation. See *Sect. 15.2.3* for proper stage-specific base-medium formulations and Table 15.1 for additional factors to be added on each day of differentiation.
- Supplements and factors used for endocrine differentiation:
  - Sodium bicarbonate ($NaHCO_3$) (Sigma, Cat. No. S6297).
  - Recombinant Activin A (AA) (R&D Systems, Cat. No. 338-AC/CF).
  - Recombinant Wnt-3a (R&D Systems, Cat. No. 1324-WN/CF).
  - GlutaMAX (Life Technologies, Cat. No. 35050061).
  - BSA (Fisher, Cat. No. 7500804).
  - L-Ascorbic Acid (Vitamin C, VIT-C) (Sigma, Cat. No. A4544).
  - Recombinant KGF/FGF-7 (R&D Systems, Cat. No. 251-KG/CF).
  - D-Glucose (Fisher Scientific, Cat. No. D161).
  - Retinoic Acid (RA) (Sigma, Cat. No. R2625).
  - LDN193189 (Stemgent, Cat. No. 04-0074).
  - ITS-X (Life Technologies, Cat. No. 51500056).
  - SANT-1 (Sigma, Cat. No. S4572).
  - TPB (EMD Millipore, Cat. No. 565740).
- Cell fixation solution for immunofluorescence (IF) analysis: 4 % paraformaldehyde (PFA) (96 % extra pure, Acros Organics, Cat. No. 41678) in DPBS.
- Cell permeabilization and blocking buffer for IF analysis: 0.15 % Triton X-100 (Fisher, Cat. No. BP151) and 1 % donkey serum (Gemini Bio-Products, Cat. No. 100-151) in DPBS.
- Superfrost Plus microscope slides (Fisher Scientific, Cat. No. 12-550-15).
- VECTASHIELD Antifade mounting medium (Vector Laboratories, Cat. No. H-1000).
- FACS buffer: 0.2 % (w/v) BSA (Lampire, Cat. No. 7500804) in DPBS, keep at 2–8 °C.
- BD fixation/permeabilization solution (BD Biosciences, Cat. No. 554714) for flow cytometry analysis, keep at 2–8 °C.

**Table 15.1** Medium formulations used for each day of differentiation. See *Sect. 15.2.3* for stage-specific base-medium formulations

| Day | Base medium | Added factors |
|---|---|---|
| 1 | Stage 1 medium | 100 ng/ml activin A<br>25 ng/ml Wnt-3a |
| 2–3 | Stage 1 medium | 100 ng/ml activin A |
| 4–5 | Stage 2 medium | 0.25 mM VIT-C<br>50 ng/ml FGF7 |
| 6–7 | Stage 3 medium | 0.25 mM VIT-C<br>50 ng/ml FGF7<br>0.25 μM SANT-1<br>1 μM retinoic acid<br>100 nM LDN193189<br>1:200 ITS-X<br>200 nM TPB |
| 8–10 | Stage 4 medium | 0.25 mM VIT-C<br>2 ng/μl FGF7<br>0.25 μM SANT-1<br>0.1 μM retinoic acid<br>200 nM LDN193189<br>1:200 ITS-X<br>100 nM TPB |

The same base media formulations and factors are used for either adherent cultures or suspension cultures

- BD perm/wash buffer (BD Biosciences, Cat. No. 554723) for flow cytometry analysis, keep at 2–8 °C.

### 15.2.3 Stage-Specific Medium Formulations

The supplements listed below are stable in the MCDB 131 medium for 1 month if kept at 2–8 °C. Because sodium bicarbonate ($NaHCO_3$) and BSA are supplied as a non-sterile powder, the medium must be sterile filtered (0.22 μm vacuum filter) after $NaHCO_3$ and BSA are added.

**Stage 1 Medium**

- MCDB 131 medium.
- 1.5 g/l $NaHCO_3$.
- 1× GlutaMAX.
- 10 mM D-Glucose (final concentration; MCDB 131 already contains 5.55 mM).
- 0.5 % BSA.

**Stage 2 Medium**

- MCDB 131 medium.
- 1.5 g/l $NaHCO_3$.
- 1× GlutaMAX.
- 10 mM D-Glucose (final concentration; MCDB 131 already contains 5.55 mM).
- 0.5 % BSA.

**Stage 3/4 Medium**

- MCDB 131 medium.
- 2.5 g/l $NaHCO_3$.
- 1× GlutaMAX.
- 10 mM D-glucose (final concentration; MCDB 131 already contains 5.55 mM).
- 2 % BSA.

## 15.2.4 Antibodies Used for Cell Characterization

**Primary Antibodies for Flow Cytometry**

- Mouse anti-SOX17-PE (1:20, BD Biosciences, Cat. No. 561591).
- Mouse anti-PDX1-PE (1:20, BD Biosciences, Cat. No. 562161).
- Mouse anti-NKX6.1-Alexa Fluor 647 (1:20, BD Biosciences, Cat. No. 563338).
- Rabbit anti-SOX9 (1:50, Millipore, Cat. No. AB5535).
- Mouse anti-IgG1, κ-PE isotype control (1:20, BD Biosciences, Cat. No. 556650).
- Mouse anti-IgG1, κ-Alexa Fluor 647 isotype control (1:20, BD Biosciences, Cat. No. 557732).

**Secondary Antibodies for Flow Cytometry**

- Donkey anti-rabbit IgG-PE (1:20, eBioscience, Cat. No. 12-4739-81).

**Primary Antibodies for Immunofluorescence**

- Goat anti-SOX17 (1:1000, R&D Systems, Cat. No. AF1924).
- Rabbit anti-OCT4 (1:1000, Cell Signaling Technology, Cat. No. 2840).
- Rabbit anti-HNF4A (1:500, Santa Cruz, Cat. No. SC-6556).
- Rabbit anti-SOX9 (1:1000, Millipore, Cat. No. AB5535).
- Goat anti-PDX1 (1:500, Abcam, Cat. No. ab47383).
- Mouse anti-NKX6.1 (1:300, Developmental Studies Hybridoma Bank, Cat. No. F55A10).

**Secondary Antibodies for Immunofluorescence**

- Donkey anti-goat Cy3 (1:1000, Jackson Immuno, Cat. No. 705-165-147).
- Donkey anti-rabbit Cy3 (1:1000, Jackson Immuno, Cat. No. 711-165-152).
- Donkey anti-goat Alexa Fluor 488 (1:1000, Jackson Immuno, Cat. No. 705-545-147).

## 15.3 Methods

**Cell culture and passaging of hESCs**

Proper handling and preparation of hESCs is important to ensure successful and efficient differentiation. The protocol is optimized for H1 cells, which can be obtained from WiCell Research Institute (WA01). Below are the recommended methods for proper culture and expansion of H1 cells prior to differentiation.

### *15.3.1 Preparation of Matrigel-Coated Tissue Culture Dishes*

Expansion is performed in 6-well plates and differentiation in 12-well plates. If different sizes of culture dishes are desired, adjust volumes as necessary.

1. Thaw Matrigel on ice and keep cold at all times;
2. Following the manufacturer's instructions dilute Matrigel in the appropriate volume of cold DMEM/F12 medium and keep on ice. The dilution used in this protocol is 1:100, but may vary from lot to lot. Unused diluted Matrigel can be stored at 2–8 °C, but should be used within 2 weeks from the time of dilution;
3. Load diluted Matrigel into each well of the tissue culture plate to be used. Ensure there is sufficient volume to completely cover the entire well surface (~2 ml per well of a 6-well plate and ~1 ml per well of a 12-well plate);
4. Incubate the plate at 37 °C for 30 min before use.

### *15.3.2 Culturing hESCs*

1. Aspirate excess Matrigel from pre-coated plates before seeding cells;
2. Thaw frozen cells in 37 °C water bath until just a small amount of ice is left in the vial;
3. Wash cells by resuspending in ~5 ml of E8 medium and spin at $200 \times g$ for 4 min at room temperature;
4. Aspirate supernatant carefully to not disrupt the cell pellet;
5. Resuspend cells in E8 medium (~$5 \times 10^5$ cells/ml) containing freshly added ROCK inhibitor (10 μM);
6. Add 2 ml cell suspension per well of the Matrigel-coated 6-well plate and place in a 37 °C incubator at 5 % $CO_2$;
7. Replace medium with fresh E8 medium (w/o ROCK inhibitor) every 24 h;
8. Cells are passaged at ~80 % confluency.

### *15.3.3 Passaging hESCs with TrypLE*

1. Aspirate medium from wells and wash with DPBS without $Ca^{2+}$ and $Mg^{2+}$;
2. Aspirate DPBS and load 0.5–1 ml of room temperature TrypLE in each well and incubate at 37 °C for 1 min;

3. After 1 min examine cells under a microscope to ensure sufficient detachment from plate. Cells should appear balled up at the edges of colonies, but not free-floating. If more time is required for cell detachment, place cells back in 37 °C and examine under microscope every 1 min until cells are sufficiently detached (see *Note 1*);
4. Stop enzymatic reaction by adding 4 volumes of E8 medium;
5. Use a cell scraper to lift cell clusters off plate;
6. Use a serological pipette to transfer the cell suspension into sterile 15 ml conical tubes;
7. Centrifuge for 4 min at 200 × *g* to pellet cells;
8. Aspirate media and resuspend cells in appropriate volume of E8 medium containing 10 μM ROCK inhibitor (Y-27632) to achieve desired dilution (see *Note 2*);
9. Add 2 ml of cell suspension per well of a Matrigel-coated 6-well plate;
10. Replace medium with fresh E8 medium (w/o ROCK inhibitor) every 24 h.

## *15.3.4 Cell Differentiation*

We describe a step-wise protocol for the directed differentiation of H1 hESCs into pancreatic progenitor cells, which we refer to as pancreatic endoderm (Fig. 15.1). During *in vitro* differentiation, the cells progress through different lineage intermediates, resembling cell populations found in the developing embryo. The first step is the differentiation to definitive endoderm, from which endodermal organs, such as liver, lungs, thymus, intestine, and pancreas are derived (Wells and Melton 1999). In subsequent steps, definitive endoderm cells are directed to acquire primitive gut tube, posterior foregut, and finally pancreatic endoderm identity. Expression of early pancreatic transcription factors, such as PDX1 and SOX9, is observed as early as the end of the

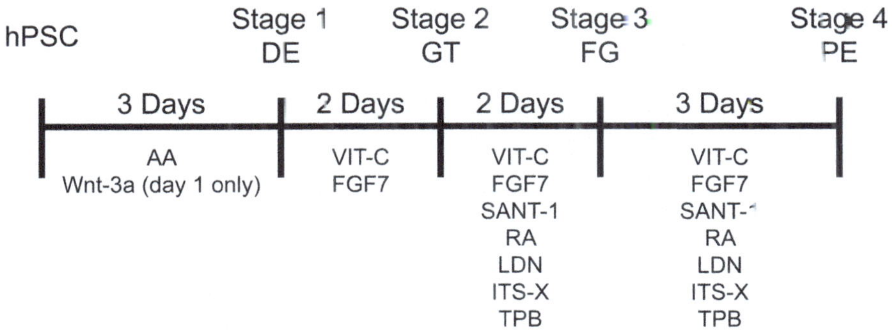


**Fig. 15.1** Schematic outlining the protocol for differentiation of human pluripotent stem cells into pancreatic progenitor cells (pancreatic endoderm). The duration of each differentiation step and the required factors are depicted. Exact concentrations of factors and medium composition can be found in *Sects. 15.2.2* and *15.2.3*. *hPSC* human pluripotent stem cell, *DE* definitive endoderm, *GT* primitive gut tube, *FG* posterior foregut, *PE* pancreatic endoderm, *AA* activin A, *VIT-C* vitamin C, *RA* retinoic acid

posterior foregut stage. When implanted subcutaneously or under the kidney capsule into mice, pancreatic endoderm differentiates into fully functional beta cells, other endocrine cell types, as well as a small percentage of pancreatic ductal cells (Kelly et al. 2011; Kroon et al. 2008; Rezania et al. 2013; Schulz et al. 2012; Xie et al. 2013). The *in vivo* differentiation and maturation process of implanted pancreatic endoderm takes ~16 weeks, regardless of the protocol and cell line used to generate pancreatic endoderm. Although the application of the protocol we describe is not limited to H1 hESCs, when other hPSC lines are used, the protocol needs to be adapted to achieve similar differentiation efficiencies as observed with H1 hESCs. Key variables include the concentration of the factors and the duration of each differentiation step.

#### 15.3.4.1 Preparation of Coverslips for Culturing hESCs

*Optional: If analysis will be performed later, it is necessary to load coverslips into the culture dish wells prior to coating with Matrigel. To prepare coverslips for culturing hESCs follow the procedure below, otherwise continue with the next Section.*

1. Put coverslips into a sterile 10 cm petri dish with lid;
2. Pour enough 70 % ethanol into the dish to completely submerge all coverslips and place the lid on the petri dish;
3. Allow coverslips to soak in 70 % ethanol overnight;
4. The following day, aspirate 70 % ethanol and replace it with 100 % ethanol, ensuring all coverslips are submerged;
5. Allow coverslips to soak for 5 min;
6. Aspirate 100 % ethanol and replace it with fresh 100 % ethanol, again submerging all coverslips;
7. Allow coverslips to soak for 5 min;
8. Place petri dish lid upside down inside the biological safety cabinet;
9. Using sterilized tweezers, remove coverslips from 100 % ethanol and place inside the inverted lid;
10. Lean each coverslip against the inside wall of the lid at an angle so that both sides of the coverslip are exposed to air;
11. Allow the coverslips to dry completely;
12. Once coverslips have dried, using sterilized tweezers, place one coverslip on the bottom of each well of the 12-well culture dish;
13. Proceed to coat the now sterilized coverslips with Matrigel as previously explained.

#### 15.3.4.2 Seeding hESCs for Differentiation

H1 cells must be plated at an appropriate density before beginning *in vitro* differentiation. Proper starting density will ensure efficient and reproducible differentiations. After expanding cells to reach the desired cell numbers, the cells are passaged and plated on Matrigel-coated 12-well plates. To allow for accurate counting, the cells must be dissociated into a single cell suspension before seeding wells to prepare for the *in vitro* differentiation.

1. Aspirate medium from wells and rinse with DPBS without $Ca^{2+}$ and $Mg^{2+}$ (~2 ml);
2. Place 0.5–1 ml of room temperature TrypLE in each well and incubate at 37 °C for 3–5 min;
3. After the first 2 min, examine cells under a microscope to ensure sufficient detachment from plate (see ***Note 3***);
4. Continue to incubate cells at 37 °C and examine under microscope every 1 min until cells appear balled and many are free-floating;
5. Stop enzymatic reaction by adding 4 volumes of E8 medium;
6. Use a 1 ml pipette to triturate the detached cells several times to break up remaining clumps, leaving a single-cell suspension (see ***Note 4***);
7. Use a serological pipette to transfer the cell suspension into sterile 15 ml conical tubes;
8. Reserve a small aliquot of cell suspension for cell counting;
9. Centrifuge for 4 min at 200×*g* to pellet cells and count cells during centrifugation;
10. Aspirate media and resuspend cells in E8 medium (~3×$10^5$ cells/ml) containing freshly added ROCK inhibitor (10 μM);
11. Add 1.5 ml of cell suspension per well of a Matrigel-coated 12-well plate;
12. After 24–48 h cells should be roughly 90% confluent and ready to begin *in vitro* differentiation.

### *15.3.5 Differentiation to Pancreatic Endoderm (Adherent Culture)*

Figure 15.1 outlines the step-wise differentiation process and key factors contained in the differentiation medium at each stage of differentiation. Once cells have reached 90% confluency (Fig. 15.2a), the differentiation process can be started. Figure 15.2b shows an example of cells that need to be expanded further before differentiation can be initiated. Stage-specific base-media are stable at 2–8 °C for up to 1 month and can therefore be mixed ahead of time. The formulations for stage-specific base-media are listed in *Sect. 15.2.3*. Table 15.1 shows the final concentrations of the added factors at each day of differentiation. *These factors are not stable at 2–8 °C for long periods of time and must be added to stage-specific base-medium immediately prior to warming the medium on each day of differentiation.*

Described here is a method in which the cells are adhered to 12-well plates in planar culture. However, this protocol can also be performed in suspension culture (Fig. 15.2c), as previously described in detail (Schulz et al. 2012). Suspension culture requires scaling up medium volumes and cell numbers, as well as orbital rotation (see ***Notes 5*** and ***6***).

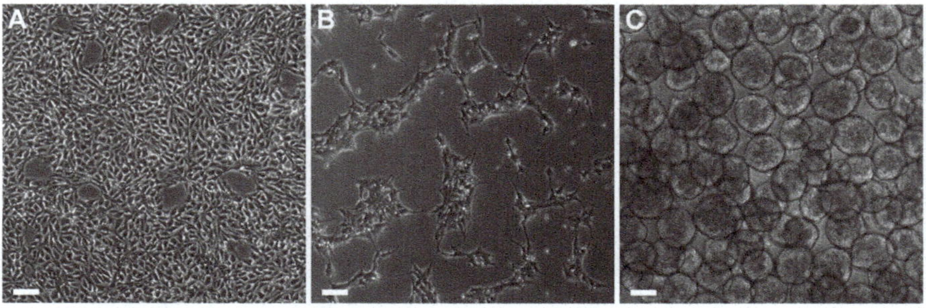


**Fig. 15.2** Brightfield images of H1 human embryonic stem cells. H1 cells cultured under adherent conditions (**A**, **B**) or as suspension aggregates (**C**). (**A**) Image showing adherent H1 cells at the correct density to initiate *in vitro* differentiation. (**B**) Image showing cells that are of insufficient density to initiate differentiation. (**C**) Image showing aggregated H1 cells prior to induction of directed differentiation. Scale bars = 50 μM in **A**, **B** and 100 μM in **C**

- **Day 1:** Aspirate E8 medium and wash cells with DPBS without $Ca^{2+}$ and $Mg^{2+}$.
- To each well add 1.5 ml of prewarmed Stage 1 Medium with freshly added Activin A (100 ng/ml) and Wnt-3a (25 ng/ml);
- **Day 2:** Aspirate medium and replace with 1.5 ml of prewarmed Stage 1 Medium with freshly added Activin A (100 ng/ml);
- **Day 3:** Repeat the previous step. At this point (or the following day) it is recommended to analyze the cells for differentiation efficiency to the definitive endoderm (DE) stage. The success of downstream differentiation is largely dependent upon highly efficient DE induction. Figure 15.3a provides an example of efficient DE induction (data generated using protocols in *Sect. 15.3.6.1*);
- **Day 4:** Aspirate Stage 1 Medium and wash cells with MCDB 131 base medium without added supplements or factors. To each well add 1.5 ml of prewarmed Stage 2 Medium with freshly added L-Ascorbic Acid (VIT-C) (0.25 mM) and FGF7 (50 ng/ml);
- **Day 5:** Repeat the previous step;
- **Day 6:** Aspirate medium and replace with 1.5 ml of fresh, prewarmed Stage 3 Medium with freshly added L-Ascorbic Acid (VIT-C) (0.25 mM), FGF7 (50 ng/ml), SANT-1 (0.25 μM), Retinoic Acid (1 μM), LDN193189 (100 nM), ITS-X (1:200) and TPB (200 nM);
- **Day 7:** Repeat the previous step;
- **Day 8:** Aspirate medium and replace with 1.5 ml of fresh, prewarmed Stage 4 Medium with freshly added L-Ascorbic Acid (VIT-C) (0.25 mM), FGF7 (2 ng/ml), SANT-1 (0.25 μM), Retinoic Acid (0.1 μM), LDN193189 (200 nM), ITS-X (1:200) and TPB (100 nM);
- **Day 9:** Repeat the previous step;
- **Day 10:** Repeat the previous step (same as *Day 8*).

At the end of the differentiation protocol the cells have adopted characteristics of pancreatic progenitor cells. A recent report by Rezania et al. describes a method to further differentiate these H1-derived pancreatic endoderm cells into beta-like cells *in vitro* (Rezania et al. 2014). The method involves culturing the cells in a liquid-air interface after the pancreatic endoderm stage. An alternative method to the liquid-air interface culture is to perform the differentiation into beta-like cells in suspension culture. However, the differentiation efficiency is reported to be lower in suspension culture than in liquid-air interface culture (Rezania et al. 2014). Two other recent reports utilized suspension culture methods to derive beta-like cells, similar to the ones reported by Rezania et al. from hPSC-derived pancreatic endoderm (Pagliuca et al. 2014; Russ et al. 2015).

## *15.3.6 Cell Characterization*

### 15.3.6.1 Characterization of Cells by Flow Cytometry and Immunofluorescence Analysis

To ensure efficient differentiation, it is important to characterize the cells at various stages throughout the differentiation. Flow cytometry analysis and IF staining for stage-specific proteins are the hallmark methods to assess differentiation efficiencies. Figure 15.3 shows example images of IF analysis as well as flow cytometry plots of stage-specific markers from a successful differentiation. For a complete list of antibodies used for flow cytometry and IF analysis see *Sect. 15.2.4*. Below are recommended protocols for both flow cytometry and IF analysis.

**Flow cytometry analysis protocol**

1. Thaw Accutase at room temperature or 2–8 °C prior to use;
2. Aspirate medium from each well to be analyzed;
3. Wash cells with DPBS (~1.5 ml per well of 12-well plate) and aspirate;
4. Add enough Accutase® to cover the surface of each well (0.3–0.5 ml per well of 12-well plate);
5. Incubate at 37 °C for 5 min, then check briefly under microscope to ensure cells appear balled up and many are detached. If needed, place back at 37 °C and check every 1 min for good detachment (see ***Note 7***);
6. Add ~1 ml cold FACS buffer (0.2 % BSA in DPBS) to each well (final volume ~1.5 ml) to stop Accutase activity;
7. Using a 1000 µl pipette triturate cells to break up any clumps and obtain a single cell suspension;
8. Label a 5 ml polystyrene tube with cell strainer cap for each sample and add ~250 µl FACS buffer to each cap to pre-wet the cell strainer. Ensure the buffer flows through the cap into the tube;
9. Load each sample into the cap of the appropriately labeled tube and allow the entire suspension to flow through the cell strainer. This will ensure good cell

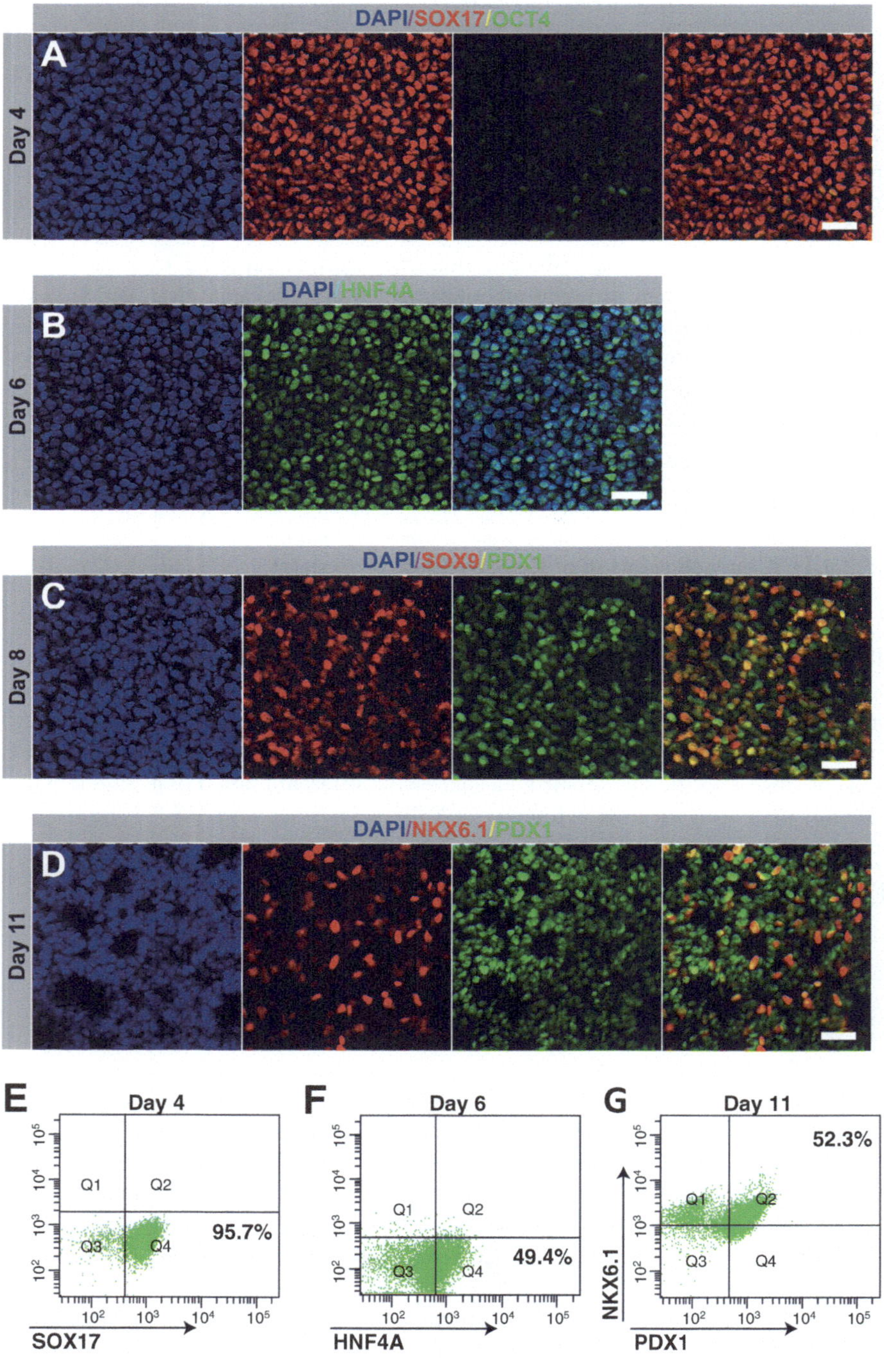


**Fig. 15.3** Immunofluorescence staining and flow cytometry analysis of H1 cells throughout pancreatic differentiation. Immunofluorescent staining (**A–D**) or flow cytometry analysis (**E–G**) for stage-restricted transcription factors in H1 cells at different days of differentiation toward pancreatic endoderm. Day 4 represents definitive endoderm; day 6 primitive gut tube; day 8 posterior foregut; and day 11 pancreatic endoderm. Scale bars = 50 μm

separation, which will provide better flow cytometry results. If the suspension does not easily flow through the cell strainer it may be necessary to pull the sample through. To do so, place the tip of a 1000 μl pipette on the underside of the cell strainer cap containing the sample. Slowly release the pipette plunger to gently pull the sample through the filter top. Carefully eject the sample from the 1000 μl pipette into the bottom of the tube;

10. Transfer the strained cells into new 1.5 ml microcentrifuge tubes and centrifuge at room temperature and 200 × *g* for 5 min;
11. Resuspend each cell pellet with cold BD fixation/permeabilization solution (300 μl per tube);
12. Incubate cells for 20 min at 2–8 °C (i.e. refrigerate—*do not place on ice*);
13. Wash cells by adding ~1.1 ml cold 1× BD Perm/Wash™ Buffer to each tube following fixation/permeabilization (1.5 ml per well of 12-well plate) and centrifuge at 10 °C and 200 × *g* for 5 min;
14. Aspirate supernatant and repeat *step 11*;
15. Aspirate supernatant and resuspend cells in 50 μl cold 1× BD Perm/Wash Buffer for each staining and isotype control to be used (e.g. add 150 μl total for two different staining samples and one isotype control sample);
16. Aliquot 50 μl of cell suspension into separate microcentrifuge tubes for each individual reaction. Co-staining with PE and AlexaFluor 647 conjugated antibodies can be performed together in one 50 μl aliquot;
17. For analysis of SOX17 expression (DE marker) load 2.5 μl SOX17-PE antibody into one 50 μl cell suspension aliquot. Load 2.5 μl PE isotype into a separate 50 μl cell suspension aliquot to serve as a control;
18. For analysis of PDX1 expression (pancreatic endoderm marker) load 2.5 μl PDX1-PE antibody into one 50 μl cell suspension aliquot. Load 2.5 μl PE isotype into a separate 50 μl cell suspension aliquot to serve as a control;
19. For analysis of NKX6.1 expression (pancreatic endoderm marker) load 2.5 μl NKX6.1-AlexaFluor 647 antibody into one 50 μl cell suspension aliquot. Load 2.5 μl AlexaFluor 647 isotype into a separate 50 μl cell suspension aliquot to serve as a control;
20. After the antibodies and matching isotypes have been loaded into cell suspensions, incubate cells in the dark for 1 h at 2–8 °C (i.e. refrigerate—*do not place on ice*);
21. Wash cells by adding 1.25 ml cold 1× BD Wash Buffer to each tube and centrifuge at room temperature and 200 × *g* for 5 min;
22. Aspirate supernatant and resuspend each sample in 300 μl cold FACS buffer;
23. The samples are now stained and ready for analysis on a flow cytometer, such as the FACSCanto used to generate the plots in Fig. 15.3.

**Immunofluorescence analysis protocol**

1. Aspirate medium from each well to be analyzed;
2. Wash cells with DPBS (~1.5 ml per well of 12-well plate) and aspirate;
3. Repeat *step 2*;
4. Add 1 ml of 4 % PFA to each well and incubate at 2–8 °C overnight;
5. The following day, aspirate PFA and wash with DPBS (~1.5 ml per well of 12-well plate);

6. Aspirate DPBS and add 1 ml of blocking buffer (0.15 % Triton X-100, 1 % normal donkey serum in DPBS) and incubate at room temperature for 1 h;
7. Prepare primary antibody solutions by mixing appropriate volumes of each antibody in blocking buffer (1 ml per well of a 12 well plate) and vortex gently. See *Sect. 15.2.4* for appropriate working dilutions. For example, to analyze SOX17 expression, add 1 µl of goat anti-SOX17 antibody to 1 ml of blocking buffer;
8. Aspirate buffer from each well and load 1 ml of pre-mixed primary antibody solution to the appropriate wells and incubate at 2–8 °C overnight;
9. The following day, aspirate antibody solution and wash with DPBS (~1.5 ml per well of 12-well plate);
10. Aspirate DPBS and repeat *step 9* three times;
11. Prepare secondary antibody solutions by mixing appropriate volumes of each antibody in blocking buffer (1 ml per well of 12 well plate) and vortex gently. See *Sect. 15.2.4* for appropriate working dilutions;
12. Aspirate DPBS from each well and add 1 ml of pre-mixed secondary antibody solution to the appropriate wells and incubate at room temperature for 1 h in the dark;
13. Aspirate antibody solution and wash with DPBS (~1.5 ml per well of 12-well plate);
14. Prepare nuclear staining solution by mixing Hoescht 33342 in DPBS (1:3000, 1 ml per well of 12-well plate);
15. Aspirate DPBS from each well and add 1 ml of nuclear staining solution and incubate at room temperature for 5 min in the dark;
16. Aspirate nuclear staining solution and wash with DPBS (~1.5 ml per well of 12-well plate);
17. Aspirate DPBS and wash once more with DPBS (~1.5 ml per well of 12-well plate);
18. The coverslips containing stained cells are now ready to be mounted on slides. Gently remove coverslips with tweezers and rinse by briefly dipping each coverslip into distilled water;
19. Allow fluid to run off the coverslip and remove excess fluid by blotting the edge of the coverslip with a paper towel. Capillary action should help remove the excess;
20. Place one drop of VECTASHIELD Antifade Mounting Medium (or desired mounting solution) on the coverslip on the side containing the cells;
21. Gently invert coverslip while placing it cell side down on a slide. Be careful to avoid any air bubbles between the slide and coverslip;
22. Seal the edges of the coverslip to the slide with clear nail polish and allow it to fully dry in the dark before analyzing the slides. The nail polish prevents the slides from drying out if stored for extended periods of time;
23. The slide containing the coverslip with the stained cells is now ready to be analyzed on an inverted fluorescent microscope, such as the Zeiss Axio Observer used to generate the images in Fig. 15.3;
24. Slides can be stored at 4 °C for short periods of time and at −20 °C for extended periods of time.

## 15.4 Notes

1. During expansion of H1 cells it is better to passage cells as clusters rather than as a single cell suspension. This will enhance health and survival of the cells.
2. Typical passaging dilutions for expanding H1 cells is 1:6 (i.e. 1 well split into 6 wells).
3. Avoid incubating with TrypLE for longer than necessary as this can reduce cell viability.
4. Avoid excessive pipetting as this can reduce cell viability.
5. *Performing differentiation in suspension culture*—To culture cells in suspension, use ultra-low attachment 6-well plates and rotate on an orbital shaker at 100 rpm. This ensures proper cell aggregation, which is necessary for cell survival and differentiation (Fig. 15.2c). The culture volume is 5.5 ml/well and base media formulations and concentrations of factors are the same as used for differentiation in adherent cultures (Table 15.1). Following expansion and dissociation of hESCs with TrypLE, seed wells with $1 \times 10^6$ cells/ml, which is a total of $5.5 \times 10^6$ cells/well. Add 10 μM ROCK inhibitor to medium when seeding cells to promote cell survival and aggregation. 48 h following cell seeding the now aggregated cells (Fig. 15.2c) are ready to initiate the *in vitro* differentiation. The differentiation protocol is the same as described for adherent culture, but the volume of culture medium has to be adjusted.
6. *Handling cell aggregates in suspension culture*—A solution of 0.2 % BSA in DPBS should be used to coat pipettes in order to prevent aggregates from adhering to plastic surfaces during medium exchanges. To properly exchange culture medium, draw PBS/BSA solution into the pipette and eject the solution. Swirl the 6-well plate in a circular motion to bring all aggregates to the center. Tilt the plate and draw medium off the side of each well with the pre-coated pipette, removing most of the medium. Leave a very small amount of medium to ensure no aggregates are removed. Load each well with 5.5 ml of the appropriate culture medium for that day. Place the plate back on the orbital shaker at 100 rpm inside a 37 °C incubator.
7. Avoid prolonged exposure to Accutase. Detachment of monolayer H1 cells should occur after 5 min, however if the Accutase has started to become inactive, more time may be necessary. To ensure swift cell detachment use freshly thawed Accutase.

**Acknowledgements** The authors would like to thank Andrea Carrano and Allen Wang for comments. This work was supported by National Institutes of Health grants (U01-DK089567 and UC4-DK104202), California Institute for Regenerative Medicine grants (RB5-07236 and RB4-06144), and Helmsley Charitable Trust grant 2012PG-T1D074 to M.S. N.V. was supported by the National Cancer Institute Cancer Cell Biology training grant 5T32 CA067754.

## References

Caicedo A (2013) Paracrine and autocrine interactions in the human islet: more than meets the eye. Semin Cell Dev Biol 24:11–21

Chen G, Gulbranson DR, Hou Z, Bolin JM, Ruotti V, Probasco MD, Smuga-Otto K, Howden SE, Diol NR, Propson NE et al (2011) Chemically defined conditions for human iPSC derivation and culture. Nat Methods 8:424–429

Cnop M, Welsh N, Jonas JC, Jorns A, Lenzen S, Eizirik DL (2005) Mechanisms of pancreatic beta-cell death in type 1 and type 2 diabetes: many differences, few similarities. Diabetes 54(Suppl 2):S97–S107

D'Amour KA, Bang AG, Eliazer S, Kelly OG, Agulnick AD, Smart NG, Moorman MA, Kroon E, Carpenter MK, Baetge EE (2006) Production of pancreatic hormone-expressing endocrine cells from human embryonic stem cells. Nat Biotechnol 24:1392–1401

Haumaitre C, Fabre M, Cormier S, Baumann C, Delezoide AL, Cereghini S (2006) Severe pancreas hypoplasia and multicystic renal dysplasia in two human fetuses carrying novel HNF1beta/MODY5 mutations. Hum Mol Genet 15:2363–2375

Johnson JD, Luciani DS (2010) Mechanisms of pancreatic beta-cell apoptosis in diabetes and its therapies. Adv Exp Med Biol 654:447–462

Kahn SE, Hull RL, Utzschneider KM (2006) Mechanisms linking obesity to insulin resistance and type 2 diabetes. Nature 444:840–846

Kelly OG, Chan MY, Martinson LA, Kadoya K, Ostertag TM, Ross KG, Richardson M, Carpenter MK, D'Amour KA, Kroon E et al (2011) Cell-surface markers for the isolation of pancreatic cell types derived from human embryonic stem cells. Nat Biotechnol 29:750–756

Kroon E, Martinson LA, Kadoya K, Bang AG, Kelly OG, Eliazer S, Young H, Richardson M, Smart NG, Cunningham J et al (2008) Pancreatic endoderm derived from human embryonic stem cells generates glucose-responsive insulin-secreting cells in vivo. Nat Biotechnol 26:443–452

Lee YH, Sauer B, Gonzalez FJ (1998) Laron dwarfism and non-insulin-dependent diabetes mellitus in the Hnf-1alpha knockout mouse. Mol Cell Biol 18:3059–3068

Mayer C, Bottcher Y, Kovacs P, Halbritter J, Stumvoll M (2008) Phenotype of a patient with a de novo mutation in the hepatocyte nuclear factor 1beta/maturity-onset diabetes of the young type 5 gene. Metabolism 57:416–420

Meloche RM (2007) Transplantation for the treatment of type 1 diabetes. World J Gastroenterol 13:6347–6355

Merkle FT, Eggan K (2013) Modeling human disease with pluripotent stem cells: from genome association to function. Cell Stem Cell 12:656–668

Mfopou JK, Chen B, Sui L, Sermon K, Bouwens L (2010) Recent advances and prospects in the differentiation of pancreatic cells from human embryonic stem cells. Diabetes 59:2094–2101

Nathan DM (2015) Diabetes: advances in diagnosis and treatment. JAMA 314:1052–1062

Nostro MC, Sarangi F, Ogawa S, Holtzinger A, Corneo B, Li X, Micallef SJ, Park IH, Basford C, Wheeler MB et al (2011) Stage-specific signaling through TGFbeta family members and WNT regulates patterning and pancreatic specification of human pluripotent stem cells. Development 138:861–871

Nostro MC, Sarangi F, Yang C, Holland A, Elefanty AG, Stanley EG, Greiner DL, Keller G (2015) Efficient generation of NKX6-1+ pancreatic progenitors from multiple human pluripotent stem cell lines. Stem Cell Rep 4:591–604

Pagliuca FW, Melton DA (2013) How to make a functional beta-cell. Development 140:2472–2483

Pagliuca FW, Millman JR, Gurtler M, Segel M, Van Dervort A, Ryu JH, Peterson QP, Greiner D, Melton DA (2014) Generation of functional human pancreatic beta cells in vitro. Cell 159:428–439

Rezania A, Bruin JE, Riedel MJ, Mojibian M, Asadi A, Xu J, Gauvin R, Narayan K, Karanu F, O'Neil JJ et al (2012) Maturation of human embryonic stem cell-derived pancreatic progenitors into functional islets capable of treating pre-existing diabetes in mice. Diabetes 61:2016–2029

Rezania A, Bruin JE, Xu J, Narayan K, Fox JK, O'Neil JJ, Kieffer TJ (2013) Enrichment of human embryonic stem cell-derived NKX6.1-expressing pancreatic progenitor cells accelerates the maturation of insulin-secreting cells in vivo. Stem Cells 31:2432–2442

Rezania A, Bruin JE, Arora P, Rubin A, Batushansky I, Asadi A, O'Dwyer S, Quiskamp N, Mojibian M, Albrecht T et al (2014) Reversal of diabetes with insulin-producing cells derived in vitro from human pluripotent stem cells. Nat Biotechnol 32:1121–1133

Robertson RP, Davis C, Larsen J, Stratta R, Sutherland DE (2000) Pancreas and islet transplantation for patients with diabetes. Diabetes Care 23:112–116

Russ HA, Parent AV, Ringler JJ, Hennings TG, Nair GG, Shveygert M, Guo T, Puri S, Haataja L, Cirulli V et al (2015) Controlled induction of human pancreatic progenitors produces functional beta-like cells in vitro. EMBO J 34:1759–1772

Schulz TC, Young HY, Agulnick AD, Babin MJ, Baetge EE, Bang AG, Bhoumik A, Cepa I, Cesario RM, Haakmeester C et al (2012) A scalable system for production of functional pancreatic progenitors from human embryonic stem cells. PLoS One 7:e37004

Standl E, Fuchtenbusch M (2003) The role of oral antidiabetic agents: why and when to use an early-phase insulin secretion agent in Type II diabetes mellitus. Diabetologia 46(Suppl 1):M30–M36

Stumvoll M, Goldstein BJ, van Haeften TW (2005) Type 2 diabetes: principles of pathogenesis and therapy. Lancet 365:1333–1346

ViaCyte I (2014) ViaCyte, Inc. Announces FDA acceptance of IND to commence clinical trial of VC-01™ candidate cell replacement therapy for type 1 diabetes. [Press release online] http://viacyte.com/press-releases/viacyte-inc-announces-fda-acceptance-of-ind-to-commence-clinical-trial-of-vc-01-candidate-cell-replacement-therapy-for-type-1-diabetes/

Wells JM, Melton DA (1999) Vertebrate endoderm development. Annu Rev Cell Dev Biol 15:393–410

Xie R, Everett LJ, Lim HW, Patel NA, Schug J, Kroon E, Kelly OG, Wang A, D'Amour KA, Robins AJ et al (2013) Dynamic chomatin remodeling mediated by polycomb proteins orchestrates pancreatic differentiation of human embryonic stem cells. Cell Stem Cell 12:224–237

Yamagata K, Furuta H, Oda N, Kaisaki PJ, Menzel S, Cox NJ, Fajans SS, Signorini S, Stoffel M, Bell GI (1996) Mutations in the hepatocyte nuclear factor-4alpha gene in maturity-onset diabetes of the young (MODY1). Nature 384:458–460

# Chapter 16
# Efficient Generation of Skeletal Myogenic Progenitors from Human Pluripotent Stem Cells

**Jaemin Kim, Alessandro Magli, and Rita C.R. Perlingeiro**

**Abstract** The rapid progress of research involving pluripotent stem cells in recent years provides an unprecedented opportunity to develop new and efficient cell replacement therapies for the treatment of degenerative diseases, such as muscular dystrophies. Nevertheless, the directed differentiation of pluripotent stem cells into a specific lineage is challenging, particularly the skeletal myogenic lineage which is poorly recapitulated during the *in vitro* differentiation of pluripotent stem cells. As demonstrated by our previous work, this limitation can be overcome by controlled expression of PAX7, which forces the commitment and expansion of human skeletal myogenic progenitors. Since variation in paraxial mesoderm formation among different human pluripotent stem cell lines/clones can affect the ability of PAX7 to induce the skeletal myogenic lineage, here we describe the optimization of our previously published protocol for deriving skeletal myogenic progenitors from these cells. This method takes advantage of the ability of the GSK3β inhibitor to induce paraxial mesoderm commitment in differentiating cells. In addition, upon flow cytometry purification, human PAX7-induced skeletal myogenic progenitors can be propagated and differentiated into skeletal myotubes. The following protocol will enable researchers to derive skeletal myogenic progenitors from pluripotent stem cells for disease modeling as wells as therapeutic purposes.

**Keywords** Pluripotent • ES and iPS cells • Skeletal muscle • GSK3β inhibitor • PAX7 • Myogenic precursors • Muscle differentiation • Cell therapy

---

J. Kim • A. Magli • R.C.R. Perlingeiro (✉)
Department of Medicine, Lillehei Heart Institute, University of Minnesota,
4-128 CCRB, 2231 6th St. SE, Minneapolis, MN 55455, USA
e-mail: perli032@umn.edu



H. Ulrich, P. Davidson Negraes, *Working with Stem Cells*,
DOI 10.1007/978-3-319-30582-0_16

## 16.1 Introduction

Muscle homeostasis is impaired in pathologic conditions such as cachexia, sarcopenia and muscular dystrophies, and there is currently no treatment to permanently reverse or strengthen the impaired skeletal muscles of these patients. Among the potential therapeutic approaches to treat these diseases, cell therapy has the advantage to contribute both to the existing muscle fibers and the resident stem cell pool (Rinaldi and Perlingeiro 2014). Importantly, a major requirement for the development of an efficient cell therapy approach to treat such degenerative diseases in human patients is an appropriate protocol to produce large quantities of skeletal myogenic progenitors. Pluripotent stem cells (PSCs) such as human embryonic stem cells (hESCs) and induced pluripotent stem cells (hiPSCs) represent an attractive source for the development of cell replacement therapies because of their unlimited proliferation capability and ability to differentiate into all different cell types of the human body.

In recent years, a few groups, including ours, have reported protocols for the derivation of skeletal myogenic progenitors from hESCs or hiPSCs (Barberi et al. 2007; Chang et al. 2009; Darabi et al. 2012; Borchin et al. 2013; Shelton et al. 2014; Chal et al. 2015). Our group has developed the most efficient *in vitro* hESC/iPSC differentiation approach to date, which involves the formation of embryoid bodies (EBs) followed by conditional expression of PAX7 using viral vectors (Darabi and Perlingeiro 2014). Notably, these *in vitro* generated myogenic precursors are able to (1) engraft into damaged muscles of an immuno- and dystrophin-deficient mouse model (Arpke et al. 2013), (2) generate myofibers expressing human DYSTROPHIN, (3) improve muscle contractile function (Darabi et al. 2012), (4) engraft into the host satellite cell pool (Darabi et al. 2012; Darabi and Perlingeiro 2014), and (5) produce persistent engraftment several months post-transplantation (Darabi et al. 2008). In order to successfully translate this technology to human patients, consistent large numbers of myogenic progenitors need to be generated. In this regard, we identified limitations that could make this protocol less reproducible among different hiPS clones, such as: (1) differences in mesoderm formation among clones; (2) batch to batch variations in sizes of EBs for PAX7 induction and; (3) necessity of at least 3 days for seeding the EBs prior to PAX7 induction as they exhibit low adherence. Among these, we reasoned that one of the most sensitive variables in the current protocol is the presomitic mesoderm induction, which is a required step for the myogenic-specific transcriptional activity of PAX3/7 (Darabi et al. 2008; Maroto et al. 1997).

Mesoderm formation is initiated by switching from hESC-hiPSC media to serum-containing media and, due to variations among different fetal bovine serum (FBS) lots, this can result in a sub-optimal mesoderm induction. Mesodermal patterning is regulated by the gradients of bone morphogenic protein (BMP), Nodal/Activin, fibroblast growth factor (FGF) and WNT (Tam and Loebel 2007) and different combinations of these morphogens have been used to direct the commitment of both mouse and human ESCs toward a specific type of mesoderm (Mendjan et al. 2014). Since presomitic mesoderm specification requires activation of WNT and

FGF signaling, modulation of these two pathways could have a positive effect on the reproducibility of our current protocol. In particular, activation of WNT/β-Catenin pathway is essential for paraxial mesoderm commitment of the neuromesodermal progenitors in the developing embryo (Garriock et al. 2015) and treatment of differentiating ES/iPS cells with Glycogen synthase kinase 3 β (GSK3β) inhibitors, which results in β-Catenin stabilization, has been shown to recapitulate this process *in vitro* (Borchin et al. 2013; Mendjan et al. 2014; Shelton et al. 2014).

Based on this rationale, we have further refined our previously published protocol (Darabi and Perlingeiro 2014) by incorporating GSK3β inhibitor (CHIR990217) to induce the canonical WNT signaling pathway. Notably, this optimization step positively affects also other variables such as EB formation, adhesion to the culture dish, and substantially reduces the overall protocol length. For this optimized protocol, we used H9 hESC line, which has been transduced with a lentiviral vector encoding an inducible hPAX7-IRES-mRFP (monomeric Red Fluorescent Protein—mCherry) transgene. In brief, after EB formation in PSC culture media, cells are treated with GSK3β inhibitor to induce paraxial mesoderm formation. The EBs are then dissociated as single cells, supplemented with doxycycline (Dox) to induce hPAX7 expression and seeded onto a gelatinized culture flask. After 4 days of PAX7 induction, mRFP$^+$ (PAX7$^+$) cells are sorted to obtain a homogeneous population of skeletal myogenic progenitors. The mRFP$^+$ cells are then propagated in Dox supplemented culture medium on gelatinized culture dishes before induction of terminal differentiation into skeletal myotubes.

In this chapter, we describe in details the improved version of our previously published protocol (Fig. 16.1).

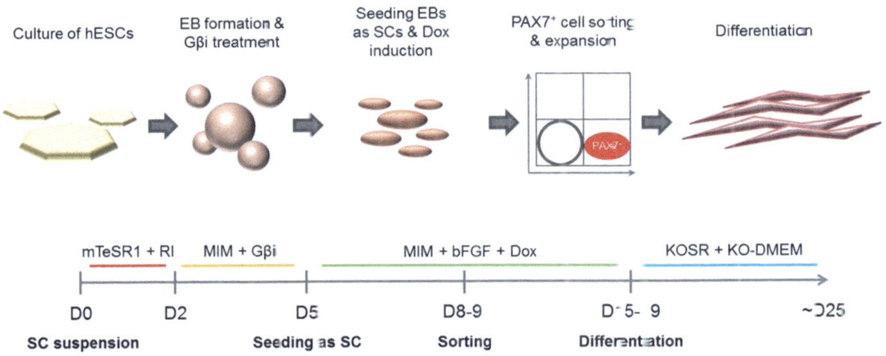


**Fig. 16.1** Schematic representation of the skeletal myogenic differentiation from hESCs protocol. The hESCs are dissociated as single cells to form EBs which is followed by treatment with GSK3β inhibitor to initiate paraxial mesoderm formation. The treated EBs are dissociated as single cells before seeding and are induced with Dox to overexpress PAX7. PAX7$^+$ myogenic precursors are isolated by FACS and terminally differentiated into skeletal myotubes. *RI* ROCK inhibitor, *Gβi* GSK3β inhibitor, *SC* single cell. *MIM* Myogenic induction medium, *bFGF* Basic fibroblast growth factor, *KOSR* Knockout serum replacement

## 16.2 Materials

### 16.2.1 Reagents for Culturing EBs

- 5 mM Rho-associated kinase (ROCK) inhibitor (Y27632) (Millipore, Cat. No. 688002).
- mTeSR1 basal medium (Stem Cell Technologies, Cat. No. 5850).
- mTeSR1 supplement (Stem Cell Technologies, Cat. No. 5850).

### 16.2.2 Reagents for the Induction of Paraxial Mesodermal Cells

- 10 mM Glycogen synthase kinase 3 beta (GSK3β) inhibitor (CHIR990217) (Millipore, Cat. No. 361571).
- Iscove's Modified Dulbecco's Medium (IMDM), glutamax supplemented (Invitrogen, Cat. No. 31980-030).
- Fetal bovine serum (FBS), qualified, USDA-approved regions (Invitrogen, Cat. No. 10437-028).
- Horse serum (HS), heat-inactivated (Sigma-Aldrich, Cat. No. H1138).
- Chicken embryo extract (CEE), ultrafiltrate (US Biological Life Sciences, Cat. No. C3999).
- L-Ascorbic acid 2-phosphate sesquimagnesium salt hydrate (Sigma-Aldrich, Cat. No. A8960).
- Monothioglycerol (Sigma-Aldrich, Cat. No. 88640).
- Penicillin/Streptomycin (Pen/Strep) (Invitrogen, Cat. No. 15140-122).

### 16.2.3 Reagents for the Culture/Expansion of Satellite-Like Cells

- 1 mg/ml Doxycycline (Dox) hyclate (Sigma-Aldrich, Cat. No. D9891).
- 10 μg/ml Basic fibroblast growth factor (bFGF), human (Roche, Cat. No. 11120417001).
- Dimethyl sulfoxide (DMSO) (Sigma-Aldrich, Cat. No. D2650).

### 16.2.4 Reagents for the Terminal Differentiation

- KnockOut Dulbecco's Minimal Essential Medium (KO-DMEM) (Invitrogen, Cat. No. 10829-018).
- KnockOut Serum Replacement (KOSR) (Invitrogen, Cat. No. 10828-028).
- GlutaMAX supplement (Invitrogen, Cat. No. 35050-061).
- Minimum Essential Medium Non-Essential Amino Acids (NEAA) Solution (100X) (Invitrogen, Cat. No. 11140-050).

### 16.2.5 *Reagents for General Tissue Culture Work*

- 24-Well Tissue culture plate (Thermo Scientific, Cat. No. 142475).
- 12-Well Tissue culture plate (Thermo Scientific, Cat. No. 150628).
- 6-Well Tissue culture plate (Thermo Scientific, Cat. No. 140675).
- T25-Tissue culture flask (Thermo Scientific, Cat. No. 156367).
- Cryogenic vial (Corning, Cat. No. 430488).
- Cell strainer (Corning, Cat. No. 352235).
- Petri dish, low adherent (60 mm × 15 mm) (Fisher Scientific, Cat. No. 351007).
- Gelatin from porcine skin (Sigma-Aldrich, Cat. No. G2500).
- Accumax (Sigma-Aldrich, Cat. No. A7089).
- Trypsin-EDTA (0.25 %), phenol red (Invitrogen, Cat. No. 25200-072).
- Dulbecco's Modified Eagle Medium/Nutrient Mixture F-12 (DMEM-F12) (Invitrogen, Cat. No. 11320-033).
- 7-AAD Viability Staining Solution (eBioscience, Cat. No. 00-6993).
- Dulbecco's Phosphate buffered saline (D-PBS) ($Mg^-$, $Cl^-$) (Invitrogen, Cat. No. 14190-144).

### 16.2.6 *Culture Media*

**Table 16.1** Table of culture media

| | mTeSR1 | Myogenic induction medium (MIM) | KOSR terminal differentiation medium |
|---|---|---|---|
| mTeSR1 basal | 158 ml | – | – |
| mTeSR1 supplement | 40 ml | – | – |
| IMDM | – | 146 ml | – |
| FBS | – | 30 ml | – |
| HS | – | 20 ml | – |
| CEE | – | 2 ml | – |
| Ascorbic acid | – | 200 μl | – |
| Monothioglycerol | – | 7.56 μl | – |
| KO-DMEM | – | – | 154 ml |
| KOSR | – | – | 40 ml |
| NEAA | – | – | 2 ml |
| GlutaMAX | – | – | 2 ml |
| Pen/Strep | 2 ml | 2 ml | 2 ml |

## 16.3 Methods

### *16.3.1 Formation of EBs*

1. From ~$5 \times 10^5$ cells of iPAX7-H9 cells, aspirate the mTeSR1 medium and wash once with D-PBS to remove the remaining medium and floating dead cells;
2. Add 1 ml of Accumax to cover the entire surface of the well and incubate at 37 °C for 3–5 min. Check under the microscope to ensure cell dispersion (agitate or tap the plate if the cells are not dispersing);
3. Add 1 ml of DMEM-F12 to the well to dilute the Accumax, collect the cells into a 15 ml tube and centrifuge at $300 \times g$ for 5 min;
4. Resuspend with 4 ml of mTeSR1 supplemented with 8 μl of ROCK inhibitor to inhibit apoptosis during the EB formation and plate in a low adherent 60 mm diameter Petri dish;
5. Transfer the dish onto an incubator shaker (orbital shaking speed of 60 rpm) and grow the cells for 2 days in the 5 % $CO_2$ incubator.

### *16.3.2 Induction of EBs into Paraxial Mesodermal Cells*

1. After 2 days, check EBs under the microscope to ensure EB formation. By this time, EBs should be formed and they should look transparent and homogeneous in terms of size;
2. Swirl the plate slowly to gather all the EBs at the center of the dish. Allow the EBs to settle down for a min;
3. Carefully remove the entire medium by pipetting from the peripheral sides of the dish without disturbing the EBs;
4. Wash the EBs once by adding 4 ml of DMEM-F12 (repeating *steps 2* and *3*) and resuspend in 4 ml of MIM supplemented with 4 μl of GSK3β inhibitor to induce paraxial mesoderm;
5. Transfer the dish onto the incubator shaker and culture the EBs for 3 days in the $CO_2$ incubator.

### *16.3.3 Induction/Expansion of Satellite-Like Cells*

1. After 3 days, discard the entire medium and add 4 ml of D-PBS;
2. Swirl the plate slowly to gather all the EBs at the center of the dish. Carefully transfer the EBs to a 15 ml tube and spin down at $300 \times g$ for 1 min;
3. Add 0.5 ml of Accumax to the pellet and incubate at 37 °C water bath for 3–5 min;

4. Gently triturate with a micropipette (P1000) to further aid in breaking the clumps into single cells and add 4.5 ml of DMEM-F12 to dilute the Accumax;
5. Count the cells and determine how many T25 flasks will be required (~$5 \times 10^5$ cells per 25 $cm^2$);
6. Spin down at $300 \times g$ for 5 min and resuspend in 4 ml of MIM with 4 μl of ROCK inhibitor, 3 μl of Dox and 2 μl of bFGF to ensure survivability of the cells and allow induction of PAX7 expression;
7. Plate the cells onto a pre-coated 0.1 % gelatinized T25 flask and gently shake the flask side to side to evenly distribute the cells all over the surface. Grow the cells for 1 day in the $CO_2$ incubator;
8. After 24 h, wash once with D-PBS and change to fresh 4 ml of MIM supplemented with 3 μl of Dox and 2 μl of bFGF and grow the cells for 2 more days in the $CO_2$ incubator;
9. After 2 days, wash once with D-PBS, add 2 ml of Accumax per T25 flask and incubate at 37 °C for 3–5 min;
10. Add 2 ml of DMEM-F12, transfer to a 15 ml tube and spin down at $300 \times g$ for 5 min;
11. Resuspend with 4 ml of D-PBS, filter the cells through a cell strainer and spin down at $300 \times g$ for 5 min;
12. Resuspend in 500 μl of Fluorescence activated cell sorting (FACS) buffer (D-PBS supplemented with 10 % FBS, 1 % Pen/Strep and 0.1 % 7-ADD dye) and place the tube in ice. Protect the cells from direct exposure to light as 7-AAD dye is sensitive to light;
13. Sort for mCherry$^+$ (PAX7$^+$) cells using a flow cytometry system (such as BD FACSAria II Flow Cytometer, BD Biosciences). Make sure to set the gate for mCherry$^+$ expression using the non-induced mCherry$^-$ control cells, and then run the Dox-induced cell preparation;
14. After cell sorting, spin down at $300 \times g$ for 5 min and seed them at a density of ~$2 \times 10^5$ cells/well of 0.1 % gelatinized 6-well plate with 3 ml/well of MIM supplemented with 6 μl of ROCK inhibitor, 2.25 μl of Dox and 1.5 μl of bFGF. Grow the cells for 1 day in the $CO_2$ incubator;
15. After 24 h, check the cells for cell quality/survival and morphology. If the quality of cell sorting is good, you should see minimal floating dead cells. Rinse once with D-PBS and replace with 2 ml/well of MIM supplemented with 1.5 μl of Dox and 1 μl of bFGF. Culture the cells in the $CO_2$ incubator to expand the progenitor cells;
16. After initial purification, myogenic progenitors can be passaged every 3–4 days or frozen for future uses (Fig. 16.2). Cells are harvested by incubation with 0.25 % trypsin for 1–2 min followed by trypsin inactivation using MIM and centrifuge ($300 \times g$, 5 min). For expansion, cells can be seeded in gelatinized dishes as described above. Alternatively, cells can be resuspended in a mixture of 60 % MIM, 30 % FBS and 10 % DMSO (Dox and bFGF supplementation is not necessary), aliquoted into cryogenic vials (0.5 ml/~$5 \times 10^5$ cells/vial), stored at −80 °C and ultimately transferred in liquid nitrogen tank for long term storage.

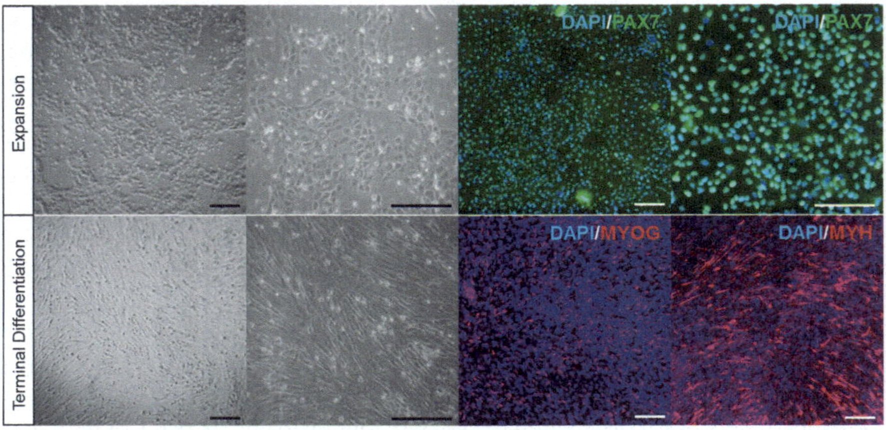


**Fig. 16.2** Generation of skeletal myogenic progenitors and skeletal myotubes from hESCs. The differentiated iPAX7-H9 cells are sorted and expanded prior to generation of skeletal myotubes, showing PAX7 expression (*top panels*). The expanded PAX7⁺ myogenic precursors are further differentiated into skeletal myotubes which expressed MYOG and MYH (*bottom panels*). *MYOG* MYOGENIN, *MYH* MYOSIN HEAVY CHAIN. Scale = 100 μm

### *16.3.4 Terminal Differentiation of mCherry⁺ Cells into Myotubes*

1. Grow the mCherry⁺ cells in the $CO_2$ incubator until the cells reach 90–95 % confluency;
2. Wash the Dox-induced mCherry⁺ (PAX7⁺) myogenic progenitors once with D-PBS and replace with KOSR terminal differentiation medium;
3. Culture the cells at 37 °C for 7–10 days and analyze for gene and protein expressions of skeletal myotubes such as MYOGENIN (1:25, DSHB, Cat. No. F5D-s) and MYOSIN HEAVY CHAIN (1:50, DSHB, Cat. No. MF20-s) (Fig. 16.2).

## 16.4 Conclusion

The described protocol will allow researchers to derive and expand skeletal myogenic progenitors from PSCs for disease modeling as wells as potential development of cell replacement therapies for treating human patients who are suffering from muscular dystrophies.

## References

Arpke RW, Darabi R, Mader TL, Zhang Y, Toyama A, Lonetree CL, Nash N, Lowe DA, Perlingeiro RC, Kyba M (2013) A new immuno-, dystrophin-deficient model, the NSG-mdx(4Cv) mouse, provides evidence for functional improvement following allogeneic satellite cell transplantation. Stem Cells 31(8):1611–1620

Barberi T, Bradbury M, Dincer Z, Panagiotakos G, Socci ND, Studer L (2007) Derivation of engraftable skeletal myoblasts from human embryonic stem cells. Nat Med 13(5):642–648

Borchin B, Chen J, Barberi T (2013) Derivation and FACS-mediated purification of PAX3+/PAX7+ skeletal muscle precursors from human pluripotent stem cells. Stem Cell Rep 1(6):620–631

Chal J, Oginuma M, Al Tanoury Z, Gobert B, Sumara O, Hick A, Bousson F, Zidouni Y, Mursch C, Moncuquet P, Tassy O, Vincent S, Miyanari A, Bera A, Garnier JM, Guevara G, Hestin M, Kennedy L, Hayashi S, Drayton B, Cherrier T, Gayraud-Morel B, Gussoni E, Relaix F, Tajbakhsh S, Pourquie O (2015) Differentiation of pluripotent stem cells to muscle fiber to model Duchenne muscular dystrophy. Nat Biotechnol 33(9):962–969

Chang H, Yoshimoto M, Umeda K, Iwasa T, Mizuno Y, Fukada S-I, Yamamoto H, Motohashi N, Miyagoe-Suzuki Y, Takeda SI, Heike T, Nakahata T (2009) Generation of transplantable, functional satellite-like cells from mouse embryonic stem cells. FASEB J 23(6):1907–1919

Darabi R, Perlingeiro RC (2016) Derivation of skeletal myogenic precursors from human pluripotent stem cells using conditional expression of PAX7. In: Turksen K, Nagy A (eds) Induced Pluripotent Stem (iPS) Cells: Methods and Protocols, Springer, New York, p 423–439

Darabi R, Gehlbach K, Bachoo RM, Kamath S, Osawa M, Kamm KE, Kyba M, Perlingeiro RC (2008) Functional skeletal muscle regeneration from differentiating embryonic stem cells. Nat Med 14(2):134–143

Darabi R, Arpke RW, Irion S, Dimos JT, Grskovic M, Kyba M, Perlingeiro RC (2012) Human ES- and iPS-derived myogenic progenitors restore DYSTROPHIN and improve contractility upon transplantation in dystrophic mice. Cell Stem Cell 10(5):610–619

Garriock RJ, Chalamalasetty RB, Kennedy MW, Canizales LC, Lewandoski M, Yamaguchi TP (2015) Lineage tracing of neuromesodermal progenitors reveals novel Wnt-dependent roles in trunk progenitor cell maintenance and differentiation. Development 142(9):1628–1638

Maroto M, Reshef R, Munsterberg AE, Koester S, Goulding M, Lassar AB (1997) Ectopic Pax-3 activates MyoD and Myf-5 expression in embryonic mesoderm and neural tissue. Cell 89(1):139–148

Mendjan S, Mascetti VL, Ortmann D, Ortiz M, Karjosukarso DW, Ng Y, Moreau T, Pedersen RA (2014) NANOG and CDX2 pattern distinct subtypes of human mesoderm during exit from pluripotency. Cell Stem Cell 15(3):310–325

Rinaldi F, Perlingeiro RC (2014) Stem cells for skeletal muscle regeneration: therapeutic potential and roadblocks. Transl Res 163(4):409–417

Shelton M, Metz J, Liu J, Carpenedo RL, Demers SP, Stanford WL, Skerjanc IS (2014) Derivation and expansion of PAX7-positive muscle progenitors from human and mouse embryonic stem cells. Stem Cell Rep 3(3):516–529

Tam PP, Loebel DA (2007) Gene function in mouse embryogenesis: get set for gastrulation. Nat Rev Genet 8(5):368–381

# Chapter 17
# Genome Editing in Stem Cells

**Leon Tejwani, Cleber A. Trujillo, Charles A. Thomas, and Alysson R. Muotri**

**Abstract** Monumental advances in molecular biology over the past few decades have provided us with tools that have revolutionized the study of genetics. With the expansive amount of information gained by the sequencing of the human genome, scientists have been charged with the task of attributing cellular phenomena to the information encoded by genes. A "reverse genetics" approach has been employed by introducing specific modifications in the expression of a particular gene to examine the resulting phenotype. While initial techniques revolved around interference at the post-transcriptional level, targeting the genome directly has enhanced our ability to study gene function. In this chapter, we will discuss the history of genome editing, the various technologies to modify the genome and, in particular, highlight the CRISPR/Cas9 system. Due to its versatility and ease of programmability, CRISPR/Cas9 captivates the attention of scientists and is crucial to elucidate genetic function. Now, scientists can emulate monogenic diseases to study pathogenesis on the molecular and cellular levels. In addition, medical personnel could revert a

L. Tejwani • C.A. Thomas
Department of Pediatrics, Stem Cell Program, School of Medicine, University of California San Diego, La Jolla, CA 92093-0695, USA

C.A. Trujillo
Department of Pediatrics, Stem Cell Program, School of Medicine, University of California San Diego, La Jolla, CA 92093-0695, USA

Sanford Consortium for Regenerative Medicine, La Jolla, CA, USA

A.R. Muotri (✉)
Department of Pediatrics, Stem Cell Program, School of Medicine, University of California San Diego, La Jolla, CA 92093-0695, USA

Department of Pediatrics and Cellular and Molecular Medicine, School of Medicine, University of California San Diego, La Jolla, CA 92093-0695, USA

Sanford Consortium for Regenerative Medicine,
2880 Torrey Pines Scenic Dr. – Room 3005, La Jolla 92093, CA, USA
e-mail: muotri@ucsd.edu



H. Ulrich, P. Davidson Negraes, *Working with Stem Cells*,
DOI 10.1007/978-3-319-30582-0_17

deleterious mutation to its wild-type form as a potential genetic therapy. Genome editing has already advanced and continues to advance academic research and personalized medicine to unprecedented heights.

**Keywords** Stem cells • Genome editing • CRISPR • Programmable nucleases

## 17.1 Introduction

Since the initial description of the accepted structure of deoxyribonucleic acid (DNA) (Watson and Crick 1953), extensive attention has been dedicated to correlating the nucleotide sequences that comprise the genome and the encoded cellular functionalities. Several hallmark concepts and techniques that have revolutionized the field of molecular biology and redefined our understanding of modern genetics have emerged from the interest in understanding DNA. Notable breakthroughs include the description of the Central Dogma of Biology (Crick 1970), the isolation of DNA restriction enzymes from *Escherichia coli* (Meselson and Yuan 1968), the development of the polymerase chain reaction (Saiki et al. 1985), and the sequencing of the human genome (Venter et al. 2001; Lander et al. 2001). The knowledge emerged from development of these tools has been seminal in enhancing the understanding of the finest intricacies of human genome function.

The completion of the Human Genome Project (Lander et al. 2001; Venter et al. 2001) marked the beginning of a new era in science, known as the "post-genomic era". At the transition to this era, the vast amount of newly obtained genomic knowledge opened doors for a genetic revolution to correlate genes with their respective functions, a field known as "functional genomics" (Eisenberg et al. 2000; Vukmirovic and Tilghman 2000; Fields et al. 1999). To thoroughly study the stepwise pathway that comprises the process of a particular gene's function, the traditional "forward genetics" approach of random genome mutagenesis and tracing the alterations in cellular phenotype back to changes in the genome was too slow and indeterminate. The resulting drawbacks render it less effective in precisely articulating a gene's function. In order to bridge the huge disparity in our understanding of genes and their functions, the direct "reverse genetics" approach emerged as a more efficient method to decipher a particular gene's function (Ueda 2001; Pickart et al. 2006; Hardy et al. 2010). By specifically manipulating expression of the gene of interest, we gain insight on the normal function of that particular gene in the cell. Along the way, we have the ability to probe and assess the status of the all of the "omics" (Horgan and Kenny 2011) that results from perturbations to the normal gene expression, providing a more complete view of cellular biology.

Advances in "reverse genetics" made since the advent of the post-genomic era have markedly increased our ability to study a gene's function and determine causality between genetics and biological phenotypes. Techniques have been developed that target the transcriptome to generate hypomorphic phenotypes, such as RNA interference (RNAi) (Fire et al. 1998; Castanotto and Rossi 2009) and morpholino anti-sense oligonucleotides (Ekker 2000), which reduce translation of a particular transcript. Additionally, scientists have developed methods that directly target the genome to create permanent genetic changes, such as protein-guided endonucleases, which include zinc-finger nucleases (ZFNs) and transcription activator-like effector nucleases (TALENs), and RNA-guided endonucleases (RGENs) like the clustered, regularly interspersed, short palindromic repeats (CRISPR)-guided CRISPR-associated (Cas9) endonucleases have been developed. In recent years, these exciting breakthroughs in the field of genome editing have provided targeted methods to manipulate the genome with relative ease. Whether it is mutating a gene to create a model to study its cellular function or role in pathogenesis of various diseases or restoring a mutated allele to its healthy form as a therapy for a genetic disorder, the technology is in place for application and continues to be advanced and applied to even more contexts.

### 17.1.1 History of Genome Editing

The zinc-finger nuclease, the first sequence-specific artificial nuclease system, was initially described in 1996 (Kim et al. 1996). Although similar zinc-finger technology was harnessed to alter transcriptional regulation (Beerli et al. 2000; Zhang et al. 2000; Bartsevich and Juliano 2000), ZFN technology was successfully applied to specifically target and modify higher eukaryotic genomes only after the publication of the human genome sequence (Bibikova et al. 2002; Urnov et al. 2005). As ZFNs proved useful in deciphering gene function, the use of this pioneering technology popularized, expanding the field of genome editing.

As ZFN technology became more widespread and developed, a new *in vitro* nuclease system emerged and took over the spotlight. This technology, known as TALENs, took advantage of DNA-binding proteins (TALEs) derived from plant pathogens to recognize specific sequences and induce similar genetic alterations as ZFNs (Chistian et al. 2010). Based on knowledge acquired during the previous two decades of working with zinc-finger proteins, fusing various effector domains, such as nucleases, to the TALEs was easily achieved (Gaj et al. 2013). TALENs were very successful in genetic modification of various species due to the nearly unambiguous TALE-nucleotide pairing code deciphered in 2009, which is described in a later section of this chapter (Boch et al. 2009; Moscou and Bogdanove 2009).

While ZFN and TALEN technologies were both crucial in advancing the field of reverse genetics, the introduction of a new, easy-to-use artificial nuclease system in 2012 allowed the field to really blossom. The CRISPR/Cas system, derived from an adaptive bacterial immune response against foreign DNA, is an extremely facile and versatile tool in genome modification for scientists in many disciplines (Jinek et al. 2012). Several groups have reported success in utilizing the simple CRISPR/Cas system to modify genomes of various species, including higher eukaryotes (Jinek et al. 2013; Cong et al. 2013; Mali et al. 2013; Hwang et al. 2013; Jiang et al. 2013; Cho et al. 2013). Owing its success to its ease of use and customizability, the RNA-guided CRISPR/Cas system has become the preferred endonuclease system for gene modification. Together, these advances in genome editing, which will be discussed in further detail throughout this chapter, have been monumental in enabling scientists to effectively explore the field of and continue deciphering the genetic code. A summary and comparison of these techniques are described in Fig. 17.1.

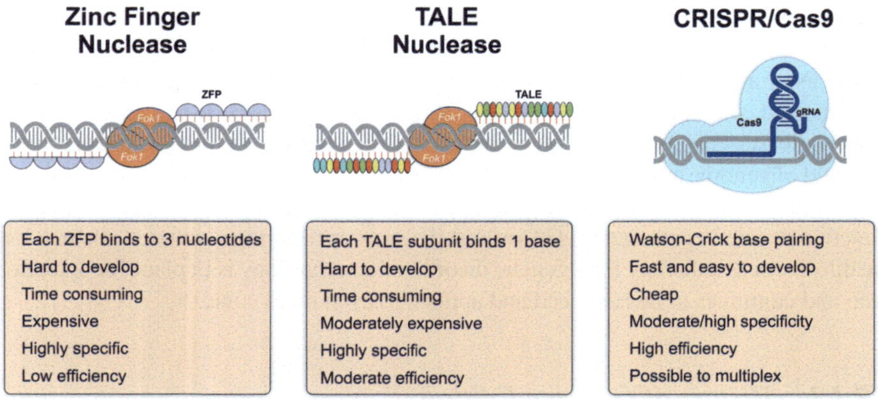


**Fig. 17.1** Schematic comparison of genome editing technologies. Zinc-Finger Nucleases are composed of a zinc-finger proteins (ZFPs) and the *FokI* nuclease. The ZFN target sequences are 18–36 bp in length. TALENs are composed of transcription activator-like effectors (TALE) and the *FokI* nuclease. Each TALE repeat is composed of 33–35 amino acids and recognizes a single base pair. CRISPR/Cas9 uses a single gRNA and Cas9 endonuclease. There are pros and cons of each method to specifically modify the genome

### 17.1.2 DNA Double-Stranded Break Repair

A cell's innate ability to repair damage to its DNA is essential for targeted genome editing to be possible. Upon introduction of a double-stranded break (DSB) in the DNA, the cell can undergo error-prone nonhomologous end-joining (NHEJ) (Moore and Haber 1996) or the more accurate homology-directed repair (HDR). NHEJ can result in new insertions and deletions (indels) that can alter the reading frame of protein translation and change the amino acid sequence and stop-codon site, and thus will likely result in a truncated, nonsensical protein and loss of function. If a donor nucleic acid sequence is present that contains homology with arms flanking the site of the DSB, the cell can employ the donor sequence for HDR (Liang et al. 1998). Because HDR incorporates the donor sequence at the DSB site, one could design the donor to modify the DSB site to incorporate a very specific deletion, insertion, or sequence substitution with high fidelity (Iliakis et al. 2004). Exploitation of the endogenous repair mechanisms has led to many successes in generating specific alterations in the human genome (Urnov et al. 2010; Durai et al. 2005).

### 17.1.3 Zinc-Finger Nucleases (ZFNs)

Native to nearly all eukaryotes, zinc-finger proteins are endogenously expressed and serve various functions, including DNA binding for direct transcriptional regulation, (Pavletich and Pabo 1991; Wolfe et al. 2000) mRNA binding to modulate protein synthesis (Mendez-Vidal et al. 2002; Lu et al. 2003), and protein binding to regulate protein interactions and functions (Tsang et al. 1997; Fox et al. 1999). The most widespread DNA-binding class of zinc-finger proteins seen in eukaryotes have a DNA-recognition domain containing a tandem array of Cysteine$_2$-Histidine$_2$ (C2H2) fingers, which each recognize 3–4 base pairs of DNA (Klug and Schwabe 1995). To introduce permanent alterations in gene expression, zinc-finger proteins have been successfully engineered to allow site-directed mutagenesis by tethering modified non-specific DNA cleavage domains of restriction endonuclease *FokI* to tandem arrays of sequence-specific C2H2 zinc-finger domains, forming zinc-finger nucleases (ZFNs) (Kim et al. 1996; Bibikova et al. 2003). To be catalytically active in facilitating (DSB) formation, the *FokI* domains of at last two ZFNs must dimerize (Bitinaite et al. 1998). The dual *FokI* endonuclease requirement necessitates one to design a pair of ZFNs to bind the preferred DNA locus for proper DNA cleavage, increasing cleavage specificity by doubling the length of recognition sites (Mani et al. 2005). Depending on the design of the DNA-binding zinc-finger domain, each ZFN is usually composed of 3–6 zinc-fingers that function virtually independently to collectively recognize between 9 and 18 nucleotides. By increasing the number of zinc-finger proteins the number of potential target sites is decreased and target specificity is further enhanced. Upon introduction of the DNA DSB by dimerized *FokI* monomers, the cell's endogenous DNA repair mechanisms described above are initiated, enabling permanent alterations in the genome.

### *17.1.4 Transcription Activator-Like Effector Nucleases (TALENs)*

TALENs provide an alternative technique for genome editing. Scientists have employed TALENs successfully in a variety of organisms, including zebrafish (Sander et al. 2011), rat (Tesson et al. 2011), fruitfly (Liu et al. 2012), yeast (Li et al. 2011), and human pluripotent and somatic cells (Hockemeyer et al. 2011; Cermak et al. 2011). Transcription activator-like effectors (TALEs) are unique to the plant pathogen *Xanthomonas* spp. and are involved in pathogenic modification of infected plant transcription (Kay and Bonas 2009). TALEs are proteins with tandem repeated 33–35 amino acid subunits with hypervariable repeat-variable di-residues (RVD) at amino acid positions 12 and 13. The di-residues recognize specific DNA nucleotides as follows: asparagine and isoleucine recognize the nucleotide adenine; histidine and aspartate recognize cytosine; asparagine and guanine recognize thymine; and two asparagines recognize guanine (Boch et al. 2009; Moscou and Bogdanove 2009). The simple amino acid-nucleotide binding code underlying TALE DNA-recognition makes for increased customizability, however, binding efficiency of the TALE is highly variable and dependent on the order of RVDs (Cermak et al. 2011).

TALENs, much like ZFNs, are fusion proteins that consist of a nonspecific *FokI* DNA cleavage domain coupled to a site-specific DNA-binding domain. Unlike the zinc-finger domains, tandem arrays of sequence-specific TALE-like repeats serve to bind DNA in TALENs. Likewise, TALENs require dimerization of *FokI* monomers for catalytic function, and thus two TALEN binding domains are necessary, enhancing the locus-recognition specificity (Chistian et al. 2010). The dimerized *FokI* endonucleases induce a DSB, initiating DNA repair mechanisms that can alter the genomic DNA by the mechanisms described above.

### *17.1.5 CRISPR/Cas Technologies*

The origins of the CRISPR/Cas system lie in an archaic adaptive immunity pathway that protects bacteria and archaea from exogenous plasmids and phages, which has most likely assisted microbial life to persist in environments dominated by foreign invaders (Carte et al. 2014). The Type II CRISPR/Cas utilizes RNA sequences to recognize foreign DNA and Cas endonucleases to cleave the exogenous DNA. The biogenesis of the system is detailed here.

In response to exogenous DNA within the cell, the bacterial Cas endonuclease cleaves the foreign DNA, resulting in cleaved DNA fragments that are integrated into the CRISPR locus of the bacterial host genome as spacers between the palindromic repeat-spacer array during the adaptation phase. Next, during CRISPR RNA (crRNA) biogenesis, these spacers serve as templates for pre-crRNA transcription (Brouns et al. 2008), and complementarily bind to a portion of another non-coding RNA, the trans-activating CRISPR RNA (tracrRNA). The resultant complex, along with the double-stranded RNA-specific endoribonuclease RNase III, facilitate the maturation of the pre-crRNA, forming a chimeric crRNA-tracrRNA complex (Deltcheva et al. 2011). The RNA heteroduplex has two main features that are integral to the functioning of the system—the crRNA that binds to complementary DNA, and the tracrRNA that interacts with the Cas endonuclease to form the crRNA-tracrRNA-Cas endonuclease complex. Finally, with the ternary crRNA-tracrRNA-Cas9 complex assembled, the crRNA recognizes invading DNA cognates and the RuvC/HNH domains of Cas9 endonuclease catalyze a double-stranded break upon the exogenous DNA (Jinek et al. 2012).

In recent years, many scientists have coerced the bacterial Type II CRISPR/Cas system to alter genomes, due to the simplicity and ease of the system. Scientists have adapted this bacterial pathway in a wide array of organisms to allow for various applications, most notably, to alter gene sequences and their expression. While this system has been repurposed to control gene expression at multiple levels, one of particular importance is the application of CRISPR/Cas technology in permanently modifying the genome directly at the DNA level. Coexpression of humanized Cas endonuclease, such as the *Streptococcus pyogenes* Cas9 (Cas9), and a synthetic chimeric guide-RNA (gRNA) that contains all sequences necessary to function analogously to the crRNA and tracrRNA seen in the bacterial system, can create double-stranded breaks at any locus within the genome containing an appropriate protospacer adjacent motif (PAM). Because the Cas9 PAM sequence (NGG) is ubiquitous throughout the genome, virtually any locus can be targeted for gene editing. Similar to the other ZF and TALE nucleases, the CRISPR/Cas9 system also relies on the host cell's endogenous repair mechanisms to give rise to permanent sequence alterations.

Due to the laborious task of synthesizing ZFNs and TALENs, which requires designing proteins with unique DNA-binding domains for each locus of interest, the more easily customizable CRISPR/Cas system has recently emerged as the preferred tool in the field of genome editing. While the ZF and TALE nuclease systems utilize protein-mediated sequence recognition to bind to the target DNA locus, the CRISPR/Cas9 system utilizes RNA-mediated recognition and localizes the endonuclease to the desired target locus. In addition, the Cas9 endonuclease is expressed independently of the gRNA, thus only the gRNA needs to be altered to target different loci. The extremely programmable CRISPR/Cas system simplifies the design process by obviating the time-consuming challenge of generating unique polypeptides for each target, as is required by ZF and TALE nucleases. The design process for approaching CRISPR/Cas9 mutagenesis will be discussed in detail throughout the following sections of this chapter. An overview of the CRISPR/Cas9 mutagenesis process, including locus-site identification, DSB generation, and subsequent repair systems provided by the host cell are depicted in Fig. 17.2.

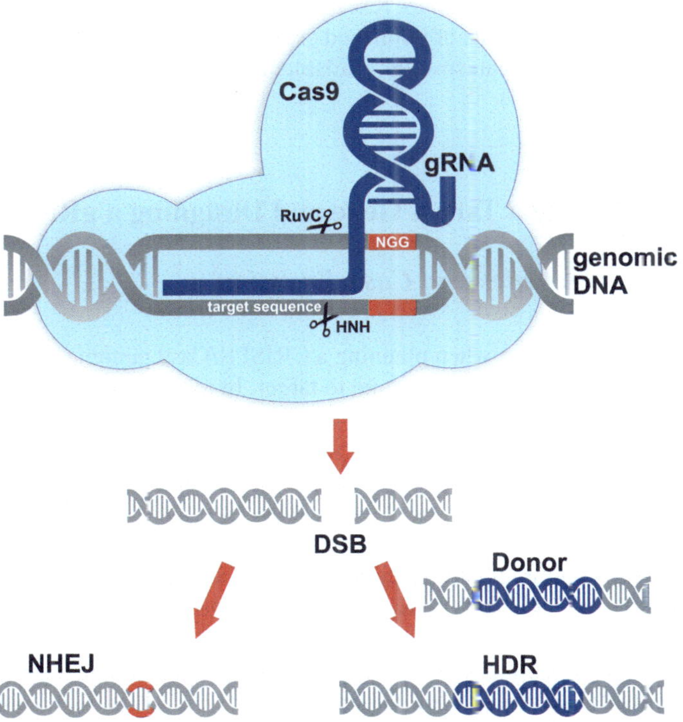


**Fig. 17.2** CRISPR and mechanisms of DNA repair. CRISPR/Cas9 genome editing system introduces double stranded breaks (DSB). The cleavage site will be repaired by non-homologous end joining (NHEJ) DNA repair pathway, an error-prone process that may result in insertions/deletions that may disrupt gene function, or homology-directed repair (HDR) that can introduce precise point mutations or insertions from a single-stranded or double-stranded DNA donor template

## 17.2 Selecting an Appropriate Cell Type to Mutagenize

### *17.2.1 Depending on the Experiment of Choice, Select a Candidate Cell Type*

CRISPR/Cas9 targeted mutagenesis in humans has been successfully realized in both transformed/cancerous cell lines, such as HEK293, K562, Jurkat, HeLa U-2-OS, UM-UC-3 and nontransformed, pluripotent stem cell (PSC) lines such as induced pluripotent stem cells (iPSCs) and embryonic stem cells (ESCs) (Sander et al. 2011). The ease of transfection and culturing immortalized cell lines make them good candidates for quick CRISPR/Cas9 screening experiments, such as assessing a gRNA's targeting efficacy, discussed in a later section. Moreover, scientists have given great attention to improving mutability of PSCs. Genetic alterations in these early stem cells provide both a renewable source of genetically modified cells and enable directed differentiation into any mature cell type of interest. The tremendous amount of utility granted by having a population of identical genetically altered stem cells allows researchers to study genetic function like never before. For reference, lists of published examples of successful CRISPR/Cas9 gene editing in different organisms are available, and will continue to expand over time (see Byrne et al. 2014).

## 17.3 Selecting a Target Gene and Designing a gRNA

### *17.3.1 Epigenetics and Cas9 Function*

The next consideration when planning a CRISPR/Cas9 genome editing experiment is selecting a location in the genome to target. In terms of selecting a specific gene to alter, the possibilities of success with CRISPR/Cas9 are seemingly limitless. Unlike with ZFNs and TALENs, CRISPR/Cas9 mediated DNA cleavage may overcome hurdles such as epigenetic hindrances that limit the gene targeting abilities of other programmable nucleases (Moreno-Mateos et al. 2015). For example, while TALENs are extremely sensitive to changes in DNA methylation, CRISPR/Cas9 has been shown to maintain its ability to cleave at loci with methylated CpG dinucleotides (Hsu et al. 2013), however, others report at a lower efficiency (Wu et al. 2014). Although scientists have yet to fully understand the effects of chomatin accessibility on recognition and cleavage by Cas9 endonucleases, it has been reported that Cas9 can achieve targeted mutagenesis at heterochomatic regions *in vitro* (Yu et al. 2013; Hsu et al. 2013; Kuscu et al. 2014) and inactivated Cas9 (dCas9) can modify transcription at sites lacking DNaseI hypersensitive regions (Perez-Pinera et al. 2013). This lifting of epigenetic restrictions grants the freedom to select virtually any gene for editing. Once a gene has been selected, the next step is to select which region of the gene to target by designing a site-specific gRNA.

## 17.4 Use an Online gRNA Design Tool to Identify and Score Potential gRNAs

Arguably, the most crucial component of designing a CRISPR/Cas9 gene editing experiment is customizing an appropriate gRNA that is both specific and efficient in targeting the gene of interest. As mentioned previously, the designed chimeric gRNA is the synthetic analog of the crRNA-tracrRNA heteroduplex in bacterial Type II system. In order to design an effective gRNA, several tools are available online that should be consulted. Some gRNA design tools include: http://tools.genome-engineering.org (Hsu et al. 2013) and www.e-crisp.org/E-CRISP/designcrispr.html (Heigwer et al. 2014). By providing a target species, gene name, and sequence, these tools computationally generate a ranked list of potential gRNAs for that locus and predict potential off-target sites. Using the ranked results, multiple gRNA sequences for a target gene should be selected to create gRNA expression constructs. These constructs could be created with primers and PCR, or alternatively, synthetic DNA can be purchased, for example GeneArt™ Strings DNA fragments from ThermoFisher (https://www.thermofisher.com/us/en/home/life-science/cloning/gene-synthesis/gene-strings-dna-fragments.html). Cloning the gRNA-containing sequences into expression plasmids are discussed in detail in a subsequent section below.

***Note****: In order to better understand the algorithm underlying the scoring of gRNAs from online design tools, two main criteria are considered: the PAM motif and gRNA design. The PAM sequence immediately secedes the gRNA sequence on the 3′ end. Depending on the species the Cas9 protein is derived from, different PAM sequences are required for proper recognition and cleavage. After selecting the target gene, one can select which Cas9 protein would be appropriate to use based on the presence or lack of specific PAM sequences in the gene. For the most commonly used* Streptococcus pyogenes *Cas9 (SpCas9), the simple 5′-NGG-3′ is the PAM sequence required for recognition (Mojica et al.* 2009*). Different Cas orthologs such as that from* Neisseiria Meningitidis *(NmCas9) may also be used for site-directed mutagenesis, however, their targeting capabilities differ from those of SpCas9 because the PAM sequence required for NmCas9 function is 5′-NNNNGATT-3′ (Hou et al.* 2013*). Because the SpCas9 only has to recognize a 3 bp 5′-NGG-3′ PAM sequence, which occurs every 8–12 bp on average (Cong et al.* 2013*), there are many more potential targets for designing suitable gRNAs, explaining why it is the most commonly used ortholog in CRISPR/Cas9 mutagenesis. The increased amount of possible targets is advantageous because it means more customizability in generating site-specific alterations; however, it also increases the likelihood of deleterious off-target effects. The 5′-NGG-3′ will be immediately downstream (3′) of the targeted sequence and the Cas9 protein is predicted to cleave 3 bp upstream (within the complementary sequence) of the PAM (Jinek et al.* 2012*). Although the PAM is required for proper function, it serves merely as a recognition site for the Cas9 protein to bind and should NOT be included in the gRNA itself.*

The other main concern is the uniqueness and specificity of the gRNA design. When targeting a single locus in the genome, it is important that the gRNA is entirely complementary to the particular target sequence while sharing as little homology as possible with all other regions of the genome. Unwanted binding of the gRNA and Cas9 to an off-target site can result in unintentional mutations that can contribute to spurious phenotypes not correlated to the target sequence. Therefore, in a well-designed CRISPR/Cas9 experiment, multiple gRNAs should be employed individually in distinct cell lines. Because the different gRNA sequences are unlikely to share off-targets, the observed phenotypes shared between the distinct cell lines ensures phenotypes are accurately attributed to the mutation for the target gene, not possible off-targets.

As it is very unlikely for two regions in the genome to share exact 23 bp homology (20 bp gRNA + 3 bp PAM sequence), the theoretical off-target effects should be very minimal, as demonstrated by various groups (Cho et al. 2014; Hsu et al. 2013; Pattanayak et al. 2014). However, some groups have contradictorily reported the ability of gRNAs to induce DSBs on off-target sequences with up to five-base pair mismatches, resulting in a high frequency of off-target mutations (Fu et al. 2013; Cradick et al. 2014). Further analysis of gRNA tolerability of mismatches have shown a seed region 6–11 nt upstream of the PAM that does not tolerate mismatching between gRNA and target well, however, the region more 5′ is able to tolerate mismatches especially well (Cho et al. 2014). Recently, whole genome sequencing was used to examine genome-wide off-target effects from RNA-guided endonucleases and CRISPR/Cas9's high specificity was confirmed, as off-target mutations consisted of less than 0.1 % of the sequencing results—near the lower limit detectable by targeted deep sequencing (Kim et al. 2015). Although findings like Kim et al. have dispelled some of the skepticism behind the specificity CRISPR/Cas9 endonucleases, an optimal gRNA design would minimize off-target effects.

### 17.4.1 If Necessary, Make Changes to Reduce Off-Target Effects

In addition to optimizing the gRNA sequence, there are many other measures to mitigate off-target effects and enhance target cleavage, including:

- *Optimizing the amount of gRNA/Cas9 delivered:* Increasing the amount of gRNA/Cas9 containing plasmid increases efficiency of cleavage, however it also decreases specificity of targeting. For plasmid delivery of components by electroporation, it is advised to use 1 μg of total DNA (equimolar amounts of individual components) for $2 \times 10^5$ cells (Ran et al. 2013), however, concentrations of transfected DNA should be titrated to satisfy this trade-off between specificity and efficiency, as necessary.
- *Changing the method of gRNA/Cas9 delivery:* Although the plasmid delivery of necessary components will be discussed in detail in this chapter, alternative

methods, such as delivering gRNA/Cas9 ribonucleoprotein (RNP) directly, can decrease the likelihood of off-target mutations (Liang et al. 2015). Since the RNP complex is more transient than plasmid DNA, it can quickly work to generate on-target specific mutations and decreases the time for potential induction of off-target changes.

- *Slight modifications to standard gRNA design:* Truncation at the 3′-end of a gRNA to yield 17–18 bp "tru-gRNAs" decreases off-target mutagenesis by up to 5000-fold at some loci without losing its ability to target intended sites. Although there is a small chance that the three fewer nucleotides can increase off-target sites, the first 3 nucleotides of the gRNA are not necessary for CRISPR function, however, they may facilitate in generating DSBs at off-target locations (Fu et al. 2014) Additionally, the addition of two guanine molecules at the 5′ end of gRNAs (ggX$_{20}$ gRNAs) reduce off-target mutations to nearly undetectable levels (by sequencing) without significantly compromising on-target editing (Kim et al. 2015).
- *Using modified forms of Cas9 to alter enzymatic cleavage function:* Because wild-type Cas9 can individually catalyze a double-stranded break by itself, its conversion to single-stranded nickase (Cas9n) by mutating either the RuvC (D10A mutant) or HNH (H840A mutant) catalytic cleavage domains significantly decreases the likelihood of off-target mutations by lowering the chance of DSB formation (Trevino and Zhang 2014). Since each gRNA/Cas9n complex would only generate a single-stranded break, two gRNAs are necessary, one for each strand, to induce two distincts SSBs to form a DSB. Since it is unlikely for two different gRNAs to share the same off-target sites, the likelihood of accidental off-target DSBs is significantly reduced. However, it is also difficult to design such a system because the presence of nearly adjacent PAM motifs on opposite strands is much more rare than if using a single gRNA (Sander and Joung 2014).
- *Using RNA-guided endonucleases that can only cut DNA upon dimerization:* Similar to the paired nickase, using inactivated Cas9 (dCas9) tethered to *FokI* would necessitate dual binding of gRNAs on opposite strands and dimerization of *FokI* monomers to catalyze the DSB. In addition to doubling the necessary recognition sites by requiring two gRNAs, this method also prevents any off-target effects that may result from single-stranded breaks induced from individuals of the paired nickases (Wyvekens et al. 2015).
- *Use a different ortholog of Cas9:* As mentioned earlier, orthologs other than the SpCas9 have longer PAM sequences required for DNA recognition and binding by the gRNAs, thereby decreasing the number of potential off-target sites in the genome (Hou et al. 2013; Esvelt et al. 2013).
- *Slightly modify gRNA design to increase transcription:* It has been shown that the U6 RNA Pol III promoter prefers to have a G as the first base of the transcript for increased transcription. If the gRNA is being cloned into a plasmid and its transcription is being driven by the U6 promoter, add a "G" nucleotide to the 5′ end of the gRNA (unless G is already the first nucleotide) (Mussolino and Cathomen 2013).

## 17.5 Preparing Plasmid for Introduction into the Cell

### *17.5.1 Clone Plasmid into Appropriate Vector for Transfection into Cell*

Due to the popular interest to utilize CRISPR/Cas9 technologies for genome editing, biotechnology companies have developed many tools that greatly facilitate the construction of appropriate vectors for delivery of the gRNA and Cas9 in various expression systems. Several different DNA-based delivery methods of gRNA and Cas9 vectors have been successfully used edit the genome. These include (1) co-transfection of dual plasmids, one containing the gRNA and the other containing Cas9, (2) co-transfection of a plasmid containing the Cas9 and a PCR amplicon of gRNA in an expression cassette (Ran et al. 2013), or (3) transfection of an all-in-one plasmid containing both the gRNA and the gene for Cas9.

Commercially produced plasmids for each of these techniques are available from many companies including, Addgene, Sigma-Aldrich, among several others. For expression of Cas9, ready-to-use plasmids already containing the humanized, codon-optimized SpCas9 driven by a highly active promoter (usually CMV, CAG, or CBh) and containing a nuclear localization signal (NLS) can be purchased; for example, hCas9 #41815 (Mali et al. 2013) or pSpCas9 #48137 (Ran et al. 2013) can be purchased on the Addgene website. While these are the simplest Cas9-expressing plasmids, there are a variety of other, more complex plasmids that enable additional applications such as Cas9 D10A nickase (Addgene, Cat. No. 51130) or GFP co-expression (Addgene, Cat. No. 44719) that are also available.

Plasmids for gRNA expression require the additional step of cloning, as the gRNA sequences vary based on target site and must be cloned into expression plasmids. Empty vectors for easy gRNA cloning are available (Addgene, Cat. Nos. 41824, 51132), and these typically use either the T7 or U6 promoters to drive expression of the gRNA. Cloning into these sites can be easily achieved by using the intact restriction sites on the plasmid to ligate sticky ends with the gRNA (Ran et al. 2013). Additionally, companies like GenScript and Sigma-Aldrich have begun to provide plasmids already containing the desired gRNA for as little as $199/clone, obviating the need for manual cloning or sequence verification after purchase.

In addition to plasmids that only express either the gRNA or Cas9 separately, more complex plasmids have been manufactured that contain both the humanized Cas9 and a cloning site for a gRNA, thus only requiring transfection of a single plasmid for simultaneous expression of Cas9 and the gRNA (Addgene, Cat. No. 42230) (Cong et al. 2013). These plasmids have also been further developed to enable additional functionality, such as fluorescence tagging and endonuclease modification.

### 17.5.2 If Necessary, Design Donor Construct for DSB-Mediated HDR

If the gene-modification goal is to introduce site-specific changes and not induce random indel formation by NHEJ, a donor repair template containing the insert must be designed as well. The donor construct can be generated as a plasmid, preferred for longer modifications, (Wu et al. 2008) (or a single-stranded oligodeoxynucleotide (ssODN), preferred for smaller insertions (Ran et al. 2013). To function successfully as a donor, plasmids should have long arms that share homology of at least 500 bp flanking each side of the target cleavage site (Hasty et al. 1991), while the sense or antisense ssODNs should share at least 40 bp homology on each side of the alteration site (Chen et al. 2011). Online tools from Sigma-Aldrich and other companies can be used to design and synthesize donor plasmids or oligos.

## 17.6 Using the Mutation Detection Kits to Check Efficacy of gRNAs

### 17.6.1 Check Efficacy of gRNAs

Once the gRNA has been designed, synthesized, and cloned into an appropriate vector for transfection, it is ready for introduction into the desired cell type. Difficult to transfect cells, such as a pluripotent stem cells or fibroblasts, can benefit from an additional step to increase the likelihood of success. Prior to performing the mutagenesis, it is useful to verify the functional quality of the gRNA in vitro, as cleavage efficiencies canvary from gRNA to gRNA and from locus to locus. In order to assess cleavage efficiency, an easy-to-culture and easy-to-transfect cell line such as HEK cells can be first used to evaluate cleavage abilities with a particular gRNA/Cas9 construct. Mutation detection kits such as the T7 Endonuclease I (T7 E1 NEB M0302S) Mutation Detection Assay (Cradick et al. 2013) or Surveyor Mutation Detection Kit (IDT, Cat. No. 706025) (Ran et al. 2013) can verify the effectiveness of a particular gRNA and Cas9 combination. The kits utilize an enzyme that cleaves mismatched base-pairs to detect mutagenesis of a desired locus (Vouillot et al. 2015). To use a mutation detection kit, genomic DNA should be extracted from cells transfected with the gRNA/Cas9 construct. The genomic DNA can be PCR amplified on the mutagenized locus. The resulting PCR amplicons are hybridized and treated with a mismatch-detecting enzyme such as T7E1 or Surveyor Nuclease. The mismatch-detecting enzymes will detect and cleave imperfectly matched hybridized mutation-containing strands. Perfectly-matched homoduplexes will remain intact. The resultant DNA after treatment with the mismatch-detecting nuclease can be analyzed by several techniques to estimate the approximate levels of error-prone NHEJ-repair, such as agarose or polyacrylamide gel electrophoresis (PAGE).

## 17.7 Delivering CRISPR/Cas9 Components

### *17.7.1 Transfect Cells with CRISPR/Cas9*

Depending on the cell type being mutated, different methods of delivery are established for successful mutagenesis. For functional validation of the gRNAs, HEK cells at 70–90 % can be transfected with 500 ng total DNA (equimolar amounts of gRNA and Cas9) using Lipofectamine 2000 (ThermoFisher, Cat. No. 11668027), Amaxa 4D Nucleofector (Lonza, Cat. No. AAF-1002B), or polyethylenimine (Sigma-Aldrich, Cat. No. 408727) (Ran et al. 2013). When generating CRISPR/Cas9 altered pluripotent stem cell lines, nucleofection with the Amaxa P3 primary cell 4D Nucleofector kit (Lonza, Cat. No. V4XP-1012) is recommended. Cells grown to 50–70 % confluency can be dissociated into single cells using cell dissociation reagent such as Accutase (Cat no. AT 104), resuspended in S1-supplemented P3 nucleofection solution, then nucleofected with 1 μg total DNA using the suggested program for the specific cell type. ROCK-inhibitor (Tocris, Cat. No. 1254) can be added to enhance post-transfection cell survival. Selection by puromycin or fluorescence, which will be discussed in the following section, can be performed after 24 h (Ran et al. 2013).

As CRISPR/Cas9 technology is being more widely applied and developed, a focus in the optimization of the system is placed on these different methods to introduce the gRNA and the Cas9 endonuclease into the target cells that increase efficiency and decrease off-target cleavage. Alternative approaches from the pure DNA-based delivery of the gRNA and Cas9 include delivery of *in vitro* transcribed gRNA and Cas9 mRNA (Kim et al. 2014) or delivery of pre-assembled Cas9/gRNA ribonucleoproteins (RNPs), which greatly increase specificity and efficiency (Liang et al. 2015). Finally, adeno-associated viral and lentiviral vectors have been successful for *in vivo* genome editing (Swiech et al. 2015; Kabadi et al. 2014).

## 17.8 How to Select Clones for Clonal Expansion and Stock Generation

### *17.8.1 Select Clones That Were Successfully Transfected*

Depending on cell type, the quality of the gRNA, and the method of delivery of components for the CRISPR/Cas9 mutagenesis, levels of transfection, generation of DSBs, and subsequent mutagenesis can vary greatly from experiment to experiment. In difficult-to-transfect cell types such as PSCs, plasmid-based introduction of the gRNA/Cas9 has been seen to only induce mutations in up to 20 % of cells (Liang et al. 2015), making it necessary to evaluate many clones for changes in their genotypes. In order to facilitate selection of an enriched population cells in which CRISPR/Cas9 could have been potentially active, it is greatly helpful to introduce a means of selection to detect cells that were actually transfected gRNA and Cas9.

If the designed gRNA or Cas9 expression vector contains a fluorescent tag or gene that confers antibiotic resistance, selection of only the successfully transfected cells is possible. For example, when using a multicistronic vector that expresses a fluorescent tag in addition to Cas9, such as the GFP-containing PX458 pSpCas9(BB)-2A-GFP plasmid available from AddGene (Plasmid 48138), the GFP marks a population of cells with high Cas9 translation thus, the GFP-positive cells have a higher probability of Cas9-induced DSB. By fluorescence-activated cell sorting (FACS) for the GFP$^+$ population, cells absent of GFP (therefore, absent of the Cas9 endonuclease) can be removed, thereby concentrating the population of cells that may have a desired alteration in their genome. Similarly, antibiotic selection with puromycin can be used to select for cells successfully transfected with the Cas9-containing plasmid conferring puromycin-resistance. However, timing must be carefully optimized to coincide with transient plasmid expression. If there is no selectable marker included in the Cas9 or gRNA plasmids, then co-transfection with a surrogate reporter GFP-containing plasmid can be used to assist in enriching transfected populations (Ramakrishna et al. 2014). If a selectable marker like a GFP-tag on the Cas9 expression cassette is present, proper isolation of cells with more highly active gRNA/Cas9 expression can be performed using routine FACS protocols 24–48 h post-transfection (Ran et al. 2013; Li et al. 2014).

## 17.9 Clonally Expand Single Cells and Make Stocks of Individual Clones

After sorting, selected cells should be seeded at a low density to allow single cells to clonally expand in isolation, away from other expanding colonies. Once colonies have formed from the single cells, individual clones should be isolated and cultured individually, as appropriate for the chosen cell line. Ideally, if the clonal derivation of cell lines was successful, all cells that make up a single clone have a uniform genotype

## 17.10 How to Screen/Genotype Clones

### *17.10.1 Screen Clones for Changes in Genotype*

After expanding each clone derived from individual CRISPR/Cas9 treated cells, it is necessary to assess the genotypic changes that may have resulted at the target site from the introduction the gRNA and Cas9 nuclease before being proceeding to use the cell lines for experimentation. While methods such as Western blotting can be used in order to check simply if a null allele was generated or gel electrophoresis based assays such as the Surveyor Nuclease assay to detect indel formation, they do not provide comprehensive information regarding the exact nature of genotypic changes that may have occurred. Therefore, methods such as Sanger sequencing or next-generation sequencing of gDNA (Bell et al. 2014) provide a better means to precisely characterize any alterations that may have been induced.

In order to prepare gDNA samples for sequencing, primers flanking the target gene can be designed and used to amplify the region of DNA of interest for examination, which can then be sequenced. By aligning the sequence of a CRISPR/Cas9 treated clone with the sequence of the parental, non-CRISPR/Cas9-treated cell line, any changes around the targeted site can be detected. If the sequencing provides an unclear result, there may be three possibilities it can be attributed to: the mutation could have been only in a single allele, therefore the heterozygosity leads to a mixed result, the colonies were not clonally derived and that the sequenced DNA originates from different clones growing together, or sequencing failed for any number of reasons. To verify, PCR amplicons can be cloned into plasmids, such as the TOPO vector (ThermoFisher, Cat. No. 450071), isolating individual alleles, and many copies of the plasmid sequenced (Geng et al. 2006). If a heterozygous mutation was introduced, the sequencing results should cleanly display approximately half wild-type and half mutant sequences. If this ratio is skewed in one direction, it is possible the clone was derived from two or more cells. Clones in which this is the case can either be discarded or clonally derived again by plating as single cells and isolating individual colonies without contamination from other cells. Once a mutation has been observed in the DNA, it can be further confirmed by sequencing cDNA produced from reverse-transcribing mRNA, as the mutation that is seen in the DNA should also be carried in its RNA.

In addition to assessing the target gene's sequence, it is critical to assess any mutagenesis that may have occurred in the potential off-target sites. Mentioned earlier, changes in off-target sites could result in specious phenotypes not associated with mutations in the intended target. By using online tools to predict off-target sites, one can design primers to specifically amplify these regions for Sanger sequencing and evaluation (Lin et al. 2014). However, as there have been findings that gRNAs may have the potential to bind with a significant number of mismatched bases to additional sites in the genome, the number and location of off-target lesions may be higher than originally predicted. In order to get a more unbiased understanding of changes in the genotype, exome or whole genome sequencing can be performed and the results compared to a parental cell line. Although the exome data neglects to analyze changes that might have taken place in intronic regions, some of which play crucial roles in transcription, it does provide sufficient insight on any mutagenesis in coding regions that may be involved in generation of null alleles (Cho et al. 2014). An overall schematic protocol of the CRISPR design process is shown in Fig. 17.3.

## 17.11 Potential Uses of CRISPR Technology

The application of CRISPR technology has not been limited to altering the human genome exclusively. In addition, successful attempts have applied this technology to a multitude of other species. Heritable germline modifications have been successfully generated using CRISPR/Cas9 in fruitfly (Bassett et al. 2013; Yu et al. 2013), mosquito (Dong et al. 2015), *C. elegans* (Friedland et al. 2013), *Xenopus* (Nakayama

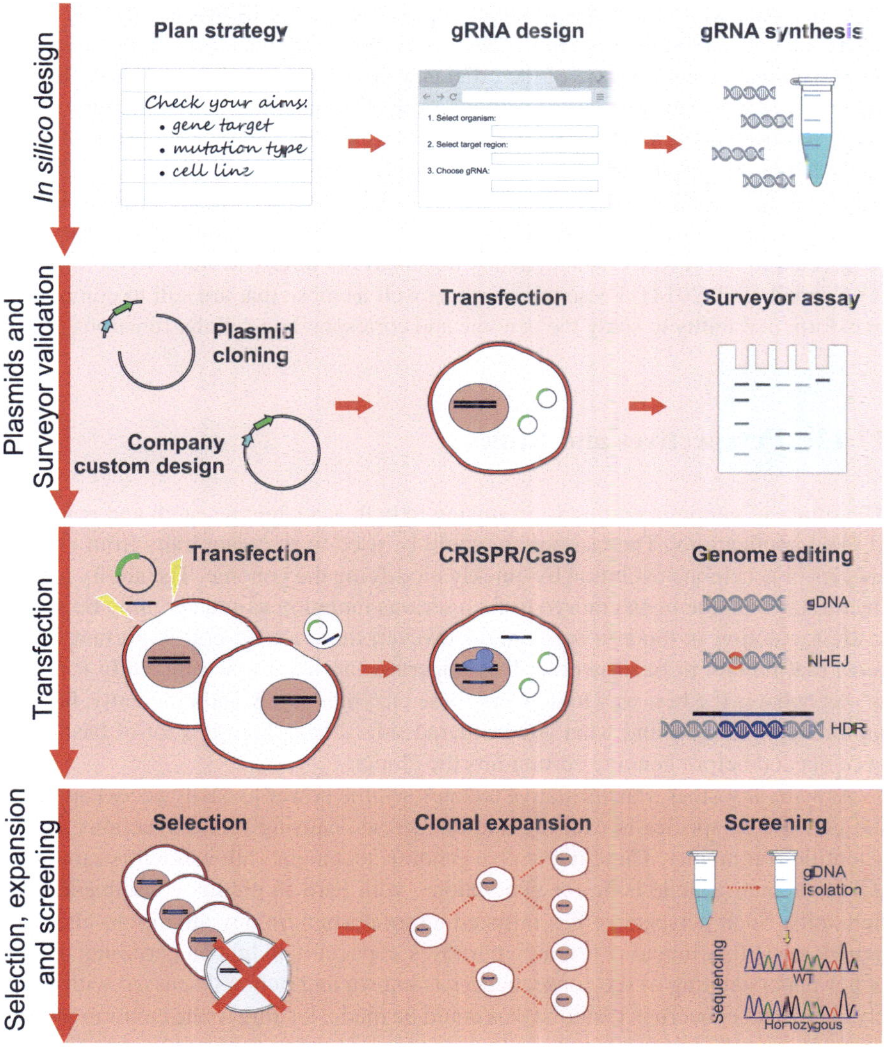


**Fig. 17.3** Overview of CRISPR/Cas9 genome editing. The guide RNA is first designed *in silico* depending on the aims, gene and cell targets. The gRNA can be cloned into an expression plasmid and transfected into a model cell in order to validate efficacy. The designed plasmid and optional repair template are transfected into the target cells. Finally, transfected cells can be clonally expanded and screened to generate isogenic cell lines with defined mutations

et al. 2013), mouse/rat (Li et al. 2013), rabbit (Yang et al. 2014), and sheep (Crispo et al. 2015), and continue to be generated in other species. In plants, CRISPR/Cas9 has been utilized to create site-specific alterations in the genomes of *Arabidopsis* (Feng et al. 2014), soybean (Jacobs et al. 2015), tobacco (Nekrasov et al. 2013), rice (Zhou et al. 2014), and wheat (Shan et al. 2013), among others.

In humans, while most of the initial applications of CRISPR/Cas9 system have been for targeted modifications of the DNA sequence, additional techniques have been developed that take advantage of the efficiency and programmability of the CRISPR system to accomplish other tasks involving nucleotide targeting. Transcriptional regulation has been achieved by coupling dCas9 to a repressive KRAB domain (Gilbert et al. 2013) or activating VP64 domain (Maeder et al. 2013; Gilbert et al. 2014). Additionally, attaching fluorescent molecules to dCas9 has allowed for dynamic imaging of genomic loci (Chen et al. 2013). CRISPR-guided Cas9 binding has also been re-directed to target RNA, which greatly enhances our ability to study RNA function (O'Connell et al. 2014). These advances, as well as those that are still to come, will transform our ability to study the genome and corresponding cellular functions.

## 17.12 Perspectives and Ethics

The future of genomic editing is promising to both academic research and potential clinical applications. The technology could be used to re-create traits from extinct species or accelerate evolution by quickly modifying the genome. The ability to edit the human genome in an embryo to fix a serious mutation would certainly be a fantastic revolution in modern medicine. However, there are several important questions that remain to be answered. It is imperative to learn how efficiently it really works, what is the best way to delivery these enzymes safely and efficiently, how to minimize off-targets and what is considered safe. Thus, there is a lot of basic science needed before genome editing hits the clinics.

As with any other transformative technology, the power to easily access and edit the genome of species is exciting and dangerous, causing mixed reactions in the scientific community. These innovative genomic techniques allow scientists to modify the human genetic code for generations, with hard to predict consequences to humanity. Such perspective has initiated a moratorium on any attempt to alter the genetic code for clinical or research purposes, especially in human germline, until a better understanding of the consequences is known and public discussed with society. The main concern is that mistakes could be made, leading to the creation of new diseases or malformations, instead of curing or enhancing the human conditions. It could also be used to design a new generation of babies, carrying unique individual features that could lead to intentionally or not selective advantage, such as hair color, resistance to disease or boosted intelligence. We just do not know enough about evolution to start to genetically engineer the human race or any other species in the planet. As for now, we do not have data to back up these initiatives, but only speculative and philosophical arguments. This will likely be a risk-reward kind of consideration for the future. Thus, a moratorium at this stage can be considered a cautious step forward rather than harmful. In the meantime, there are several issues related to intellectual property, commercialization and conflict of interest that will keep scientists and bioethicists busy.

## References


Bartsevich VV, Juliano RL (2000) Regulation of the MDR1 gene by transcriptional repressors selected using peptide combinatorial libraries. Mol Pharmacol 58(1):1–10

Bassett AR, Tibbit C, Ponting CP, Liu JL (2013) Highly efficient targeted mutagenesis of Drosophila with the CRISPR/Cas9 system. Cell Rep 4(1):220–228

Beerli RR, Dreier B, Barbas CF 3rd (2000) Positive and negative regulation of endogenous genes by designed transcription factors. Proc Natl Acad Sci U S A 97(4):1495–1500

Bell CC, Magor GW, Gillinder KR, Perkins AC (2014) A high-throughput screening strategy for detecting CRISPR-Cas9 induced mutations using next-generation sequencing. BMC Genomics 15:1002

Bibikova M, Golic M, Golic K, Carroll D (2002) Targeted chomosomal cleavage and mutagenesis in Drosophila using zinc-finger nucleases. Genetics 161(3):1169–1175

Bibikova M, Beumer K, Trautman JK, Carroll D (2003) Enhancing gene targeting with designed zinc finger nucleases. Science 300(5620):764

Bitinaite J, Wah DA, Aggarwal AK, Schildkraut I (1998) FokI dimerization is required for DNA cleavage. Proc Natl Acad Sci U S A 95(18):10570–10575

Boch J, Scholze H, Schornack S, Landgraf A, Hahn S, Kay S, Lahaye T, Nickstadt A, Bonas U (2009) Breaking the code of DNA binding specificity of TAL-type III effectors. Science 326(5959):1509–1512

Brouns SJ, Jore MM, Lundgren M, Westra ER, Slijkhuis RJ, Snijders AP, Dickman MJ, Makarova KS, Koonin EV, van der Oost J (2008) Small CRISPR RNAs guide antiviral defense in prokaryotes. Science 321(5891):960–964

Byrne SM, Mali P, Church GM (2014) Genome editing in human stem cells. Methods Enzymol 546:119–138

Carte J, Chistopher RT, Smith JT, Olson S, Barrangou R, Moineau S, Glover CV 3rd, Graveley BR, Terns RM, Terns MP (2014) The three major types of CRISPR-Cas systems function independently in CRISPR RNA biogenesis in Streptococcus thermophilus. Mol Microbiol 93(1):98–112

Castanotto D, Rossi JJ (2009) The promises and pitfalls of RNA-interference-based therapeutics. Nature 457(7228):426–433

Cermak T, Doyle EL, Chistian M, Wang L, Zhang Y, Schmidt C, Baller JA, Somia NV, Bogdanove AJ, Voytas DF (2011) Efficient design and assembly of custom TALEN and other TAL effector-based constructs for DNA targeting. Nucleic Acids Res 39(12), e82

Chen F, Pruett-Miller SM, Huang Y, Gjoka M, Duda K, Taunton J, Collingwood TN, Frodin M, Davis GD (2011) High-frequency genome editing using ssDNA oligonucleotides with zinc-finger nucleases. Nat Methods 8(9):753–755

Chen B, Gilbert LA, Cimini BA, Schnitzbauer J, Zhang W, Li GW, Park J, Blackburn EH, Weissman JS, Qi LS, Huang B (2013) Dynamic imaging of genomic loci in living human cells by an optimized CRISPR/Cas system. Cell 155(7):1479–1491

Chistian M, Cermak T, Doyle EL, Schmidt C, Zhang F, Hummel A, Bogdanove AJ, Voytas DF (2010) Targeting DNA double-strand breaks with TAL effector nucleases. Genetics 186(2):757–761

Cho SW, Kim S, Kim JM, Kim JS (2013) Targeted genome engineering in human cells with the Cas9 RNA-guided endonuclease. Nat Biotechnol 31(3):230–232

Cho SW, Kim S, Kim Y, Kweon J, Kim HS, Bae S, Kim JS (2014) Analysis of off-target effects of CRISPR/Cas-derived RNA-guided endonucleases and nickases. Genome Res 24(1):132–141

Cong L, Ran FA, Cox D, Lin SL, Barretto R, Habib N, Hsu PD, Wu XB, Jiang WY, Marraffini LA, Zhang F (2013) Multiplex genome engineering using CRISPR/Cas systems. Science 339(6121):819–823

Cradick TJ, Fine EJ, Antico CJ, Bao G (2013) CRISPR/Cas9 systems targeting beta-globin and CCR5 genes have substantial off-target activity. Nucleic Acids Res 41(20):9584–9592

Cradick TJ, Qiu P, Lee CM, Fine EJ, Bao G (2014) COSMID: a web-based tool for identifying and validating CRISPR/Cas off-target sites. Mol Ther Nucleic Acids 3:e214

Crick F (1970) Central dogma of molecular biology. Nature 227(5258):561–563

Crispo M, Mulet A, Tesson L, Barrera N, Cuadro F, dos Santos-Neto P, Nguyen T, Crénéguy A, Brusselle L, Anegón I, Menchaca A (2015) Efficient generation of myostatin knock-out sheep using CRISPR/Cas9 technology and microinjection into zygotes. PLoS One 10(8):e0136690

Deltcheva E, Chylinski K, Sharma CM, Gonzales K, Chao Y, Pirzada ZA, Eckert MR, Vogel J, Charpentier E (2011) CRISPR RNA maturation by trans-encoded small RNA and host factor RNase III. Nature 471(7340):602–607

Dong S, Lin J, Held NL, Clem RJ, Passarelli AL, Franz AW (2015) Heritable CRISPR/Cas9-mediated genome editing in the yellow fever mosquito, Aedes aegypti. PLoS One 10(3):e0122353

Durai S, Mani M, Kandavelou K, Wu J, Porteus MH, Chandrasegaran S (2005) Zinc finger nucleases: custom-designed molecular scissors for genome engineering of plant and mammalian cells. Nucleic Acids Res 33(18):5978–5990

Eisenberg D, Marcotte EM, Xenarios I, Yeates TO (2000) Protein function in the post-genomic era. Nature 405(6788):823–826

Ekker SC (2000) Morphants: a new systematic vertebrate functional genomics approach. Yeast 17(4):302–306

Esvelt KM, Mali P, Braff JL, Moosburner M, Yaung SJ, Church GM (2013) Orthogonal Cas9 proteins for RNA-guided gene regulation and editing. Nat Methods 10(11):1116–1121

Feng Z, Mao Y, Xu N, Zhang B, Wei P, Yang DL, Wang Z, Zhang Z, Zheng R, Yang L, Zeng L, Liu X, Zhu JK (2014) Multigeneration analysis reveals the inheritance, specificity, and patterns of CRISPR/Cas-induced gene modifications in Arabidopsis. Proc Natl Acad Sci U S A 111(12):4632–4637

Fields S, Kohara Y, Lockhart DJ (1999) Functional genomics. Proc Natl Acad Sci 96(16):8825–8826

Fire A, Xu S, Montgomery MK, Kostas SA, Driver SE, Mello CC (1998) Potent and specific genetic interference by double-stranded RNA in Caenorhabditis elegans. Nature 391(6669):806–811

Fox AH, Liew C, Holmes M, Kowalski K, Mackay J, Crossley M (1999) Transcriptional cofactors of the FOG family interact with GATA proteins by means of multiple zinc fingers. EMBO J 18(10):2812–2822

Friedland AE, Tzur YB, Esvelt KM, Colaiacovo MP, Church GM, Calarco JA (2013) Heritable genome editing in C. elegans via a CRISPR-Cas9 system. Nat Methods 10(8):741–743

Fu Y, Foden JA, Khayter C, Maeder ML, Reyon D, Joung JK, Sander JD (2013) High-frequency off-target mutagenesis induced by CRISPR-Cas nucleases in human cells. Nat Biotechnol 31(9):822–826

Fu Y, Sander JD, Reyon D, Cascio VM, Joung JK (2014) Improving CRISPR-Cas nuclease specificity using truncated guide RNAs. Nat Biotechnol 32(3):279–284

Gaj T, Gersbach CA, Barbas CF 3rd (2013) ZFN, TALEN, and CRISPR/Cas-based methods for genome engineering. Trends Biotechnol 31(7):397–405

Geng LA, Xin W, Huang DW, Feng G (2006) A universal cloning vector using vaccinia topoisomerase I. Mol Biotechnol 33(1):23–28

Gilbert LA, Larson MH, Morsut L, Liu Z, Brar GA, Torres SE, Stern-Ginossar N, Brandman O, Whitehead EH, Doudna JA, Lim WA, Weissman JS, Qi LS (2013) CRISPR-mediated modular RNA-guided regulation of transcription in eukaryotes. Cell 154(2):442–451

Gilbert LA, Horlbeck MA, Adamson B, Villalta JE, Chen Y, Whitehead EH, Guimaraes C, Panning B, Ploegh HL, Bassik MC, Qi LS, Kampmann M, Weissman JS (2014) Genome-scale CRISPR-mediated control of gene repression and activation. Cell 159(3):647–661

Hardy S, Legagneux V, Audic Y, Paillard L (2010) Reverse genetics in eukaryotes. Biol Cell 102(10):561–580

Hasty P, Rivera-Perez J, Bradley A (1991) The length of homology required for gene targeting in embryonic stem cells. Mol Cell Biol 11(11):5586–5591

Heigwer F, Kerr G, Boutros M (2014) E-CRISP: fast CRISPR target site identification. Nat Methods 11(2):122–123

Hockemeyer D, Wang HY, Kiani S, Lai CS, Gao Q, Cassady JP, Cost GJ, Zhang L, Santiago Y, Miller JC, Zeitler B, Cherone JM, Meng XD, Hinkley SJ, Rebar EJ et al (2011) Genetic engineering of human pluripotent cells using TALE nucleases. Nat Biotechnol 29(8):731–734

Horgan RP, Kenny LC (2011) 'Omic' technologies: genomics, transcriptomics, proteomics and metabolomics. Obstet Gynaecol 13(3):189–195
Hou ZG, Zhang Y, Propson NE, Howden SE, Chu LF, Sontheimer EJ, Thomson JA (2013) Efficient genome engineering in human pluripotent stem cells using Cas9 from Neisseria meningitidis. Proc Natl Acad Sci U S A 110(39):15644–15649
Hsu PD, Scott DA, Weinstein JA, Ran FA, Konermann S, Agarwala V, Li Y, Fine EJ, Wu X, Shalem O, Cradick TJ, Marraffini LA, Bao G, Zhang F (2013) DNA targeting specificity of RNA-guided Cas9 nucleases. Nat Biotechnol 31(9):827–832
Hwang WY, Fu Y, Reyon D, Maeder ML, Tsai SQ, Sander JD, Peterson RT, Yeh JR, Joung JK (2013) Efficient genome editing in zebrafish using a CRISPR-Cas system. Nat Biotechnol 31(3):227–229
Iliakis G, Wang H, Perrault AR, Boecker W, Rosidi B, Windhofer F, Wu W, Guan J, Terzoudi G, Pantelias G (2004) Mechanisms of DNA double strand break repair and chomosome aberration formation. Cytogenet Genome Res 104(1–4):14–20
Jacobs TB, LaFayette PR, Schmitz RJ, Parrott WA (2015) Targeted genome modifications in soybean with CRISPR/Cas9. BMC Biotechnol 15:16
Jiang W, Bikard D, Cox D, Zhang F, Marraffini LA (2013) RNA-guided editing of bacterial genomes using CRISPR-Cas systems. Nat Biotechnol 31(3):233–239
Jinek M, Chylinski K, Fonfara I, Hauer M, Doudna JA, Charpentier E (2012) A programmable dual-RNA-guided DNA endonuclease in adaptive bacterial immunity. Science 337(6096):816–821
Jinek M, East A, Cheng A, Lin S, Ma E, Doudna J (2013) RNA-programmed genome editing in human cells. Elife 2:e00471
Kabadi AM, Ousterout DG, Hilton IB, Gersbach CA (2014) Multiplex CRISPR/Cas9-based genome engineering from a single lentiviral vector. Nucleic Acids Res 42(19), e147
Kay S, Bonas U (2009) How Xanthomonas type III effectors manipulate the host plant. Curr Opin Microbiol 12(1):37–43
Kim YG, Cha J, Chandrasegaran S (1996) Hybrid restriction enzymes: zinc finger fusions to Fok I cleavage domain. Proc Natl Acad Sci U S A 93(3):1156–1160
Kim S, Kim D, Cho SW, Kim J, Kim JS (2014) Highly efficient RNA-guided genome editing in human cells via delivery of purified Cas9 ribonucleoproteins. Genome Res 24(6):1012–1019
Kim D, Bae S, Park J, Kim E, Kim S, Yu H, Hwang J, Kim JI, Kim JS (2015) Digenome-seq: genome-wide profiling of CRISPR-Cas9 off-target effects in human cells. Nat Methods 12(3):237–243, 231 p following 243
Klug A, Schwabe JW (1995) Protein motifs 5. Zinc fingers. FASEB J 9(8):597–604
Kuscu C, Arslan S, Singh R, Thorpe J, Adli M (2014) Genome-wide analysis reveals characteristics of off-target sites bound by the Cas9 endonuclease. Nat Biotechnol 32(7):677–683
Lander ES, Linton LM, Birren B, Nusbaum C, Zody MC, Baldwin J, Devon K, Dewar K, Doyle M, FitzHugh W, Funke R, Gage D, Harris K, Heaford A, Howland J et al (2001) Initial sequencing and analysis of the human genome. Nature 409(6822):860–921
Li T, Huang S, Zhao X, Wright DA, Carpenter S, Spalding MH, Weeks DP, Yang B (2011) Modularly assembled designer TAL effector nucleases for targeted gene knockout and gene replacement in eukaryotes. Nucleic Acids Res 39(14):6315–6325
Li DL, Qiu ZW, Shao YJ, Chen YT, Guan YT, Liu MZ, Li YM, Gao N, Wang LR, Lu XL, Zhao YX, Liu MY (2013) Heritable gene targeting in the mouse and rat using a CRISPR-Cas system. Nat Biotechnol 31(8):681–683
Li K, Wang G, Andersen T, Zhou P, Pu WT (2014) Optimization of genome engineering approaches with the CRISPR/Cas9 system. PLoS One 9(8):e105779
Liang F, Han M, Romanienko P, Jasin M (1998) Homology-directed repair is a major double-strand break repair pathway in mammalian cells. Proc Natl Acad Sci U S A 96:5172–5177
Liang X, Potter J, Kumar S, Zou Y, Quintanilla R, Sridharan M, Carte J, Chen W, Roark N, Ranganathan S, Ravinder N, Chesnut JD (2015) Rapid and highly efficient mammalian cell engineering via Cas9 protein transfection. J Biotechnol 208:44–53
Lin Y, Cradick TJ, Brown MT, Deshmukh H, Ranjan P, Sarode N, Wile BM, Vertino PM, Stewart FJ, Bao G (2014) CRISPR/Cas9 systems have off-target activity with insertions or deletions between target DNA and guide RNA sequences. Nucleic Acids Res 42(11):7473–7485

Liu J, Li C, Yu Z, Huang P, Wu H, Wei C, Zhu N, Shen Y, Chen Y, Zhang B, Deng WM, Jiao R (2012) Efficient and specific modifications of the Drosophila genome by means of an easy TALEN strategy. J Genet Genomics 39(5):209–215
Lu D, Searles MA, Klug A (2003) Crystal structure of a zinc-finger-RNA complex reveals two modes of molecular recognition. Nature 426(6962):96–100
Maeder ML, Linder SJ, Cascio VM, Fu Y, Ho QH, Joung JK (2013) CRISPR RNA-guided activation of endogenous human genes. Nat Methods 10(10):977–979
Mali P, Yang L, Esvelt KM, Aach J, Guell M, DiCarlo JE, Norville JE, Church GM (2013) RNA-guided human genome engineering via Cas9. Science 339(6121):823–826
Mani M, Smith J, Kandavelou K, Berg JM, Chandrasegaran S (2005) Binding of two zinc finger nuclease monomers to two specific sites is required for effective double-strand DNA cleavage. Biochem Biophys Res Commun 334(4):1191–1197
Mendez-Vidal C, Wilhelm MT, Hellborg F, Qian W, Wiman KG (2002) The p53-induced mouse zinc finger protein wig-1 binds double-stranded RNA with high affinity. Nucleic Acids Res 30(9):1991–1996
Meselson M, Yuan R (1968) DNA restriction enzyme from E Coli. Nature 217(5134):1110–1114
Mojica FJ, Diez-Villasenor C, Garcia-Martinez J, Almendros C (2009) Short motif sequences determine the targets of the prokaryotic CRISPR defence system. Microbiology 155(Pt 3):733–740
Moore JK, Haber JE (1996) Cell cycle and genetic requirements of two pathways of nonhomologous end-joining repair of double-strand breaks in Saccharomyces cerevisiae. Mol Cell Biol 16(5):2164–2173
Moreno-Mateos MA, Vejnar CE, Beaudoin JD, Fernandez JP, Mis EK, Khokha MK, Giraldez AJ (2015) CRISPRscan: designing highly efficient sgRNAs for CRISPR-Cas9 targeting in vivo. Nat Methods 2(10):982–988
Moscou MJ, Bogdanove AJ (2009) A simple cipher governs DNA recognition by TAL effectors. Science 326(5959):1501
Mussolino C, Cathomen T (2013) RNA guides genome engineering. Nat Biotechnol 31(3):208–209
Nakayama T, Fish MB, Fisher M, Oomen-Hajagos J, Thomsen GH, Grainger RM (2013) Simple and efficient CRISPR/Cas9-mediated targeted mutagenesis in Xenopus tropicalis. Genesis 51(12):835–843
Nekrasov V, Staskawicz B, Weigel D, Jones JDG, Kamoun S (2013) Targeted mutagenesis in the model plant Nicotiana benthamiana using Cas9 RNA-guided endonuclease. Nat Biotechnol 31(8):691–693
O'Connell MR, Oakes BL, Sternberg SH, East-Seletsky A, Kaplan M, Doudna JA (2014) Programmable RNA recognition and cleavage by CRISPR/Cas9. Nature 516(7530):263–266
Pattanayak V, Guilinger JP, Liu DR (2014) Determining the specificities of TALENs, Cas9, and other genome-editing enzymes. Methods Enzymol 546:47–78
Pavletich NP, Pabo CO (1991) Zinc finger DNA recognition—crystal-structure of a Zif268-DNA complex at 2.1-A. Science 252(5007):809–817
Perez-Pinera P, Kocak DD, Vockley CM, Adler AF, Kabadi AM, Polstein LR, Thakore PI, Glass KA, Ousterout DG, Leong KW, Guilak F, Crawford GE, Reddy TE, Gersbach CA (2013) RNA-guided gene activation by CRISPR-Cas9-based transcription factors. Nat Methods 10(10):973–976
Pickart MA, Klee EW, Nielsen AL, Sivasubbu S, Mendenhall EM, Bill BR, Chen E, Eckfeldt CE, Knowlton M, Robu ME, Larson JD, Deng Y, Schimmenti LA, Ellis LB, Verfaillie CM et al (2006) Genome-wide reverse genetics framework to identify novel functions of the vertebrate secretome. PLoS One 1:e104
Ramakrishna S, Cho SW, Kim S, Song M, Gopalappa R, Kim JS, Kim H (2014) Surrogate reporter-based enrichment of cells containing RNA-guided Cas9 nuclease-induced mutations. Nat Commun 5:3378
Ran FA, Hsu PD, Wright J, Agarwala V, Scott DA, Zhang F (2013) Genome engineering using the CRISPR-Cas9 system. Nat Protoc 8(11):2281–2308
Saiki RK, Scharf S, Faloona F, Mullis KB, Horn GT, Erlich HA, Arnheim N (1985) Enzymatic amplification of beta-globin genomic sequences and restriction site analysis for diagnosis of sickle-cell anemia. Science 230(4732):1350–1354

Sander JD, Joung JK (2014) CRISPR-Cas systems for editing, regulating and targeting genomes. Nat Biotechnol 32(4):347–355
Sander JD, Cade L, Khayter C, Reyon D, Peterson RT, Joung JK, Yeh JRJ (2011) Targeted gene disruption in somatic zebrafish cells using engineered TALENs. Nat Biotechnol 29(8):697–698
Shan QW, Wang YP, Li J, Zhang Y, Chen KL, Liang Z, Zhang K, Liu JX, Xi JJ, Qiu JL, Gao CX (2013) Targeted genome modification of crop plants using a CRISPR-Cas system. Nat Biotechnol 31(8):686–688
Swiech L, Heidenreich M, Banerjee A, Habib N, Li Y, Trombetta J, Sur M, Zhang F (2015) In vivo interrogation of gene function in the mammalian brain using CRISPR-Cas9. Nat Biotechnol 33(1):102–106
Tesson L, Usal C, Menoret S, Leung E, Niles BJ, Remy S, Santiago Y, Vincent AI, Meng X, Zhang L, Gregory PD, Anegon I, Cost GJ (2011) Knockout rats generated by embryo microinjection of TALENs. Nat Biotechnol 29(8):695–696
Trevino AE, Zhang F (2014) Genome editing using Cas9 nickases. Methods Enzymol 546:161–174
Tsang AP, Visvader JE, Turner CA, Fujiwara Y, Yu CN, Weiss MJ, Crossley M, Orkin SH (1997) FOG, a multitype zinc finger protein, acts as a cofactor for transcription factor GATA-1 in erythoid and megakaryocytic differentiation. Cell 90(1):109–119
Ueda R (2001) Rnai: a new technology in the post-genomic sequencing era. J Neurogenet 15(3–4):193–204
Urnov FD, Miller JC, Lee YL, Beausejour CM, Rock JM, Augustus S, Jamieson AC, Porteus MH, Gregory PD, Holmes MC (2005) Highly efficient endogenous human gene correction using designed zinc-finger nucleases. Nature 435(7042):646–651
Urnov FD, Rebar EJ, Holmes MC, Zhang HS, Gregory PD (2010) Genome editing with engineered zinc finger nucleases. Nat Rev Genet 11(9):636–646
Venter JC, Adams MD, Myers EW, Li PW, Mural RJ, Sutton GG, Smith HO, Yandell M, Evans CA, Holt RA, Gocayne JD, Amanatides P, Ballew RM, Huson DH, Wortman JR et al (2001) The sequence of the human genome. Science 291(5507):1304–1351
Vouillot L, Thelie A, Pollet N (2015) Comparison of T7E1 and surveyor mismatch cleavage assays to detect mutations triggered by engineered nucleases. G3 (Bethesda) 5(3):407–415
Vukmirovic OG, Tilghman SM (2000) Exploring genome space. Nature 405(6788):820–822
Watson JD, Crick FH (1953) Molecular structure of nucleic acids; a structure for deoxyribose nucleic acid. Nature 171(4356):737–738
Wolfe SA, Nekludova L, Pabo CO (2000) DNA recognition by Cys2His2 zinc finger proteins. Annu Rev Biophys Biomol Struct 29:183–212
Wu S, Ying G, Wu Q, Capecchi MR (2008) A protocol for constructing gene targeting vectors: generating knockout mice for the cadherin family and beyond. Nat Protoc 3(6):1056–1076
Wu X, Scott DA, Kriz AJ, Chiu AC, Hsu PD, Dadon DB, Cheng AW, Trevino AE, Konermann S, Chen S, Jaenisch R, Zhang F, Sharp PA (2014) Genome-wide binding of the CRISPR endonuclease Cas9 in mammalian cells. Nat Biotechnol 32(7):670–676
Wyvekens N, Topkar VV, Khayter C, Joung JK, Tsai SQ (2015) Dimeric CRISPR RNA-guided FokI-dCas9 nucleases directed by truncated gRNAs for highly specific genome editing. Hum Gene Ther 26(7):425–431
Yang DS, Xu J, Zhu TQ, Fan JL, Lai LX, Zhang JF, Chen YE (2014) Effective gene targeting in rabbits using RNA-guided Cas9 nucleases. J Mol Cell Biol 6(1):97–99
Yu Z, Ren M, Wang Z, Zhang B, Rong YS, Jiao R, Gao G (2013) Highly efficient genome modifications mediated by CRISPR/Cas9 in Drosophila. Genetics 195(1):289–291
Zhang L, Spratt SK, Liu Q, Johnstone B, Qi H, Raschke EE, Jamieson AC, Rebar EJ, Wolffe AP, Case CC (2000) Synthetic zinc finger transcription factor action at an endogenous chromosomal site—activation of the human erythopoietin gene. J Biol Chem 275(43):33850–33850
Zhou H, Liu B, Weeks DP, Spalding MH, Yang B (2014) Large chomosomal deletions and heritable small genetic changes induced by CRISPR/Cas9 in rice. Nucleic Acids Res 42(17):10903–10914

# Chapter 18
# Neural Stem Cells: Functional Multipotency and Spinal Cord Injury Research Protocols

**Yang D. Teng, Xiang Zeng, Inbo Han, and Jaime E. Anderson**

**Abstract** Human neural stem cells (hNSCs) have been the focus in basic science and translational research as well as in investigative clinical applications. The endeavors partly aim to devise hNSCs into stem cell-based therapeutic modalities, disease models, and investigative tools for treating neurological abnormalities, understanding pathophysiology, and advancing fundamental neurobiology, respectively. Therefore, the capability to perform reliable derivation, effective expansion, and long term maintenance of primordial and uncommitted NSCs *in vitro* is essential for growing the capacities of stem cell biology and regenerative medicine. In this chapter, we systematically describe a set of protocols and unique procedures that have been developed in the principal investigator's laboratories. These regimens have been, over years, productively used to derive, propagate, maintain, and differentiate operationally defined human somatic NSCs or NSCs of different origins.

---

Y.D. Teng (✉)
Department of Neurosurgery, Harvard Medical School and Brigham and Women's Hospital, 75 Francis Street, Boston, MA 02115, USA

Departments of Physical Medicine and Rehabilitation, Harvard Medical School and Spaulding Rehabilitation Hospital, Boston, MA, USA

Division of SCI Research, VA Boston Healthcare System, Boston, MA, USA
e-mail: yang_teng@hms.harvard.edu

X. Zeng • J.E. Anderson
Department of Neurosurgery, Harvard Medical School and Brigham and Women's Hospital, 75 Francis Street, Boston, MA 02115, USA

Division of SCI Research, VA Boston Healthcare System, Boston, MA, USA

I. Han
Department of Neurosurgery, Harvard Medical School and Brigham and Women's Hospital, 75 Francis Street, Boston, MA 02115, USA

Division of SCI Research, VA Boston Healthcare System, Boston, MA, USA

Department of Neurosurgery, Cha University, Seoul, South Korea



H. Ulrich, P. Davidson Negraes, *Working with Stem Cells*,
DOI 10.1007/978-3-319-30582-0_18

Moreover, we summarize our established multimodal methodologies for characterizing the functional multipotency of NSCs and their value in basic as well as translational studies.

**Keywords** Stem cell • Neural stem cell • Mesenchymal stromal stem cell • Polymer • Biodegradation • Functional multipotency • Spinal cord injury • Stroke • Transplantation

## 18.1 Functional Multipotency of NSCs

Neural stem cells (NSCs) are developmentally primordial and uncommitted cells that could putatively give rise to the full array of specialized cells of the central nervous system (CNS) (Gage and Temple 2013; Adepoju et al. 2014; Lim and Alvarez-Buylla 2014; Lukaszewicz and Anderson 2011; Teng et al. 2011; Emborg et al. 2013; Ferrari et al. 2010). Conventionally, NSCs are operationally defined by their capabilities (1) to differentiate into cells of all neural lineages (i.e. neurons of multiple subtypes, oligodendroglia, and astroglia) in multiple regional and developmental contexts (i.e. to be phenotypically multipotent); (2) to self-renew (i.e. to also produce new NSCs with the same potential); and (3) to migrate and populate developing and/or degenerating CNS regions. Based on recently validated data, we have proposed an updated and novel concept of the biology of stem cells (SCs) and NSCs in particular. Specifically, in addition to the conventional view which has touched mainly on the essential property of lineage pluri- or multi-potency, *the emerging evidence of SC's innate capacity and capability to mediate focal or systemic homeostasis in interactions with their microenvironmental niche signals to survive and develop defines the 'Functional Multipotency of Stem Cells'*. This, in NSCs, is exemplified by their inherent ability to secrete neurotrophic factors and other cytokines, disperse exosomes, form gap junctions and cell fusions, perform specific cue molecule-mediated migration, induce neurogenesis or activities of other types of (endogenous or donor) SCs, and develop gene- and niche-determined phenotypic differentiation (i.e. the conventionally defined "lineage differentiation puri- or multi-potency of stem cells") (Teng et al. 2008, 2009, 2011; Jia et al. 2014; Imitola et al. 2006; Jäderstad et al. 2010).

In order to investigate NSCs in laboratory settings, an unambiguous demonstration of monoclonal derivation of progeny is obligatory to the definition. That is, a single cell must possess all the attributes mentioned above when assessed under appropriate conditions. In the past 20 years, it has been recognized that cells isolated from the CNS, from a variety of structures and at different developmental stages including adulthood, possess stem-like properties (Ferrari et al. 2010; Gage and Temple 2013; Lim and Alvarez-Buylla 2014). Such NSCs can be propagated in culture, and transplanted into both rodent and higher mammalian (including primate) brains where they can reintegrate appropriately and steadily express their own and engineered foreign genes (Redmond et al. 2007; Wakeman et al. 2014). Some studies, including ours, provided hope that the use of NSCs might circumvent some limitations of presently available graft material and gene transfer vehicles and make feasible a variety of therapeutic strategies (Teng et al. 2002b; Neri et al. 2010; Teng et al. 2012; Chen et al. 2015). This significant advancement has led neurobiologists to speculate how such phenomena might be harnessed for both therapeutic advantage and for better understanding developmental, physiological and pathophysiological mechanisms (Teng et al. 2011; Snyder and Teng 2012).

The true feasibility of obtaining large quantities of NSCs derived from the human brain, the newly devised induced pluripotent stem cells (iPSCs), and conditionally reprogramed NSCs holds key importance and interest for basic and applied stem cell biology and neuroscience, including therapeutic cell and gene transfer following transplantation (Takahashi and Yamanaka 2006; Takahashi et al. 2007a, b; Kim et al. 2011; Lee et al. 2015). Here, we have summarized the combination of epigenetic and genetic procedures for perpetuating rodent (in particular, murine) and human neural stem cell lines. We also included our routine procedures for *in vitro* maintenance of both kinds of NSCs as well as methods of their preparation for surgical transplantation. Lastly, we succinctly describe methods that have been effectively used to assess the biological functional multipotency of NSCs for potential roles that NSCs are capable to play for basic and translational research, tissue engineering, and therapeutic applications.

## 18.2 Murine NSCs

Cells from the CNS with stem-like properties have been successfully isolated from the embryonic, neonatal, and adult murine CNS. They can be propagated *in vitro* by many effective and safe means that comprise both epigenetic and genetic strategies. The epigenetic approach includes mitogens such as epidermal growth factor (EGF) (Deleyrolle and Reynolds 2009) or basic fibroblast growth factor (bFGF or FGF2) (Qian et al. 1997). The genetic method consists of gene transfer with propagating genes such as vmyc, large T-antigen (T-Ag), or telomerase (Cepko 1989; Dottori et al. 2011; Roy at al. 2007).

Importantly, maintaining murine NSCs (mNSCs) in a proliferative state in culture does not appear to subvert their ability to respond to normal developmental cues *in vivo* following transplantation. Upon entering the *in vivo* environment, NSCs withdraw from the cell cycle, interact with host cells, and differentiate to express cellular markers for neuronal and glial lineages (Park et al. 2002; Redmond et al. 2007). Often, a small subpopulation will survive in long terms and remain as quiescent and undifferentiated cells intermixed seamlessly among host cells (Teng et al. 2002a). These extremely plastic cells from both rodents and human intend to migrate and differentiate in temporal and regional manners in response to cue molecules, which emulates endogenous NSCs following implantation into germinal zones, as well as acts like white blood cells to migrate into lesion/inflammatory areas based on chemotaxic mechanisms (e.g. SDF-1α and CXCR4, VEGF and VEGFR-1, etc.) (Wittko et al. 1999; Schmidt et al. 2005; Imitola et al. 2006; Teng et al. 2002a, 2012). In addition, they can express developmental cytokines including trophic or anti-inflammatory factors and engineered foreign genes that are homeostatic/therapeutic, shed exosomes, form gap junctions, and are capable of neural cell replacement (i.e. functional multipotency of NSCs) (Teng et al. 2002a, 2011; Redmond et al. 2007; Park et al. 2002). For example, multipotent, clonal mNSC lines generated from neonatal murine cerebellum can participate in normal development along the rodent neuraxis, largely independent of the initial region from which they were isolated, attesting to their stemness. Donor mNSCs can integrate not only back into the developing cerebellum but throughout the entire neuroaxis of the immature and adult rodent central and peripheral nervous system in a non tumorigenic, cytoarchitecturally matching manner (Snyder et al. 1992). Differentiation fate appears to be determined by site-specific microenvironmental factors in addition to neural progenitor lineage; cells from the same clonal line differentiated into neurons or glia based on their site of engraftment. Transplanted NSCs that differentiated into ultrastructurally identifiable neurons appeared, for example, to participate in synaptogenesis, suggesting not only functional but also anatomic integration (Snyder et al. 1992; Taylor et al. 2006). Similar phenomena have been reported for adult NSCs obtained from rodent hippocampus (Kempermann et al. 2015), spinal cord (Shihabuddin et al. 2000), ventricular-subventricular zone (V-SVZ) (Lim and Alvarez-Buylla 2014), and cortex (Florio and Huttner 2014; De Juan Romero and Borrell 2015).

Therefore, NSCs, as modeled by these various mNSC lines, can be used as informative investigative probes and have the capability for repair of, or transport of genes into, the CNS. Furthermore, such cellular level plasticity provides a unique mechanism for a developmental strategy whereby multipotent progenitors migrate with commitment to cell phenotype occurring amid interaction with their microenvironment. While mouse and rat generally seem to show similar biological behaviors under most circumstances, the accumulated data suggest that mouse NSCs are mostly more practical than those of rat to propagate and maintain for prolonged periods for as-yet undetermined reasons. However, most technological approaches and biological concepts derived from mouse NSCs can be extrapolated to rat NSCs.

## 18.3 Maintenance and Propagation of Murine NSCs *In Vitro*

***Note:*** *All cell culture medium, trophic factors, cytokines, genetic molecules, and other experimental reagents can be ordered from American Type Culture Collection (Manassas, VA, USA), Atlanta Biologicals (Flowery Branch, GA, USA), EMD Millipore Life Science (Billerica, MA, USA), Innovative Cell Technologies, Inc. (San Diego, CA, USA), Life Technologies (Grand Island, NY, USA), Invitrogen (Carlsbad, CA, USA), Thermo Fisher Scientific Inc. (Hudson, NH, USA), Sigma-Aldrich Corporation (St. Louis, MO, USA), Promega (Madison, WI, USA), or other vendors with validated product reputation.*

The most widely used, genetically engineered, mNSC line in our laboratories and that of our collaborators' is C17.2 which was initially developed by Dr. E.Y. Snyder and colleagues (Snyder et al. 1992). C17.2 cells, a prototypical NSC clone, are originally derived from neonatal mouse cerebellum as previously described (Snyder et al. 1992).

### *18.3.1 Methods and Materials*

1. Derivation of NSC lines can be most readily accomplished by dissecting the primary structure of interest (e.g. cerebellum, olfactory bulb, cortex, hippocampus, spinal cord) from immature (fetal or newborn) mice and incubating the tissue in 0.5 % trypsin in phosphate buffered saline. If the tissue is abstracted from adult animals, dissociation will require collagenase and/or Accutase as well as longer incubation times;
2. Tissue is triturated and washed twice before cultures are plated on uncoated tissue culture dishes in serum-containing medium. In this step, serum both inactivates the trypsin and insures the health of the dissociated cells;
3. At this stage, cells can then be propagated in serum-free medium with mitogens—EGF (10–20 ng/ml) plus FGF2 (10–20 ng/ml) plus LIF (10 ng/ml)—or they can be transduced with a gene that enhances propagation by acting downstream of these mitogens. For the growth factor selection process of NSCs, please see section on human NSCs below;
4. If one elects to use the genetic approach to propagate NSC clones, then the primary culture described above is infected 24–48 h after plating, by incubation with the avian myc or viral myc (vmyc) vector PK VM 2. pneoMLV is the parent retrovirus vector in which vmyc DNA was inserted (Kaplan et al. 1987). Cells are then cultured in DME + 10 % FCS + 5 % horse serum (HS) + 2 mM glutamine for 3–7 days, until cultures appear to have undergone at least two doublings;
5. Cultures are then trypsinized (or accutased) and seeded at 3–10 % confluence in DME + 10 % FCS + 0.3 mg/ml G418. Neo-resistant colonies are typically observed within 1–2 weeks;
6. Chosen colonies are then isolated by brief exposure to trypsin within plastic cloning cylinders after ~2 weeks. Colonies are then re-plated and expanded on uncoated 24-well CoStar plates. At confluence, these cultures are further passaged and expanded to first 35 mm and then 60 mm, and ultimately 100 mm uncoated Corning tissue culture dishes. Cells infected with control vectors without propagating genes did not survive beyond this passage in serum-containing medium;
7. Early passages of the expanded colonies are frozen in Nunc cryostat tubes at $1 \times 10^6$ cells/vial in DME + 30 % FCS + 12 % DMSO (*This method of freezing is also effective for most other types of murine NSC lines*).

### *18.3.2 Genetic Modification*

Immortalized NSC lines such as C17.2 can then be transfected with additional genes, such as *lacZ*, the gene encoding *E. coli* β galactosidase (βGal) (Price et al. 1987)

1. This procedure is done by plating a recent 1:10 split of the cell line of interest onto 60 mm tissue culture plates. Twenty-four to forty-eight hours after plating, the cells are incubated with a lacZ-encoding retroviral vector (e.g., BAG) plus 8 μg/ml Polybrene for 1–3 h. Cells are then cultured in fresh feeding medium (DME + 10 % FCS + 5 % HS + 20 mM glutamate) for ~3 days until they will have undergone through at least two doublings;
2. The cultures are then trypsinized (or accutased) and seeded at low density (50–5000 cells on a 100 mm tissue culture dish). If the newly inserted transgene contains a selection marker distinct from your first selection marker, e.g. non-neomycin-based selection marker such as hygromycin or pyromycin, then one can select for the new-infectants as described above;
3. After ~3 days, well-separated colonies are isolated by brief exposure to trypsin (or Accutase) within plastic cloning cylinders, as described above. If the additional transgene does not have a new selection marker or also employs *neo*, identification of infectants is still possible. Select colonies as described and proceed to plate them in 24-well plates. At confluence, these cultures can be passaged to and expanded in increasingly larger surface area tissue culture dishes;
4. A representative dish from each clone can then be stained histochemically for your new gene of interest, e.g. for βGal expression via Xgal histochemistry (Snyder et al. 1992; Teng et al. 2002a; Redmond et al. 2007), directly in the culture dish. The percentage of Xgal + blue cells is then assessed microscopically. The clones with the highest percentage of blue cells are then used for future studies.

The ability to detect the *lacZ* gene product by Xgal histochemistry *in vivo* and/or by anti-βGal immunohistochemistry makes such lacZ-expressing cells effective for transplantation studies, especially when it is used under cross verification of immunocytochemistry deploying antibody specifically against βGal (Redmond et al. 2007; Teng et al. 2002a, b).

Preferred murine cerebellar or other type of NSC lines are grown in DME + 10 % FCS + 5 % HS + 2 mM glutamine on tissue culture dishes coated with poly-L-lysine (PLL). They are either fed weekly with 1/2 conditioned medium from confluent cultures + 1/2 fresh medium, or split (1:10) weekly into fresh medium. Lines should not routinely be split more dilutely than 1:10 although one may potentially have the ability to do so. We recommend that, for important transplant studies, a given thaw of cells should not have been passaged for longer than 4–8 weeks. It is better thaw out an earlier vial and use one of the two or three post-expansion vials for the planned *in vivo* study. Therefore, a stock-pile of as many as possible of early freezes is advisable.

Subclones can be abstracted by seeding 50–5000 cells on 10 cm dishes in the typical medium, or in 1/3 conditioned medium from confluent cultures. Well separated colonies can again be isolated using cloning cylinders. One should be aware that, for reasons that still remain not specified, changes in brands of tissue culture plastic, type of substrate coat, and type, concentration, and lot of serum can change the phenotypes displayed by NSC lines.

## 18.4 Preparation of Murine NSCs In Vitro for In Vivo Transplantation

For preparation of cells for transplantation in, for example, the injured spinal cord of rodents, here is a succinct description of procedure workflow.

### *18.4.1 Spinal Cord Injury Models*

1. Young adult or adult Sprague-Dawley (or a different strain) rats (body weight: 225–250 g) are randomly assigned for receiving different pre- and/or post-injury treatments with group size carrying adequate power of statistics (Teng et al. 2004). All *in vivo* protocols should be in accordance with the principal investigator's institutional animal care and use committee (IACUC) guidelines and policies issued by federal and local regulatory agencies.
2. For each SCI (spinal cord injury) site, midline contusion (Teng et al. 2004) or compression (Ropper et al. 2015) or, hemicontusion (Choi et al. 2005), or hemisection (Teng et al. 2002a) is created using a standardized injury device or size 11 surgical scalpel, under IACUC-approved sufficient anesthesia.
3. Hemostasis is normally achieved by using Gelfoam (Pfizer, New York, NY). *An independent observer who is blinded for the experimental group design should confirm the adequacy and consistency of the lesion severity or length and breadth of the lesion.* Only at that time is the surgeon informed of the treatment (e.g. cell suspension solution injection, implanting of NSC-seeded polymer construct, etc.) to be administered to the lesion.
4. Following either the full or control treatment, the musculature is sutured, skin closed, and the rodent recovered in a cage with clean bedding materials on a heating pad till it fully awakes (usually about 4 h).
5. Ringer's lactate solution (0.5–1.0 ml/mouse; 5.0–10.0 ml/rat) should be given daily for 5–7 days post-operation. *When injury affects spontaneous micturition reflex, bladders were gently evacuated twice daily until a so called "reflex bladder" function is established (Teng et al.* 1999, 2004*).*

### *18.4.2 Preparation and Quality Control of Donor NSCs*

1. <u>NSC preparation:</u> a subconfluent (75–85 %, not >90 %) 10 cm culture dish of adherent mNSCs (or cells of the line of interest) is washed with PBS (1–3 times);

   ***Cautions:*** If cells are allowed to become too confluent for more than 24–48 h prior to transplantation, they begin to elaborate an extracellular matrix that causes them to become clumpy and renders poor engraftment and often autonomous clusters of cells. Also, the cells begin to exit the cell cycle and differentiate, which also predisposes to poor engraftment. The more immature and proliferative and well isolated the cells, the better the engraftment.

***Recommendations:*** Of course, it is reassuring to purposely allow sister plates to proceed (in parallel assessments *in vitro)* to confluence and differentiation to insure that the cells still do possess the ability to differentiate into mature neural phenotypes. It is better to use many dishes of 75–85 % confluent cells than a fewer number of dishes with more cells that have become heavily confluent.

2. Cells that have been prepared appropriately for transplantation (we usually do a split 48 h before the intended grafting procedure) are then gently trypsinized (or accutased), smoothly but thoroughly triturated into well-dispersed single-cell suspensions;
3. Ressuspending the cells in PBS (pH 7.4) or in feeding medium plus ~0.05 % w/v trypan blue (which will aid with localization of the injected material) to give a high cellular concentration ($3–6 \times 10^4$ cells/μl), which is nevertheless free-flowing and will not obstruct the glass pipette (Φ: 70–100 μm) or needle used for implantation;
4. Although we employ finely drawn glass micropipettes using Flaming/Brown Micropipette Puller (Model: P-97, Sutter Instrument Co. Novato, CA, USA), one can also employ a Hamilton syringe though we recommend connection of a glass micropipette, instead of a conventional metal needle, with it for cell injections. *It is important to minimize cell settlement out of suspension (which can occur rather quickly if the suspension solution is not prepared correctly) and to ensure that there is no clump formation that is usually detected by sharply increased resistance to injection advancement;*
5. Hence, keep cells on ice and gently triturate them regularly. One should be careful not to fall into the trap of implanting cell carrying vehicle from which the cells have settled out yet believing that one has implanted cells. Interpretation of data, as one might imagine, will be entirely erroneous.

### *18.4.3 Donor Cell Tracking*

For NSCs engineered via retrovirus to express a transgene, e.g. for βGal transduced cell lines, the "helper virus-free" status needs to be demonstrated prior to transplantation. Otherwise, host-derived cells may be falsely interpreted as being donor-derived. The test can be done by confirmation of the failure of supernatants from the cell lines to produce neomycin-resistant colonies or X-gal positive colonies in naive cells (e.g. 3T3 fibroblasts). In addition, transgene+cells should be systematically validated to exclude possibilities of cell fusion-mediated contamination before confirming their true identity.

While the most efficient method for detecting cells *in vivo* is via the presence of a transgene, the risk of down-regulation of a transgene, for unspecified/unclear reasons and thus usually unpredictably, presents enough of a problem that multiple simultaneous methods of donor cell detection *in vivo* should also be put in place prior to an important transplantation study. For example,

- FISH (fluorescence *in situ* hybridization) against the *lacZ* gene itself can be effective. Using NSCs derived from a male animal and implanting them into a female animal allows one to do FISH against the Y-chromosome;
- Most easily, one can pre-incubate the NSCs in BrdU or EdU ~48 h (depending on cell doubling time) prior to transplantation and subsequently identify the cells *in vivo* through the use of an anti-BrdU antibody (which requires DNA denaturation by typically using HCl, heat, or digestion with DNase to expose the BrdU for immuno-detection);
- Or, a simple EdU detection assay (which is based on a click reaction via a copper-catalyzed reaction between an azide-containing detection reagent and an EdU-possessing alkyne). The small size of azide allows the use of mild conditions to readily detecting EdU incorporated into the DNA.

Either assay will reveal donor-derived cells via their immunopositive nucleus (Redmond et al. 2007). Most NSCs only undergo 0–2 cell divisions *in vivo* following transplantation, usually not enough to dramatically 'dilute" the BrdU marker (e.g., it takes 5 symmetric divisions to get rid of ~97 % BrDU in each cell).

## 18.5 Human NSCs

Similar to their rodent counterparts, stable clones of NSCs have been isolated from the human fetal telencephalon (Flax et al. 1998; Eriksson et al. 1998; Vescovi et al. 1999; Roy et al. 2000; Redmond et al. 2007; Teng et al. 2002a, b). These self-renewing clones *in vitro* give rise to all fundamental neural lineages. Following transplantation into germinal zones of the brain or the spinal cord of new born rodents, they participate in aspects of normal development, including migration along established migratory pathways to disseminated CNS regions, intended differentiation or differentiation into multiple developmentally- and regionally-appropriate cell types, and non-disruptive interspersion with host progenitors and their progeny. Prototype or genetically engineered *ex vivo*, human NSCs (hNSCs) are capable of expressing innate or foreign transgenes *in vivo* in these disseminated locations (Flax et al. 1998; Eriksson et al. 1998; Roy et al. 2000). The soluble products secreted from and/or gap junctions formed by these hNSCs can cross-correct an inherited genetic metabolic defect in neurons and glia *in vitro* or *in vivo*, further supporting their gene therapy potential. Finally, human NSCs have the potential to replace specific deficient neuronal populations (Brustle et al. 1998; Flax et al. 1998; Eriksson et al. 1998; Roy et al. 2000; Redmond et al. 2007; Jäderstad et al. 2010).

For research and potential clinical applications, hNSCs have been studied and found feasible for cryopreservation. In addition, these cells can be propagated by both epigenetic and genetic means that are comparably safe and effective. The observations encourage investigations of NSC transplantation for a range of disorders. Thus, we have tested various culture conditions and genes for those that optimally allow for the continuous, efficient expansion and passaging of prototype hNSCs (Redmond et al. 2007; Yu et al. 2009; Teng et al. 2012). For genetic engineering modifications, *vmyc* (e.g. the p110 gag-myc fusion protein derived from the avian retroviral genome) seems to be one of the most effective genes (Flax et al. 1998; Kim et al. 2011). In either case, we have also identified a strict requirement for the presence of mitogens (FGF2 and EGF) in the growth medium, in effect constituting a conditional perpetuality or immortalization (LIF has been effective for blunting potential senescence) (Redmond et al. 2007; Yu et al. 2009; Teng et al. 2012; Haragopal et al. 2015).

### *18.5.1 hNSC Protocols*

- In general, our monoclonal, nestin-positive, human neural stem cell lines perpetuated in this way divide every ~40 h and stop dividing upon mitogen removal, undergoing spontaneous morphological differentiation and upregulating markers of the three fundamental lineages in the CNS (neurons, astrocytes and oligodendrocytes) (Wakeman et al. 2014; Haragopal et al. 2015);
- Therefore, hNSC lines (e.g. HFB2050) retain basic features of epigenetically expanded human neural stem cells (Redmond et al. 2007; Teng et al. 2012; Wakeman et al. 2014; Haragopal et al. 2015). Clonal analysis confirmed the stability, multipotency, and self-renewability of the cell lines;
- Finally, hNSC lines can be transfected and transduced using a variety of procedures and genes encoding proteins for marking purposes (e.g. lacZ) and of therapeutic interest (e.g. BDNF) (Snyder et al. 1992, 1995; Flax et al. 1998);
- The isolation, propagation, characterization, cloning, and transplantation of NSCs from the human CNS appeared to share a similar tack established by prior experience with the successful murine NSC clone C17.2 (propagated following transduction of a constitutively downregulated *vmyc* Snyder et al. 1992, 1995) and with growth factor-expanded murine NSC clones (Weiss et al. 1996).

## 18.6 Induced Conditional Self-Renewing Human NSCs

For the current field of induced pluripotency or potency and fate reprogramming, an interesting question is that to what extent a cell should be reprogramed back to a more primitive state for translational purposes. The standard scenario requires reprogramming a terminally committed cell back to pluripotence before instructing it toward a particular specialized phenotype. The process is in general demanding and increases risks of developing neoplasia and undesired cell types. Since precursor/progenitor cells derived for therapeutic goals typically lack only one critical attribute—the capacity for sustained (and/or controllable) self-renewal, we deemed that the capacity could be induced in a regulatable manner such that cells proliferate only *in vitro* (or, for most cases, in short terms following transplantation) and differentiate *in vivo* without common dangers related to promoting pluripotence or specifying lineage identity.

As proof-of-concept, our collaboration team theorized and tested the efficiency, safety, engraftability, and therapeutic utility of "induced conditional self-renewing progenitor cells" derived from the human CNS. In the study, the self-renewal capability was efficiently induced within neural progenitors solely by introducing *vmyc* tightly regulated by a tetracycline (Tet)-on gene expression system (details of methods in Kim et al. 2011). Tet in the culture medium activated *myc* transcription and translation, allowing efficient expansion of homogeneous, clonal, karyotypically normal human CNS precursors *ex vivo*. After transplantation *in vivo*, when Tet was not administered, *myc* expression was absent, which impeded self-renewal and its tumorigenic potential. Cell proliferation ceased, and differentiation into electrophysiologically active neurons and other CNS cell types *in vivo* was facilitated upon transplantation into rats.

Besides showing the efficacy in both developmental rats and adult rats post ischemia injury (e.g. functional improvement and without neoplasia, donor cell overgrowth, emergence of non-neural cell types, phenotypic or genomic instability, etc.), this approach also mitigated the need for immunosuppression. We postulate that the strategy of "induced conditional self-renewal (ICSR)" might be applied to progenitors of other lineages. It may ultimately prove to be a safe, effective, efficient, and practical method for optimizing investigative or therapeutic potential gained from the ability to genetically reprogram stem or progenitor cells (Kim et al. 2011).

## 18.7 Isolation, Selection, Maintenance and Propagation of Human NSCs In Vitro

1. Prepare a suspension of primary dissociated neural cells ($5 \times 10^5$ cells/ml) from the telencephalon (particularly the periventricular region) of an early second trimester human fetus (e.g. 13 or 15 weeks) (Flax et al. 1998; Vescovi et al. 1999; Teng et al. 2002b);
2. Plate the cells on uncoated tissue culture dishes (Corning) in serum-containing medium, as described for mouse, for ~24–48 h;
3. Switch to "growth medium": Dulbecco's Modified Eagles Medium (DMEM)+F12 medium (1:1) supplemented with N2 medium (Gibco) plus FGF2 (10–20 μg/ml)+heparin (8 μg/ml) and EGF (10–20 μg/ml) and LIF (10 ng/ml);
4. Next, cultures are put through the following "growth factor" selection process: transfer cells to FGF2-containing serum-free medium for 2–3 weeks then; culture the cells in EGF-containing serum-free medium for another 2–3 weeks and; returne cells to FGF2-containing serum-free medium for 2–3 more weeks;
5. Finally, cells are maintained in serum-free medium containing FGF2 plus LIF (EGF is optional);

***Note:*** *At each stage of selection, large numbers of cells would die or fail to survive passaging. What would be left following the above selection process are passageable, immature, proliferative cells that are responsive to both FGF2 and EGF (i.e. co-expression of both receptors), qualities essential, in our view, for operationally defining an NSC.*

6. Medium is changed every 5–7 days. Cell aggregates are dissociated in trypsin/EDTA (0.05 %) or Accutase when ≥10–20 cell diameters in cluster or sphere sizes and replated in growth medium at $5 \times 10^5$ cells/ml;
7. To verify differentiation potential, plate some dissociated hNSCs on poly-L-lysine (PLL)-coated slides (Nunc) in DMEM (devoid of growth factors) + 10 % fetal bovine serum (FBS; one may add 2 mM L-glutamine and 1000 U/ml penicillin/1000 μg/ml streptomycin). Proceed to weekly rounds of immunocytochemistry (ICC). In most cases, differentiation occurres spontaneously or under enhanced promotion of selected trophic factors (Haragopal et al. 2015);
8. For astrocytic maturation, clones can be co-cultured with primary dissociated embryonic CD-1 mouse brain (Ryder et al. 1990). As a new approach, neural cell (neuron, in particular) differentiation can be induced by culturing NSCs in nanofibers fabricated in unique patterns (Jia et al. 2014);
9. Next, separate polyclonal populations of hNSCs into single clonal lines either by serial dilution alone (i.e. one cell per well—the process often has poor yields for prototypic hNSCs), or by first infecting the cells with a retrovirus (allowing there to be an engineered mitotic gene and/or molecular marker of clonality, i.e. the proviral integration site) and then performing serial dilution. The transgenes transduced via retrovirus are either (non-transforming) propagation enhancement genes, such as *vmyc* or human telomerase reverse transcriptase (hTERT) and/or purely reporter genes such as *lacZ* (Flax et al. 1998; Roy et al. 2007);
10. For retrovirus-mediated gene transfer, two xenotropic, replication-incompetent retroviral vectors have been used to infect hNSCs. A vector encoding *lacZ* is similar to BAG (Snyder et al. 1992) except for bearing a PG13 xenotropic envelope. An amphotropic vector encoding *vmyc* was generated using the ecotropic vector described for generating murine NSC clone C17.2 (Ryder et al. 1990) to infect the GP+envAM12 amphotropic packaging line (Markowitz et al. 1988). No helper virus needs to be produced. Infection of FGF2-maintained human neural cells with either vector (titer: $4 \times 10^5$ CFUs) can follow similar, previously detailed procedures (Markowitz et al. 1988; Snyder et al. 1992);
11. For cloning hNSCs, cells are dissociated as above, diluted to 1 cell/15 μl and plated at 15 μl/well of a Terasaki or 96-well dish. Wells with single cells are immediately noted. Single cell clones can be expanded and maintained in FGF2-containing growth medium. Single cells grow best when conditioned medium from dense hNSC cultures is included as at least 20–50 % of the growth medium;

12. Monoclonality is confirmed by identifying in all progeny a single and identical genomic insertion site by Southern analysis for, as an example, either the lacZ- or the vmyc-encoding provirus, as previously detailed (Ryder et al. 1990). The *vmyc* probe is generated by nick translation labeling with 32P dCTP; a probe to the *neo* sequence of the lacZ-encoding vector is generated by PCR utilizing 32P dCTP;
13. Lastly, for cryopreservation, resuspend post-trypsinized/accutased hNSCs in a freezing solution composed of 10 % DMSO, 50 % FBS, 40 % FGF2-containing growth medium. Decrease the temperature of cells slowly, first to 4 °C for 1 h, then to −80 °C for 24 h and, then to −140 °C (Teng et al. 2002b).

### *18.7.1 Preparation of Human NSCs* In Vitro *for* In Vivo *Transplantation*

Prepare *in vivo* model properly as per the SCI modeling system described above.

1. Cultured cells grow as a combination of adherent cells and floating clusters within T25 flasks. Collect all cells and homogeneously dispersed them into a suspension of individual cells, so they will engraft efficiently;
2. To accomplish this, transfer all medium and cells (including those adherent which are mechanically dislodged) to a 15 ml conical tube, centrifuged at 1000 rpm for 3 min;
3. Removel the supernatant and add 0.7 ml of trypsin/EDTA (0.05 % Trypsin, 0.53 mM EDTA or Accutase, typically 2.5–5 ml for a T25 flask depending upon confluency and density of the cell culture) to the centrifuge tube: briefly triturate the cells (again) and incubate at 37 °C for 5 min to facilitate cell dissociation;

   ***Note:*** *Further gentle triturating is required to break up pellets and reach a true single cell suspension status. Although Trypsin can also be added to the original flask in order to retrieve cells that may have still been adherent to the flask, it must be used with caution as over-digestion could trigger irreversible cell clunking. Trypsinization is then terminated by adding 0.7 ml trypsin inhibitor (0.25 mg/ml in PBS) into the tube (or the flask), and triturating the mixture thoroughly. After centrifugation at 0.2 RCF for 3 min and removal of supernatant, cells are washed 1–3 times by resuspending them in 10 ml PBS.*
4. At this stage, additional labeling of cells with trackers such as DiI or Hoechst can be performed according to protocols suggested by manufacturers. Then cells can be resuspended with small volume of PBS (or with a sufficient amount of trypan blue, e.g. ~0.05 % w/v, to permit viability assay or localization of the injected suspension);
5. Count the cells using a hemocytometer; adjust the final volume to get an ideal concentration of cells for injection (i.e. 5–10 × $10^4$ cells/μl). The remaining transplantation procedures are similar to those employed for murine NSCs.

## 18.8 Multimodal Applications of NSC-Based Investigative and Translational Approaches

As discussed in the previous sections, the transplantation of NSCs into areas of injury can be useful for cell replacement and/or for delivery of therapeutic genes. Some of the most impressive examples of this, in platform technology establishing studies, were observed in rat models of traumatic SCI, and mouse models of hypoxic-ischemic (HI) cerebral injury (Teng et al. 2002a; Park et al. 2002). Although NSCs appear to have the potential to repopulate severely injured spinal cord or HI-injured brain, their ability to survive, reconstitute neural tissue, and reform neural connections is often limited by the vast amount of parenchyma loss and the consequent secondary injury cellular and molecular milieu. Since the epicenter of the primary lesion changes rapidly into a necrotic syrinx (the so called "cystic cavity" or "lesion cavity"), even the most vital NSC may need an organization template that partially serves as extracellular matrix (ECM) skeletons to support survival and guide restructuring. In addition, large volumes of cells will not survive if located greater than a few 100 μm from the nearest capillary (Park et al. 2002). Hence, *we first hypothesized that 3-dimensional highly porous (or, purposely "structurally patterned") scaffolds composed of biodegradable (natural or synthetic) copolymers (or polymers) such as poly(lactic-co-glycolic acid) (or polyglycolic acid, PGA) that are also well tolerated by live biological systems, if pre-seeded with NSCs (or other types of cells) for co-transplantation into the infarction cavity or space in solid organs, might facilitate donor cell survival, migration, differentiation, and reformation of structural and functional circuits (Teng et al.* 2002a*; Park et al.* 2002*).* Since PLGA or PGA is a synthetic biodegradable polymer with Food and Drug Administration (FDA) approval for a variety of clinical applications, this approach has since become a primary platform technology for devising research and clinical applications of cells, especially stem cells. Highly hydrophilic, PLGA or PGA loses its mechanical strength rapidly over 2–4 weeks in the body; the scaffold can initially provide a matrix to guide cellular organization and growth, allowing diffusion of oxygen/nutrients to the transplanted cells, become vascularized, and then disappear, obviating concerns over long-term biocompatibility.

To test this hypothesis for spinal cord repair, a multi-component, biodegradable, synthetic PLGA scaffold of specified architecture and seeded with NSC's was designed to support and structure neural repair, including possibilities of neural regeneration, direct cell replacement, impediment of glial scar formation, stabilization of blood-spinal cord-barrier, and mitigation of secondary injury events (Teng et al. 2002a). The implantation of the scaffold seeded with NSCs in an adult rat segmental hemisection model of spinal cord injury (T9-10; initial length: 4 mm) led to robust long-term improvement in function relative to the lesion control group. At 70 days post injury, the scaffold with cells group exhibited coordinated weight-bearing stepping as compared with movement of two to three hindlimb joints in the lesion control group. Importantly, transplantation of scaffold alone also showed significantly benefit for locomotion improvement, with the treated rats demonstrating body weight bearing stepping in the hindlimbs. In contrast with conventional data, tract tracing revealed no corticospinal tract axons passing through the injury epicenter to the caudal side of the cord; however, there were very few BDA (Biotinylated dextran amine)-labeled CST fibers spotted along the interface between the injured side and the contralateral intact spinal cord, suggesting that they were likely spared by the NSC-seeded polymer or polymer only treatment. Histological and immunocytochemical analysis including antibodies against donor specific or general NSC markers (e.g. nestin), neurofilament (H, M and L), GFAP, and GAP-43 indicated that functional recovery was the product of a significant mitigation of secondary tissue loss resulting in an increase in preserved tissue and reconstitution of host neural tissue of the damaged spinal cord. Most importantly, we observed no donor-derived mature neuronal replacement in the lesion epicenter, suggesting an overall suppressive environment for neurogenesis in the adult mammalian spinal cord; the few long surviving donor cells (~1–3 %) remained to be progenitors showing nestin positive immunoreactivity. Therefore, we for the first time concluded that NSCs, in addition to their lineage differentiation capacity, could produce neurotrophic and anti-inflammatory factors/cytokines in responses to lesion niche demands to promote tissue and functional repair through rebuilding homeostasis (Teng et al. 2002a). This evidence, together with the finding that NSCs can nurture functioning neural networks via forming gap junctions with their surrounding cells (Jäderstad et al. 2010; Li et al. 2010), enabled us to engender a novel concept of stem cell biology—the functional multipotency of stem cells (Teng et al. 2011). Our work showed that the broadened biology spectrum of stem cells including the essential capability of terminal phenotypic differentiation can be enhanced by polymer-mediated multimodal applications. For example, when NSC-PGA polymer combination was transplanted into the cerebral infarction cavity via a glass micropipette 4–7 days following induction of an experimental HI insult, the NSCs were observed to completely impregnate the PGA matrix and the NSC-PGA unit refilled the infarction cavity. Part of the reconstituted tissue became vascularized by the host. The polymer-scaffolded NSCs displayed robust engraftment, foreign gene expression, and differentiation into neurons and glia within the region of HI injury, which in contrast with the spinal cord showed a neurogenesis permissive environment in the adult mammalian brain. Neuronal tracing assessment with DiI and BDA suggested that the long-distance neuronal circuitry between donor-derived and host neurons in both cerebral hemispheres might have been reformed through the corpus callosum in some cases (Park et al. 2002).

These groundbreaking findings suggest that multimodal applications of stem cells (e.g. using drug-embedding polymer to promote cell survival/guide cell fate or stage "biological reactors") may facilitate even further the repair potential, function, and differentiation of donor and host neurons, and augment the ingrowth/outgrowth of such cells to help promote reformation of structural/functional spinal cord and brain neural circuits. Together, the studies performed by our teams and others suggest that organ or iPSC-derived NSCs can play a fundamental role in neural repair via mechanistically oriented strategies of neurobiological investigation (Teng et al. 2002a, 2012; Haragopal et al. 2015), tissue engineering (Jia et al. 2014), controlled drug release (Yu et al. 2009), conditional reprogramming (Kim et al. 2011), cell replacement (Park et al. 2002; Redmond et al. 2007), and gene directed enzyme prodrug therapy (GDEPT) (Ropper et al. 2015). Indeed, the data further reinforce the idea that, for CNS and PNS repair, NSCs may serve not only as an individual investigative or therapeutic vehicle but also as an anchor that holds concomitant approaches together: molecular and cell interaction, gene therapy, drug delivery, biomaterial, tissue engineering, niche modification, and cell replacement.

**Acknowledgments** The work at Teng Laboratories has been supported by Project ALS, NIH, VA, DoD, BSF, the Gordon Project to Treat and Cure Clinical Paralysis, and CASIS-NASA. We thank Dr. E. Snyder, Dr. R. Sidman, and Dr. R. Langer for their mentorship and support to our research work. Author contributions: YDT wrote and finalized the manuscript. XZ, IH and JEA actively participated in stem cell culture, cryoprotection, implantation preparation, and protocol updates in working with Dr. Teng. The authors hold no COI for this manuscript.

## References

Adepoju A, Micali N, Ogawa K, Hoeppner DJ, McKay RD (2014) FGF2 and insulin signaling converge to regulate cyclin D expression in multipotent neural stem cells. Stem Cells 32:770–778

Brustle O, Choudhary K, Karram K, Huttner A, Murray K, Dubois-Dalcq M, McKay RD (1998) Chimeric brains generated by intraventricular transplantation of fetal human brain cells into embryonic rats. Nat Biotechnol 16:1040–1044

Cepko CL (1989) Immortalization of neural cells via retrovirus-mediated oncogene transduction. Annu Rev Neurosci 12:47–65

Chen H, Qian K, Chen W, Hu B, Blackbourn LW 4th, Du Z, Ma L, Liu H, Knobel KM, Ayala M, Zhang SC (2015) Human-derived neural progenitors functionally replace astrocytes in adult mice. J Clin Invest 125:1033–1342

Choi H, Liao WL, Newton KM, Onario RC, King AM, Desilets FC, Woodard EJ, Eichler ME, Frontera WR, Sabharwal S, Teng YD (2005) Respiratory abnormalities resulting from midcervical spinal cord injury and their reversal by serotonin 1A agonists in conscious rats. J Neurosci 25:4550–4559

De Juan Romero C, Borrell V (2015) Coevolution of radial glial cells and the cerebral cortex. Glia 63:1303–1319

Deleyrolle LP, Reynolds BA (2009) Isolation, expansion, and differentiation of adult mammalian neural stem and progenitor cells using the neurosphere assay. Methods Mol Biol 549:91–101

Dottori M, Tay C, Hughes SM (2011) Neural development in human embryonic stem cells-applications of lentiviral vectors. J Cell Biochem 112:1955–1962

Emborg ME, Liu Y, Xi J, Zhang X, Yin Y, Lu J, Joers V, Swanson C, Holden JE, Zhang SC (2013) Induced pluripotent stem cell-derived neural cells survive and mature in the nonhuman primate brain. Cell Rep 3:646–650

Eriksson PS, Perfilieva E, Bjork-Eriksson T, Alborn AM, Nordborg C, Peterson DA, Gage FH (1998) Neurogenesis in the adult human hippocampus. Nat Med 4:1313–1317

Ferrari D, Binda E, De Filippis L, Vescovi AL (2010) Isolation of neural stem cells from neural tissues using the neurosphere technique. Curr Protoc Stem Cell Biol Chapter 2:Unit2D.6

Flax JD, Aurora S, Yang C, Simonin C, Wills AM, Billinghurst LL, Jendoubi M, Sidman RL, Wolfe JH, Kim SU, Snyder EY (1998) Engraftable human neural stem cells respond to developmental cues, replace neurons, and express foreign genes. Nat Biotechnol 16:1033–1039

Florio M, Huttner WB (2014) Neural progenitors, neurogenesis and the evolution of the neocortex. Development 141:2182–2194

Gage FH, Temple S (2013) Neural stem cells: generating and regenerating the brain. Neuron 80:588–601

Haragopal H, Yu D, Zeng X, Kim SW, Han IB, Ropper AE, Anderson JE, Teng YD (2015) Stemness enhancement of human neural stem cells following bone marrow MSC coculture. Cell Transplant 24:645–659
Imitola J, Raddassi K, Park KI, Mueller FJ, Nieto M, Teng YD, Frenkel D, Li J, Sidman RL, Walsh CA, Snyder EY, Khoury SJ (2006) Directed migration of neural stem cells to sites of CNS injury by the stromal cell-derived factor 1alpha/CXC chemokine receptor 4 pathway. Proc Natl Acad Sci U S A 101:18117–18122
Jäderstad J, Jäderstad LM, Li J, Chintawar S, Salto C, Pandolfo M, Ourednik V, Teng YD, Sidman RL, Arenas E, Snyder EY, Herlenius E (2010) Communication via gap junctions underlies early functional and beneficial interactions between grafted neural stem cells and the host. Proc Natl Acad Sci U S A 107:5184–5189
Jia C, Yu D, Lamarre M, Leopold PL, Teng YD, Wang H (2014) Patterned electrospun nanofiber matrices via localized dissolution: potential for guided tissue formation. Adv Mater 26:8192–8197
Kaplan PL, Simon S, Cartwright CA, Eckhart W (1987) cDNA cloning with a retrovirus expression vector: generation of a pp60c-src cDNA clone. J Virol 61:1731–1734
Kempermann G, Song H, Gage FH (2015) Neurogenesis in the adult hippocampus. Cold Spring Harb Perspect Biol 7:a018812
Kim KS, Lee HJ, Jeong HS, Li J, Teng YD, Sidman RL, Snyder EY, Kim SU (2011) Self-renewal induced efficiently, safely, and effective therapeutically with one regulatable gene in a human somatic progenitor cell. Proc Natl Acad Sci U S A 108:4876–4881
Lee JH, Mitchell RR, McNicol JD, Shapovalova Z, Laronde S, Tanasijevic B, Milsom C, Casado F, Fiebig-Comyn A, Collins TJ, Singh KK, Bhatia M (2015) Single transcription factor conversion of human blood fate to NPCs with CNS and PNS developmental capacity. Cell Rep 11:1367–1376
Li J, Gu X, Ma Y, Calicchio ML, Kong D, Teng YD, Yu L, Crain AM, Vartanian TK, Pasqualini R, Arap W, Libermann TA, Snyder EY, Sidman RL (2010) Nna1 mediates Purkinje cell dendritic development via lysyl oxidase propeptide and NF-κB signaling. Neuron 68:45–60
Lim DA, Alvarez-Buylla A (2014) Adult neural stem cells stake their ground. Trends Neurosci 37:563–571
Lukaszewicz AI, Anderson DJ (2011) Cyclin D1 promotes neurogenesis in the developing spinal cord in a cell cycle-independent manner. Proc Natl Acad Sci U S A 108:11632–11637
Markowitz D, Goff S, Bank A (1988) Construction and use of a safe and efficient amphotropic packaging cell line. Virology 167:400–406
Neri M, Maderna C, Ferrari D, Cavazzin C, Vescovi AL, Gritti A (2010) Robust generation of oligodendrocyte progenitors from human neural stem cells and engraftment in experimental demyelination models in mice. PLoS One 5:e10145
Park KI, Teng YD, Snyder EY (2002) The injured brain interacts reciprocally with neural stem cells supported by scaffolds to reconstitute lost tissue. Nat Biotechnol 20:1111–1117
Price J, Turner D, Cepko CL (1987) Lineage analysis in the vertebrate nervous system by retrovirous-mediated gene transfer. Proc Natl Acad Sci U S A 84(156):160
Qian X, Davis AA, Goderie SK, Temple S (1997) FGF2 concentration regulated the generation of neurons and glia from multipotent cortical stem cells. Neuron 18:81–93
Redmond DE Jr, Bjugstad KB, Teng YD, Ourednik V, Ourednik J, Wakeman DR, Parsons XH, Gonzalez R, Blanchard BC, Kim SU, Gu Z, Lipton SA, Markakis EA, Roth RH, Elsworth JD, Sladek JR Jr, Sidman RL, Snyder EY (2007) Behavioral improvement in a primate Parkinson's model is associated with multiple homeostatic effects of human neural stem cells. Proc Natl Acad Sci U S A 104:12175–12180
Ropper AE, Zeng X, Haragopal H, Anderson JE, Aljuboori, Z, Han Abd-El-Barr M, Lee HJ, Sidman RL, Snyder EY, Viapiano MS, Kim SU, Chi JH, Teng YD (2015) Targeted treatment of experimental spinal cord glioma with dual gene-engineered human neural stem cells. Neurosurgery. [Epub ahead of print]. PMID: 26671631
Roy NS, Wang S, Jiang L, Kang J, Benraiss A, Harrison-Restelli C, Fraser RA, Couldwell WT, Kawaguchi A, Okano H, Nedergaard M, Goldman SA (2000) In vitro neurogenesis by progenitor cells isolated from the adult human hippocampus. Nat Med 6:271–277

Roy NS, Chandler-Militello D, Lu G, Wang S, Goldman SA (2007) Retrovirally mediated telomerase immortalization of neural progenitor cells. Nat Protoc 2:2815–2825
Ryder EF, Snyder EY, Cepko CL (1990) Establishment and characterization of multipotent neural cell lines using retrovirus vector mediated oncogene transfer. J Neurobiol 21:356–375
Schmidt NO, Przylecki W, Yang W, Ziu M, Teng YD, Kim SU, Black PM, Aboody KS, Carroll RS (2005) Brain tumor tropism of transplanted human neural stem cells is induced by vascular endothelial growth factor. Neoplasia 7:623–629
Shihabuddin LS, Horner PJ, Ray J, Gage FH (2000) Adult spinal cord stem cells generate neurons after transplantation in the adult dentate gyrus. J Neurosci 20:8727–8735
Snyder EY, Teng YD (2012) Stem cells and spinal cord repair. N Engl J Med 366:1940–1942
Snyder EY, Deitcher DL, Walsh C, Arnold-Aldea S, Hartwieg EA, Cepko CL (1992) Multipotent neural cell lines can engraft and participate in development of mouse cerebellum. Cell 68:33–51
Snyder EY, Taylor RM, Wolfe JH (1995) Neural progenitor cell engraftment corrects lysosomal storage throughout the MPS VII mouse brain. Nature 374:367–370
Takahashi K, Yamanaka S (2006) Induction of pluripotent stem cells from mouse embryonic and adult fibroblast cultures by defined factors. Cell 126:663–676
Takahashi K, Okita K, Nakagawa M, Yamanaka S (2007a) Induction of pluripotent stem cells from fibroblast cultures. Nat Protoc 2:3081–3089
Takahashi K, Tanabe K, Ohnuki M, Narita M, Ichisaka T, Tomoda K, Yamanaka S (2007b) Induction of pluripotent stem cells from adult human fibroblasts by defined factors. Cell 131(5):861–872
Taylor RM, Lee JP, Palacino JJ, Bower KA, Li J, Vanier MT, Wenger DA, Sidman RL, Snyder EY (2006) Intrinsic resistance of neural stem cells to toxic metabolites may make them well suited for cell non-autonomous disorders: evidence from a mouse model of Krabbe leukodystrophy. J Neurochem 97:1585–1599
Teng YD, Mocchetti I, Taveira-DaSilva AM, Gillis RA, Wrathall JR (1999) Basic fibroblast growth factor increases long-term survival of spinal motor neurons and improves respiratory function after experimental spinal cord injury. J Neurosci 19:7037–7047
Teng YD, Lavik EB, Qu X, Park KI, Ourednik J, Zurakowski D, Langer R, Snyder EY (2002a) Functional recovery following traumatic spinal cord injury mediated by a unique polymer scaffold seeded with neural stem cells. Proc Natl Acad Sci U S A 99:3024–3029
Teng YD, Park KI, Larvik EB, Langer R, Snyder EY (2002b) Stem cell culture: neural stem cells. In: Atala A, Lanzer RP (eds) Methods of tissue engineering. Academic Press, San Diego, pp 421–437
Teng YD, Choi H, Onario RC, Zhu S, Desilets FC, Lan S, Woodard EJ, Snyder EY, Eichler ME, Friedlander RM (2004) Minocycline inhibits contusion-triggered mitochondrial cytochrome c release and mitigates functional deficits after spinal cord injury. Proc Natl Acad Sci U S A 101:3071–3076
Teng YD, Santos FN, Black PM, Konya D, Park KI, Sidman RL, Snyder EY (2008) Chapter 18: neural stem cells: multipotency beyond self-renewal and phenotypic differentiation. In: Atala A, Lanza R, Thomson JA, Nerem RM (eds) Principles of regenerative medicine. Elsevier, Inc., San Diego, pp 300–317
Teng YD, Kabatas S, Wakeman DR, Li J, Snyder EY, Sidman RL (2009) Chapter 16: functional multipotency of neural stem cells and its therapeutic implications. In: Ulrich H (ed) Perspectives of stem cells: from tools for studying mechanisms of neuronal differentiation towards therapy. Springer-Verlag, Inc., San Diego, pp 255–270
Teng YD, Yu D, Ropper AE, Li J, Kabatas S, Wakeman DR, Wang J, Sullivan MP, Redmond DE Jr, Langer R, Snyder EY, Sidman RL (2011) Functional multipotency of stem cells: a conceptual review of neurotrophic factor-based evidence and its role in translational research. Curr Neuropharmacol 9:574–585
Teng YD, Benn SC, Kalkanis SN, Shefner JM, Onario RC, Cheng B, Lachyankar MB, Marconi M, Li J, Yu D, Han I, Maragakis NJ, Lládo J, Erkmen K, Redmond DE Jr, Sidman RL, Przedborski S, Rothstein JD, Brown RH Jr, Snyder EY (2012) Multimodal actions of neural stem cells in a mouse model of ALS: a meta-analysis. Sci Transl Med 4:165ra164

Vescovi AL, Parati EA, Gritti A, Poulin P, Ferrario M, Wanke E, Frolichsthal-Schoeller P, Cova L, Arcellana-Panlilio M, Colombo A, Galli R (1999) Isolation and cloning of multipotential stem cells from the embryonic human CNS and establishment of transplantable human neural stem cell lines by epigenetic stimulation. Exp Neurol 156:71–83

Wakeman DR, Redmond DE Jr, Dodiya HB, Sladek JR Jr, Leranth C, Teng YD, Samulski RJ, Snyder EY (2014) Human neural stem cells survive long term in the midbrain of dopamine-depleted monkeys after GDNF overexpression and project neurites toward an appropriate target. Stem Cells Transl Med 3:692–701

Weiss S, Reynolds BA, Vescovi AL, Morshead C, Craig C, van der Kooy D (1996) Is there a neural stem cell in the mammalian forebrain. Trends Neurosci 19:387–393

Wittko IM, Schänzer A, Kuzmichev A, Schneider FT, Shibuya M, Raab S, Plate KH (1999) VEGFR-1 regulates adult olfactory bulb neurogenesis and migration of neural progenitors in the rostral migratory stream in vivo. J Neurosci 29:8704–8714

Yu D, Neeley WL, Pritchard CD, Slotkin JR, Woodard EJ, Langer R, Teng YD (2009) Blockade of peroxynitrite-induced neural stem cell death in the acutely injured spinal cord by drug-releasing polymer. Stem Cells 27:1212–1222

# Chapter 19
# Stem Cells and Tissue Engineering

**Fernanda Maria Policarpo Tonelli, Nicole de Cássia Oliveira Paiva, Rebecca Vasconcellos Botelho de Medeiros, Mauro Cunha Xavier Pinto, Flávia Cristina Policarpo Tonelli, and Rodrigo Ribeiro Resende**

**Abstract** Tissue engineering is a research field that aims to offer efficient strategies for tissue repair in an effective way, and it has been developing rapidly in the last years. In this context, embryonic stem cells (ESCs) and adult stem cells (ASCs) are the most promising tools to repair tissue damage and loss of function. New protocols using stem cells and different approaches for scaffolds, production are now available for scientific experimentation. In this chapter, we review the use of stem cells in regenerative medicine and also discuss some important strategies to verify stem cell's differentiation in functional cells for efficient tissue repair. Additionally, we present a protocol to induce pluripotent stem cells to differentiate into bone cells, in order to generate an *in vivo* material for bone repair.

**Keywords** Stem cells • Differentiation • Bone cells

## 19.1 Introduction

### *19.1.1 Tissue Engineering*

Tissue engineering is a multidisciplinary field focused in creating biological substitutes that mimic tissues (Langer and Vacanti 1993). These substitutes are used in research, diagnostics and mainly in regeneration of injured or diseased tissues. The basis of modern tissue engineering were consolidated in 1998 with the development of the first stem cell lines (Amit et al. 2000). However, there are several historical scientific breakthroughs events which contributed to the development of this area (Lal et al. 2003).

F.M.P. Tonelli (✉) • N. de Cássia Oliveira Paiva • R.V.B. de Medeiros • M.C.X. Pinto
F.C. Policarpo Tonelli • R.R. Resende (✉)
Cell Signaling and Nanobiotechnology Laboratory, Department of Biochemistry and Immunology, Universidade Federal de Minas Gerais and Instituto Nanocell,
Av. Antônio Carlos 6627, Bloco N4 112, Belo Horizonte, MG 31270-901, Brazil
e-mail: tonellinanda@gmail.com; resende@icb.ufmg.br



H. Ulrich, P. Davidson Negraes, *Working with Stem Cells*,
DOI 10.1007/978-3-319-30582-0_19

The tissue engineering tools are cells, scaffolds (extracellular matrix), growth factors and bioreactors (Chan and Leong 2008). Cells can be obtained from a donor and cultured in artificial media. Scaffolds or extracellular matrix (ECM) have the function to offer adequate support for cell growth, including nutrients availability, influencing on cell attachment and mechanic stiffness. Each tissue contains a specific ECM that is specialized for its function and contributes to cellular differentiation. Growth factors are molecules that stimulate cell growth, proliferation and differentiation (Lee et al. 2011). There are specific receptors localized in the cells, which activate cell signaling and trigger several cellular processes. Lastly, bioreactors are equipments that reproduce physiologic conditions for cell culture maintenance and stimulation (Bilodeau and Mantovani 2006). The culture conditions, like temperature, pH and biochemical gradient need to be controlled during the maturation period.

Stem cells represent the main pillar of tissue engineering due to their high proliferative capacity and their ability to differentiate into several types of tissue. These cells are classified according to their source and can be embryonic or adult. Ethical concerns about the use of human embryonic stem cells (ESC) are a significant limitation for their large scale utilization. Therefore, recent advances in using adult stem cells (ASC) (such as induced pluripotent cells (iPSC), placenta and umbilical cord stem cells) are enabling not only the replacement of embryonic stem cells but also the progress of stem cell application in tissue engineering. In this context, stem cell differentiation can be guided to a purpose: for example, human ESC can be differentiated into neural tissue, cartilage, liver or blood vessels by putting them into different 3D scaffolds' structures (Levenberg et al. 2003).

The use of human ASC in research and therapy is not as controversial as the use of human ESC because the production of ASC does not require destruction of an embryo. Thus, ASC are the promise for regenerative medicine due to its ease cultivation *in vitro*, high proliferation rate, and its versatility to differentiate into many cell types including osteoblasts, chondrocytes, adipocytes, hepatocytes and neurons (Pittenger et al. 1999; Amaya et al. 2013; Andrade et al. 2012; Resende et al. 2010a, b, 2008; 2013; Lee et al. 2004; Black and Woodbury 2001). One of the most successful example of ASC use is shown by the treatment of leukemia and blood cell cancer with hematopoietic stem cells, which has been occuring for over 30 years (Karanes et al. 2008). Human ASC regenerative therapies are made with intravenous administration of hematopoietic stem cells to treat various diseases such as spinal cord injury (Srivastava et al. 2010) and hepatic cirrhosis (Terai et al. 2006). Additionally, in 2007, orthopedic applications demonstrated promising results of bone marrow stem cells treatment to repair articular cartilage defects of the knee (Wakitani et al. 2007). In 2008, a trachea section was the first human organ grown from stem cells to be transplanted (Macchiarini et al. 2008).

### *19.1.2 Stem Cells*

Embryonic, adult and induced pluripotent stem cells are extremely important when it comes to tissue engineering. ESC are derived from the Inner Cell Mass (ICM), which can be found in embryos during the blastocyst stage (4–5 days after fecundation), and were isolated for the first time in 1981 from mice and, in 1998 from humans (Evans and Kaufman 1981; Thomson et al. 1998). Stem cells can maintain themselves undifferentiated during years, even after they suffer several passages (Hoffman and Carpenter 2005) and, at the same time, these cells display high potential of differentiation. Human ESC can differentiate into neurons, pancreatic beta cells, hepatocytes (Carpenter et al. 2001; Zhang et al. 2001), cardiomyocytes (He et al. 2003; Mummery et al. 2003; Assady et al. 2001; Lavon et al. 2004), among others (as described in other chapter from this volume). However, human ESCs' obtainment involves important ethical issues and their use in research is prohibited in some countries.

ASCs are undifferentiated cells that are resident in differentiated tissues. They are found virtually in all tissues (being most commonly isolated from bone marrow, placenta blood, umbilical cord (Sousa et al. 2014) and adipose tissue (Strem et al. 2005; Sousa et al. 2014)) and have an important role in the maintenance of tissue homeostasis in case of injury or illness, besides supporting cellular renewal (Tuan et al. 2003). ASCs are called multipotent because they are less versatile than ESC; as an example, hematopoietic stem cells (Till and Mc 1961; Becker et al. 1963) can give rise to all blood cells, while bone marrow stem cell (Friedenstein et al. 1970) originate bone, cartilage and adipose tissue.

It is also possible to use stem cells in regenerative medicine approaches. Mature differentiated cells can be induced to un-differentiated into stem cells by cellular reprogramming, generating the so called iPSCs. These are cells similar to ESC when it comes to morphology and molecular signature (Bock et al. 2011). After being generated, iPSCs can also be induced to differentiate into a cell lines of interest for tissue engineering purposes. But, how is this done? New protocols to generated iPSCs and to control their differentiation for tissue engineering purposes have constantly been developed. To generate iPSCs, for example, is possible to perform techniques of cell reprogramming not involving integration of genes into cell's genome (Chou et al. 2011), or to use viral vector to achieve integration (Boulting et al. 2011). However, once iPSCs are available, how to induce their differentiation into a desired cell line? The protocol described below was proposed by de Peppo and coworkers for bone tissue engineering (de Peppo et al. 2013) using human iPSCs (hiPSCs) or human ESCs (hESCs) as starting cells (Fig. 19.1).

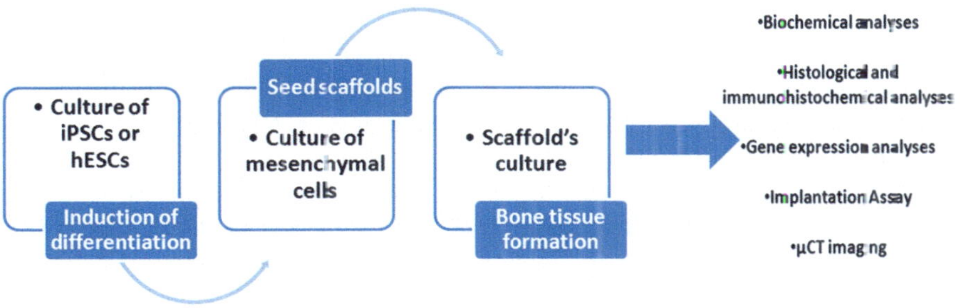


**Fig. 19.1** Experimental steps to transform an hiPSC or hESCs into bone cells and confirm the success of the methodology (based on Marolt et al. 2012)

## 19.2 Materials

### *19.2.1 Solutions and Chemicals*

- 70 % Ethanol solution.
- Mesenchymal lineage differentiation media: All components are supplied from Life Technologies, except when specified. Knockout Dulbecco's Modified Eagle Medium (DMEM) (Cat. No. 10829-018) should be supplemented with 20 % (vol/vol) HyClone fetal bovine serum (GE Life Sciences, Cat. No. SH30071.01), 1 mM of L-Glutamine (Cat. No. 25030-081), 0.1 mM of β-mercaptoethanol (Cat. No. 31350-010), 0.1 mM of MEM Non-essential amino acids (Cat. No. 11140-035), and 1 % (vol/vol) of Penicillin-Streptomycin (10,000 U/ml, Thermo Fisher, Cat. No. 15140-122). Store at 4 °C for up to 4 weeks.
- Gelatin-coated plates: Add 0.2 % (w/vol) of gelatin (Sigma-Aldrich, Cat. No. G9136) to dd$H_2O$. Autoclave the solution and add the solution to culture dishes. Incubate at room temperature for 1 h.
- Dulbecco's Phosphate-Buffered Saline (DPBS) (Thermo Fisher, Cat. No. 14040141).
- 0.25 % Trypsin/EDTA solution (Sigma-Aldrich, Cat. No. T4049).
- Antibodies:
  - Alexa Fluor 488 anti-human SSEA-1 (1:10, BD Biosciences, Cat. No. 560172).
  - Alexa Fluor 488 anti-human SSEA-4 (1:10, BD Biosciences, Cat. No. 560308).
  - Fluorescein isothiocyanate anti-human CD44 (ready to use, BD Biosciences, Cat. No. 555478).
  - Phycoerythin anti-human CD73 (ready to use, BD Biosciences, Cat. No. 550257).
  - Fluorescein isothiocyanate anti-human CD90 (ready to use, BD Biosciences, Cat. No. 555595).
  - Phycoerythin anti-human CD166 (ready to use, BD Biosciences, Cat. No. 559263).
  - Alexa Fluor 488 anti-human CD105 (BioLegend, Cat. No. 323209).
- PBS pH 7.4 (phosphate-buffered saline) (Thermo Fisher, Cat. No. 10010023).
- Buffered Hypotonic Solution (Genial Genetics, Cat. No. GGS-JL006a).
- Detergent solution: 10 mM Tris (Sigma-Aldrich, Cat. No. 252859) and 0.5 % sodium dodecyl sulfate (SDS) (Sigma-Aldrich, Cat. No. L3771).
- Low-glucose DMEM (Thermo Fisher, Cat. No. 11885-084) supplemented with 10 % HyClone™ fetal bovine serum (GE Life Sciences, Cat. No. SH30071.01) and 1 % (vol/vol) of Penicillin-Streptomycin (10,000 U/mL, Thermo Fisher, Cat. No. 15140-122).
- Osteogenic medium: Low-glucose Dulbecco's Modified Eagle Medium (DMEM) (Thermo Fisher, Cat. No. 11885-084) supplemented with 10 % HyClone™ fetal

bovine serum (GE Life Sciences, Cat. No. SH30071.01), 1 % (vol/vol) of Penicillin-Streptomycin (10,000 U/ml, Thermo Fisher, Cat. No. 15140-122), 1 μM dexamethasone (Sigma-Aldrich, Cat. No. D4902), 10 mM β-glycerophosphate (Sigma-Aldrich, Cat. No. G9422), and 50 μM ascorbic acid-2-phosphate (Sigma-Aldrich, Cat. No. 49752).
- Lysis buffer: PBS (Thermo Fisher, Cat. No. 10010023) 1 % (vol/vol), Triton X-100 (Sigma-Aldrich, Cat. No. X100), 0.5 % (wt/vol) sodium deoxycholate (Sigma-Aldrich, Cat. No. D6750), 0.1 % SDS (Sigma-Aldrich, Cat. No. L3771), 0.1 mg/mL phenylmethylsulfonyl fluoride (Sigma-Aldrich, Cat. No. P7626), and 0.3 % (vol/vol) aprotinin (Sigma-Aldrich, Cat. No. A6279).

### 19.2.2 Disposables

- 6-Well tissue culture plates (treated).
- Sterile serological pipette.
- Micropipette tips.

### 19.2.3 Equipment

- Tissue culture incubator (5 % $CO_2$ at 37 °C).
- Laminar flow hood.
- Cell counter.
- Flow cytometer.
- Micropipettes.

## 19.3 Methods

### 19.3.1 Inducing Mesenchymal Lineage

1. Induce human iPSCs to differentiate into mesenchymal lineage by culturing them in cell plates in the presence of mesenchymal lineage differentiation media for 7 days with two media changes. Keep them always like a monolayer.
2. Trypsinize the cells after removing culture medium and wash cells with sterile DPBS. After adding the Trypsin/EDTA solution to cover the cell, incubate the plate at room temperature for 30 s. Remove the solution and incubate the plate at room temperature for 1 min. Add culture medium to the wells and suspend the cells by gentle pipetting.

3. Count cells and subculture them in mesenchymal lineage differentiation media at 100,000 cells/cm$^2$ on tissue culture plates pre-coated with 0.1 % (wt/vol) gelatin (*passage 1*).
4. After reached 80 % confluence, trypsinize and subculture the cells at 10,000/cm$^2$ for up to four passages (3–4 weeks).
5. By flow cytometry, evaluate the expression of mesenchymal surface antigens CD44, CD73, CD90, CD105, and CD166. At this point, antigens like SSEA-4 (pluripotent stem cell marker) and SSEA-1 (early differentiation marker) should no longer be expressed. In order to perform these analyses:

   i. Suspend cells at a concentration of $1 \times 10^6$ cells/mL.
   ii. Incubate 100 μl of the cell suspension for 30 min on ice with the following antibodies: Alexa Fluor 488 anti-human SSEA-1, Alexa Fluor 488 anti-human SSEA-4, fluorescein isothiocyanate anti-human CD44, phycoerythin anti-human CD73, fluorescein isothiocyanate antihuman CD90, phycoerythin anti-human CD166 and Alexa Fluor 488 anti-human CD105.
   iii. Wash the samples and analyze on flow cytometer.

***Note:*** *At this step, it is possible to observe cell morphology and cell growth, by counting cell numbers at each passage. This protocol can also be performed using human ESCs (Marolt et al.* 2012*).*

## *19.3.2 Decellularized Scaffolds*

1. Obtain the trabecular bone from the subchondral region of carpometacarpal joints from 2-week to 4-month-old cows.
2. Generate bone plugs measuring approximately 4 mm in diameter and 4 mm in high.
3. Wash the bone plugs with high velocity stream of water to remove the marrow from the pore spaces.
4. Wash the plugs in PBS pH 7.4 at room temperature for 1 h.
5. Wash the plugs in hypotonic buffer overnight at 4 °C.
6. Wash the plugs in detergent for 24 h at room temperature.
7. Wash the plugs in enzymatic solution.
   ***Note:*** *At these wash steps are performed to remove any remaining cellular material.*
8. Rinse the plugs repeatedly in PBS.
9. Sterilize scaffolds in 70 % ethanol for at least 1 h (Fig. 19.2).
10. Incubate the scaffolds in low-glucose DMEM overnight in a culture dish prior to cell seeding (Grayson et al. 2011).

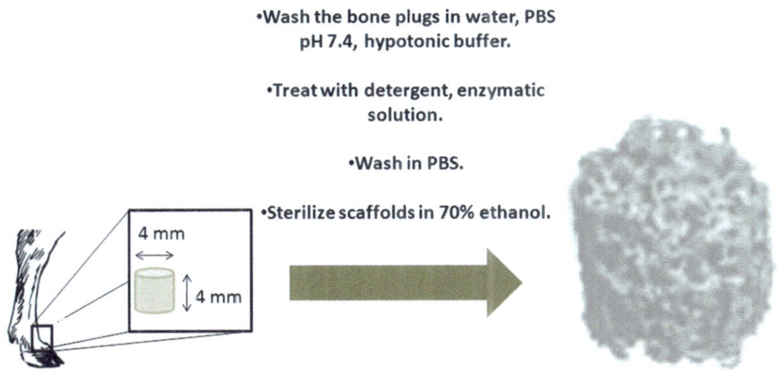


**Fig. 19.2** Procedure to obtain decellularized scaffolds

### *19.3.3 Seeding Scaffolds and Perfusion Bioreactor Culture*

1. Resuspend cells in culture medium at a density of $30 \times 10^6$ cells/mL.
2. Pipette 40 μL of the cell suspension into the dried scaffolds.
3. Add 55 μL of medium and flip the scaffolds every 15 min during 1 h to induce uniform cell distribution.
4. Culture the scaffolds in osteogenic medium for 3 days to allow cell attachment.
5. Place seeded scaffolds in either perfusion bioreactors or 6-well plates for up to 5 weeks, changing medium twice a week. The bone tissue will be generated.

***Note:*** *The bioreactor used in studies to induce bone generation (Grayson et al.* 2011*) is capable of enabling uniform perfusion of six constructs. The medium recirculates to maintain oxygen concentration of 20%* (Fig. 19.3). *The use of 6-well plates represents one important disadvantage: the amount of bone tissue generated is very lower than the amount obtained using the bioreactor (Marolt et al.* 2012*).*

## 19.4 Maintenance and Characterization of Generated Bone Tissue

The bone cells can be maintained into the scaffolds inside the bioreactor or in a 6-well plate with no problem for up to 5 weeks, only performing the medium change twice a week. After this period of time, constructs are ready to be transplanted into animals and it is possible to analyze the bone tissue generation after this procedure (as it will be mentioned ahead).

In order to observe *in vivo* bone formation and also access phenotype stability of perfused bone constructs and their safety, SCID-beige female mice from 3 to 5 months old can be used as receptors for scaffold implantation. These animals are immunodeficient to avoid immune rejection after scaffold transplantation: bone

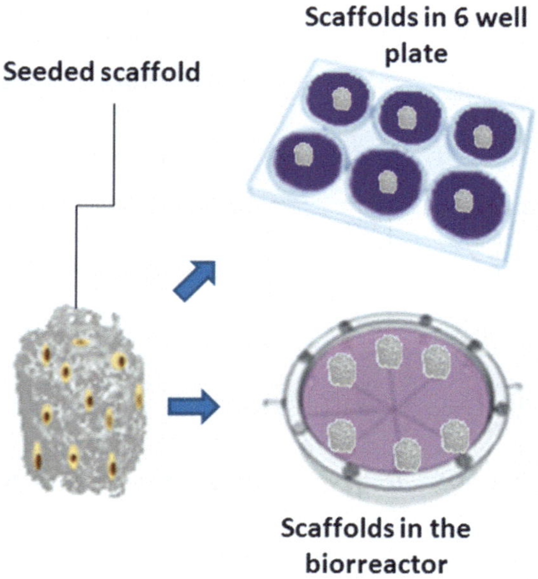


**Fig. 19.3** Culture of seeded scaffolds, which can be performed either in 6-well plates or in bioreactors

constructs developed inside the mice can be evaluated in reconstructing areas around the implanted matrix (Marolt et al. 2012), proving that the cells are able to efficiently reconstruct the bone tissue.

Among the analyses that can be performed to characterize the bone cells generated prior or after implantation, we can highlight determination of alkaline phosphatase activity, analyses of gene expression profiles and calcium content, determination of osteopontin concentration, immunohistochemical staining against osteopontin, bone sialoprotein and osteocalcin, and also microcomputed tomography (μCT) of the scaffolds.

## 19.4.1 Determination of Calcium Content

Cells can be added to trichloroacetic acid (Sigma, Cat. No. T0699) in water (0.15 mL of 5 % (vol/vol)) for 30 min and submitted to repeated up and down pipetting. The supernatants can be analyzed by using Calcium (CPC) Liquicolor assay (Stanbio, Cat. No. 0155-225), for example (Marolt et al. 2012; Pinto et al. 2015).

## 19.4.2 Determination of Alkaline Phosphatase Activity

The scaffolds should be kept in lysis buffer, on ice, and disintegrated for example by steel balls in a Minibead beater during two 15 s cycles. The extracts are then centrifugated and the supernatant incubated at 37 °C in alkaline buffer and nitrophenyl-phosphate substrate solution, until the development of a yellow color. The reactions can, at this time, be stopped by 0.5 M NaOH (Sigma, Cat. No. 221465). The absorbance at 405 nm is compared to a standard curve obtained from p-nitrophenol (Sigma, Cat. No. 73560) solutions of known concentrations to determine the alkaline phosphatase activity. Significantly higher alkaline phosphatase activity can be detected after the third week of culture in the bioreactor (Marolt et al. 2012).

### *19.4.3 Concentration of Osteopontin*

The concentration of osteopontin can be determined by using the cell culture medium and the human osteopontin ELISA kit (Abcam, Cat. No. ab100618) (Marolt et al. 2012).

### *19.4.4 Histological and Immunohistochemical Analyses*

After washing in PBS and cutting into half, constructs can be fixed, decalcified, dehydrated, paraffin (Sigma, Cat. No. 327204) or methylmethacrylate (Sigma, Cat. No. W400201) embedded (for immunohistochemical or histological analyses, respectively), sectioned, and mounted on glass slides. It is possible to stain slides by using hematoxylin/eosin (Sigma, Cat. No. H3136 and E6003) and Masson trichome (Merck, Cat. No. 100485) (for example to evaluate collagen generation—positive blue areas) or Goldner (Merck, Cat. No. 100485) (to evaluate osteoid formation—positive red areas) procedures.

To perform immunohistochemical analyses, incubate samples with primary antibodies against osteopontin (Abcam, Cat. No. ab8448), bone sialoprotein (Abcam, Cat. No. ab52128), and osteocalcin (Abcam, Cat. No. ab13420). Next, it is necessary to block the slides and incubate them with secondary antibodies (for example biotin/avidin system).

***Note:*** *Osteopontin starts to accumulate into culture medium in a significant way after 2 week of culture in the bioreactor. Bioreactor also offers denser tissue deposition than statical culture in six well plates (Marolt et al. 2012; de Peppo et al. 2013, Paschon et al. 2012, Rocha-Resende et al. 2012).*

### *19.4.5 qPCR Analyses*

Cells can be submitted to total RNA extraction (for example using TRIzol, Thermo Fisher, Cat. No. 15596-026), first strand synthesis using random hexamers and real time-PCR using probes and primers designed to alkaline phosphatase, and osteopontin mRNAs, for example (de Peppo et al. 2013).

### 19.4.6 Microarray Analyses

RNA samples can be submitted to a reverse transcriptase reaction to generate double-stranded labeled templates. Then, a microarray kit can be used, e.g. Illumina HumanHT-12 v4 Expression BeadChip Kit (Illumina, Cat. No. BD-103-0204), to analyze the expression of genes involved in osteogenesis (de Peppo et al. 2013).

### 19.4.7 In Vivo: Implantation Assay

Scaffolds can be implanted into mice to observe their development *in vivo*. Immunodeficient mice (SCID-beige female mice), can receive scaffold implants in their dorsal pockets. Some weeks later, stability of engineered bone implanted can be evaluated. These studies have already been conducted to observe bone tissue generated until 12 weeks after implantation. Given the desired period of time, mice are euthanized, tissues are harvested, and the analyses of interest can be performed (de Peppo et al. 2013; Marolt et al. 2012).

### 19.4.8 Microcomputed Tomography (μCT) Imaging

*In vitro* cultured constructs and explanted tissues can be analysed by μCT. For this purpose, it is necessary a step of fixation with glutaraldehyde (Sigma, Cat. No. G5882) before scanning the constructs. Then, parameters such as total bone volume, bone volume fraction, trabecular number, trabecular thickness, and trabecular spacing (Marolt et al. 2012) can be determined.

## 19.5 The Potential Use of Stem Cells in Tissue Engineering

Due to their pluripotency, stem cells can be used in various clinical applications. Although there are some limitations, stem cells derived from adipose tissue have been used to repair small defects in soft tissues, after mastectomy, to deal with contour defects such as wrinkles, and in plastic surgery through techniques such as lipotransfer (Kim and Jeong 2014; Kim and Heo 2014).

The analysis of MSCs and their molecular products released in the blood stream (e.g. by sequencing) can lead to the identification of potential biomarkers, which could constitute useful tools in providing (1) starting material to the development of non-invasive diagnose strategies, or (2) a way to evaluate the prognosis for a variety of diseases. For example, fetal MSCs when present in the peripheral blood of pregnant women or found together with neural cells in the amniotic fluid represent an opportunity to diagnose defects in the formation of the neural tube during the prenatal testing. It is also possible to use stem cells present in the peripheral blood of a human being as a potential biomarker to access the prognosis of patients suffering from myocardial infarction, or patients after an ischemic stroke (Valenti et al. 2015).

Damaged tissue's regeneration capacity in humans is restricted to a few specialized organs such as the liver. A patient with a damaged or diseased organ that has lost its function and is unable to regenerate alone, has the transplant as the main effective treatment option. However, the limited number of donors and the complications due to immunological rejection of the transplanted organ make transplants unfeasible for most patients. Therefore, advances in regenerative medicine bring the stem cells as an alternative to regenerate injured tissues (Cassidy 2014).

Stem cells derived from adipose tissue, after differentiation into chondrocytes, can be used to repair cartilaginous defects caused by degenerative joint diseases such as osteoarthitis (Kim and Heo 2014). In cardiovascular disease, stem cells are being used in order to enhance angiogenesis, reduce cell apoptosis rate and promote the formation of new shoots of neurons in the damaged myocardium; all of that with the intention of improving cardiac function (Kim and Heo 2014; Cutts et al. 2015).

When it comes to the treatment of neurodegenerative diseases, the stem cells are still under investigation. They are capable of differentiating into neurons and glial cells, yet it is still necessary to establish a protocol for obtaining a homogeneous population of cells to an effective therapy for neurodegenerative diseases such as Parkinson's disease, multiple sclerosis and Alzheimer's disease (Cutts et al. 2015). In eye therapy, a promising source of stem cells, which is under investigation is the primary dentition. The dental pulp cells, with their characteristic pluripotency, were capable of reconstructing the corneal epithelium in rabbits after a chemical burn in this tissue. Not only dental pulp but also hair follicle, bone marrow and umbilical cord are possible sources of stem cells to be applied in regeneration of the corneal epithelium (de Araujo and Gomes 2015).

More recently, most projects in tissue engineering involve the addition of cultures of stem cells to a matrix (scaffolds) with the intention to facilitate the differentiation of these cells and increase the efficiency in maintaining tissue-specific functions of cells after differentiation (Resende et al. 2014; Santos et al. 2013; Tonelli et al. 2012a, b). This kind of strategy became a promising alternative to deliver stem cell in damaged tissues (Cassidy 2014; Cutts et al. 2015). The scaffolds' function is to mimic the extracellular matrix of the tissue for which exists the intention to perform the stem cell differentiation process. In this context, a variety of biomaterials is being used/developed in tissue engineering to build suitable scaffolds (Cutts et al. 2015).

Commercial biomaterials (such as Matrigel and Geltrex) are based on extracellular matrix proteins isolated from animal sources, which are biocompatible and suitable to retain the biological activity exhibited by the tissue source; it favors stem cell adhesion, survival and differentiation (Sellaro et al. 2010; Cutts et al. 2015) in various cell types such as neurons (Tavakol et al. 2014; Ma et al. 2008), cardiomyocytes (Salick et al. 2014; Danoviz et al. 2010; Cutts et al. 2015), pancreatic cells (Lee et al. 2014) and liver cells (Zhang and Dong 2015). However, the rate of degradation of this material varies depending on the location and extent of the implant (Cutts et al. 2015).

One strategy that has been widely used to construct scaffolds for tissue engineering purposes is decellularized ECM (Santi and Johnson 2013). In this process, the intact organ undergoes several washings with detergent which eliminate all the parenchyma tissue, leaving only the extracellular matrix, that is used as a substrate for cell cultures (Santi and Johnson 2013; Zhang et al. 2009; Badylak et al. 2011; Zuk et al. 2001). In addition to providing a biomechanical support, these scaffolds formed from the ECM provide a microenvironment composed of a complex network of fibrillar proteins, proteoglycans and glycosaminoglycans (Badylak et al. 2011; Cassidy 2014; Hynes 2009).

When it comes to constructing scaffolds to allow the differentiation of stem cells, "natural biomaterial scaffolds" are commonly used. They may be formed by fibroin, chitosan, fibrin, alginate or agarose. These materials contain important naturally occurring proteins and polysaccharides to promote cell adhesion. Another advantage of such materials is that their physical and chemical properties can be manipulated and, after implanted *in vivo*, these materials persist long enough to promote integration with the native tissue (Zuk et al. 2001). Scaffolds based on synthetic polymers, such as poly(ethylene glycol) (PEG), poly(lactic acid) (PLA), poly(caprolactone) (PCL), poly(L-lactide-co-caprolactone) (PLCL), poly(glycolic-co-caprolactone) (PGCL), poly(glycerol sebacate) (PGS), and polyurethane (PU) are easily manufactured and adjustable, allowing a better response of stem cells to differentiate into cells from specific tissues. However, some of these materials may degrade in potentially harmful products (Zuk et al. 2001; Wang et al. 2013).

The scaffolds are often used to obtain a 3D culture, yet the rate of absorption of the biomaterial must match the rate of formation of the new tissue and, the material that form the scaffolds together with the products of biomaterial degradation must not interfere or impair tissue's regeneration (Sasaki et al. 2015). In eye therapy, the new strategy is to replace the native tissue by a matrix obtained from animal sources, including pigs, cattle and cats. After decellularization, the animal tissue turns into a biocompatible three-dimensional structure of extracellular matrix, providing adequate biomechanical stress and transparency that mimics the native cornea (de Araujo and Gomes 2015). When comparing hepatogenic differentiation of stem cells in cultures with different matrices, cells cultured in liver extracellular matrix (LEM) have a better response to differentiation, offering liver function and viability better than the ones offered by the cells grown on collagen type I matrices or without any substrate (which in turn showed less satisfactory results) (Zhang and Dong 2015).

A variety of biological acellular scaffolds can be constructed from central nervous system (CNS) tissue such as the optic nerve, the spinal cord and the brain. The cytocompatibility, the remaining growth factors of the tissue of origin, and the presence of support neural proteins make these constructs ideal for attracting neural stem cells for the CNS injury site, making them suitable for replacement of lost tissue after some accident or, to be used to treat neurodegenerative diseases. Therefore, the delivery of ECM to the sites of CNS injury (e.g. stroke or spinal cord injury area), by injection, can help the patient to "fell" the injured area: recruiting endogenous stem cells, and improving patient's regenerative response (Crapo et al. 2012).

Although there have been many advances in tissue engineering *in vitro*, some technologies applied to regenerate mainly 3D tissues suffer limitations attributed to the lack of techniques that allow the formation of blood capillaries in the tissue graft. In this context, the inside part of the engineered tissues often suffers necrosis before complete regeneration. In order to form capillary networks, and thereby increase the efficiency of human mesenchymal stem cells (hMSCs) differentiation into osteocytes, an attempt of performing a co-culture of hMSCs with human umbilical vein endothelial incorporation cells (HUVECs) was made in thermo-responsive hydrogels of poly(N)-isopropylacrylamide. The hMSCs were distributed in the peripheral region, while the HUVECs were placed in the inner region of the construct formed into hydrogels (Sasaki et al. 2015). In this study, it was observed an increased number of viable cells within the 3D culture in regions where necrosis was commonly reported due to the lack of nutrients and oxygen. The inner part of the construct showed a higher ratio of HUVECs incorporated (Sasaki et al. 2015). Furthermore, the results indicated that HUVECs were able to survive and organize themselves under the influence of hMSCs, and could also form a crosslinked structure—making them a new interesting option to be used in new strategies in the tissue engineering field.

Despite the presented research breakthroughs and promising results, much work is still needed before the new treatment strategies are available for use in a large number of patients in a secure and effective way (Cassidy 2014).

## References


Amaya MJ, Goulart VAM, Santos AK, Kihara AH, Guatimosim S, Resende RR (2013) Role of calcium signaling in stem and cancer cell proliferation. In: Resende RR, Ulrich H (eds) Trends in stem cell proliferation and cancer research. Springer, The Netherlands

Amit M, Carpenter MK, Inokuma MS, Chiu CP, Harris CP, Waknitz MA, Itskovitz-Eldor J, Thomson JA (2000) Clonally derived human embryonic stem cell lines maintain pluripotency and proliferative potential for prolonged periods of culture. Dev Biol 227:271–278

Andrade LA, Geraldo JM, Gonçalves OX, Leite MTT, Catarina AM, Guimarães MM, Leme AFP, Yokoo S, Machado CR, Rajão MA, Carvalho SM, Gomes DA, Aguiar CJ, Souza-Fagundes EM, Zani CL, Resende RR, Martins-Filho OA, Leite MF (2012) Nucleoplasmic calcium buffering sensitizes human squamous cell carcinoma to anticancer therapy. J Cancer Sci Ther 4:23–43

Assady S, Maor G, Amit M, Itskovitz-Eldor J, Skorecki KL, Tzukerman M (2001) Insulin production by human embryonic stem cells. Diabetes 50:1691–1697

Badylak SF, Taylor D, Uygun K (2011) Whole-organ tissue engineering: decellularization and recellularization of thee-dimensional matrix scaffolds. Annu Rev Biomed Eng 13:27–53

Becker AJ, Mc CE, Till JE (1963) Cytological demonstration of the clonal nature of spleen colonies derived from transplanted mouse marrow cells. Nature 197:452–454

Bilodeau K, Mantovani D (2006) Bioreactors for tissue engineering: focus on mechanical constraints. A comparative review. Tissue Eng 12:2367–2383

Black IB, Woodbury D (2001) Adult rat and human bone marrow stromal stem cells differentiate into neurons. Blood Cells Mol Dis 27:632–636

Bock C, Kiskinis E, Verstappen G, Gu H, Boulting G, Smith ZD, Ziller M, Croft GF, Amoroso MW, Oakley DH, Gnirke A, Eggan K, Meissner A (2011) Reference maps of human ES and iPS cell variation enable high-throughput characterization of pluripotent cell lines. Cell 144:439–452

Boulting GL, Kiskinis E, Croft GF, Amoroso MW, Oakley DH, Wainger BJ, Williams DJ, Kahler DJ, Yamaki M, Davidow L, Rodolfa CT, Dimos JT, Mikkilineni S, Macdermott AB, Woolf CJ,

Henderson CE, Wichterle H, Eggan K (2011) A functionally characterized test set of human induced pluripotent stem cells. Nat Biotechnol 29:279–286
Carpenter MK, Inokuma MS, Denham J, Mujtaba T, Chiu CP, Rao MS (2001) Enrichment of neurons and neural precursors from human embryonic stem cells. Exp Neurol 172:383–397
Cassidy JW (2014) Nanotechnology in the regeneration of complex tissues. Bone Tissue Regen Insights 5:25–35
Chan BP, Leong KW (2008) Scaffolding in tissue engineering: general approaches and tissue-specific considerations. Eur Spine J 17:467–479
Chou BK, Mali P, Huang X, Ye Z, Dowey SN, Resar LMS, Zou C, Zhang YA, Tong J, Cheng L (2011) Efficient human iPS cell derivation by a non-integrating plasmid from blood cells with unique epigenetic and gene expression signatures. Cell Res 21:518–529
Crapo PM, Medberry CJ, Reing JE, Tottey S, Van Der Merwe Y, Jones KE, Badylak SF (2012) Biologic scaffolds composed of central nervous system extracellular matrix. Biomaterials 33:3539–3547
Cutts J, Nikkhah M, Brafman DA (2015) Biomaterial approaches for stem cell-based myocardial tissue engineering. Biomark Insights 10:77–90
Danoviz ME, Nakamuta JS, Marques FL, Dos Santos L, Alvarenga EC, Dos Santos AA, Antonio EL, Schettert IT, Tucci PJ, Krieger JE (2010) Rat adipose tissue-derived stem cells transplantation attenuates cardiac dysfunction post infarction and biopolymers enhance cell retention. PLoS One 5:e12077
De Araujo AL, Gomes JA (2015) Corneal stem cells and tissue engineering: current advances and future perspectives. World J Stem Cell 7:806–814
De Peppo GM, Marcos-Campos I, Kahler DJ, Alsalman D, Shang L, Vunjak-Novakovic G, Marolt D (2013) Engineering bone tissue substitutes from human induced pluripotent stem cells. Proc Natl Acad Sci U S A 110:8680–8685
Evans MJ, Kaufman MH (1981) Establishment in culture of pluripotential cells from mouse embryos. Natur 292:154–156
Friedenstein AJ, Chailakhjan RK, Lalykina KS (1970) The development of fibroblast colonies in monolayer cultures of guinea-pig bone marrow and spleen cells. Cell Tissue Kinet 3:393–403
Grayson WL, Marolt D, Bhumiratana S, Fröhlich M, Guo XE, Vunjak-Novakovic G (2011) Optimizing the medium perfusion rate in bone tissue engineering bioreactors. Biotechnol Bioeng 108:1159–1170
He JQ, Ma Y, Lee Y, Thomson JA, Kamp TJ (2003) Human embryonic stem cells develop into multiple types of cardiac myocytes: action potential characterization. Circ Res 93:32–39
Hoffman LM, Carpenter MK (2005) Human embryonic stem cell stability. Stem Cell Rev 1:139–144
Hynes RO (2009) The extracellular matrix: not just pretty fibrils. Science 326:1216–1219
Karanes C, Nelson GO, Chitphakdithai P, Agura E, Ballen KK, Bolan CD, Porter DL, Uberti JP, King RJ, Confer DL (2008) Twenty years of unrelated donor hematopoietic cell transplantation for adult recipients facilitated by the National Marrow Donor Program. Biol Blood Marrow Transplant 14:8–15
Kim EH, Heo CY (2014) Current applications of adipose-derived stem cells and their future perspectives. World J Stem Cells 6:65–68
Kim YJ, Jeong JH (2014) Clinical application of adipose stem cells in plastic surgery. J Korean Med Sci 29:462–467
Lal B, Viola J, Hicks D, Grad O (2003) The emergence of tissue engeneering as a research field. In: The National Science Foundation Report. Viola J, Grad O (eds) 1st edn. Abt Associates Inc., Cambridge
Langer R, Vacanti JP (1993) Tissue engineering. Science 260:920–926
Lavon N, Yanuka O, Benvenisty N (2004) Differentiation and isolation of hepatic-like cells from human embryonic stem cells. Differentiation 72:230–238
Lee KD, Kuo TK, Whang-Peng J, Chung YF, Lin CT, Chou SH, Chen JR, Chen YP, Lee OK (2004) *In vitro* hepatic differentiation of human mesenchymal stem cells. Hepatology 40:1275–1284

Lee K, Silva EA, Mooney DJ (2011) Growth factor delivery-based tissue engineering: general approaches and a review of recent developments. J R Soc Interface 8:153–170

Lee JS, An SY, Kwon IK, Heo JS (2014) Transdifferentiation of human periodontal ligament stem cells into pancreatic cell lineage. Cell Biochem Funct 32:605–611

Levenberg S, Huang NF, Lavik E, Rogers AB, Itskovitz-Eldor J, Langer R (2003) Differentiation of human embryonic stem cells on thee-dimensional polymer scaffolds. Proc Natl Acad Sci U S A 100:12741–12746

Ma W, Tavakoli T, Derby E, Serebryakova Y, Rao MS, Mattson MP (2008) Cell-extracellular matrix interactions regulate neural differentiation of human embryonic stem cells. BMC Dev Biol 8:90–101

Macchiarini P, Jungebluth P, Go T, Asnaghi MA, Rees LE, Cogan TA, Dodson A, Martorell J, Bellini S, Parnigotto PP, Dickinson SC, Hollander AP, Mantero S, Conconi MT, Birchall MA (2008) Clinical transplantation of a tissue-engineered airway. Lancet 372:2023–2030

Marolt D, Campos IM, Bhumiratana S, Koren A, Petridis P, Zhang G, Spitalnik PF, Grayson WL, Vunjak-Novakovic G (2012) Engineering bone tissue from human embryonic stem cells. Proc Natl Acad Sci U S A 109:8705–8709

Mummery C, Ward-Van Oostwaard D, Doevendans P, Spijker R, Van Den Brink S, Hassink R, Van Der Heyden M, Opthof T, Pera M, De La Riviere AB, Passier R, Tertoolen L (2003) Differentiation of human embryonic stem cells to cardiomyocytes: role of coculture with visceral endoderm-like cells. Circulation 107:2733–2740

Paschon V, Higa GS, Resende RR, Britto LR, Kihara AH (2012) Blocking of connexin-mediated communication promotes neuroprotection during acute degeneration induced by mechanical trauma. PLoS One 7:e45449

Pinto MC, Kihara AH, Goulart VA, Tonelli FM, Gomes KN, Ulrich H, Resende RR (2015) Calcium signaling and cell proliferation. Cell Signal 27:2139–2149

Pittenger MF, Mackay AM, Beck SC, Jaiswal RK, Douglas R, Mosca JD, Moorman MA, Simonetti DW, Craig S, Marshak DR (1999) Multilineage potential of adult human mesenchymal stem cells. Science 284:143–147

Resende RR, Alves AS, Britto LR, Ulrich H (2008) Role of acetylcholine receptors in proliferation and differentiation of P19 embryonal carcinoma cells. Exp Cell Res 314:1429–1443

Resende RR, Adhikarid A, Da Costa JL, Lorençon E, Ladeira MS, Guatimosim S, Kiharag AH, Ladeira LO (2010a) Influence of spontaneous calcium events on cell-cycle progression in embryonal carcinoma and adult stem cells. Biochim Biophys Acta 1803:246–260

Resende RR, Da Costa JL, Kihara AH, Adhikari A, Lorencon E (2010b) Intracellular Ca2+ regulation during neuronal differentiation of murine embryonal carcinoma and mesenchymal stem cells. Stem Cells Dev 19:379–393

Resende RR, Andrade LM, Oliveira AG, Guimaraes ES, Guatimosim S, Leite MF (2013) Nucleoplasmic calcium signaling and cell proliferation: calcium signaling in the nucleus. Cell Commun Signal 11:14–22

Resende RR, Fonseca EA, Tonelli FMP, Sousa BR, Santos AK, Gomes KN, Guatimosim S, Kihara AH, Ladeira LO (2014) Scale/topography of substrates surface resembling extracellular matrix for tissue engineering. J Biomed Nanotechnol 10:1157–1193

Rocha-Resende C, Roy A, Resende R, Ladeira MS, Lara A, De Morais Gomes ER, Prado VF, Gros R, Guatimosim C, Prado MA, Guatimosim S (2012) Non-neuronal cholinergic machinery present in cardiomyocytes offsets hypertrophic signals. J Mol Cell Cardiol 53:206–216

Salick MR, Napiwocki BN, Sha J, Knight GT, Chindhy SA, Kamp TJ, Ashton RS, Crone WC (2014) Micropattern width dependent sarcomere development in human ESC-derived cardiomyocytes. Biomaterials 35:4454–4464

Santi PA, Johnson SB (2013) Decellularized ear tissues as scaffolds for stem cell differentiation. J Assoc Res Otolaryngol 14:3–15

Santos AK, Tonelli FMP, Silva DAF, Gomes KN, Ladeira LO, Resende RR (2013) The role of cell adhesion, cell junctions, and extracellular matrix in development and carcinogenesis. In: Resende RR, Ulrich H (eds) Trends in stem cell proliferation and cancer research. Springer, The Netherlands

Sasaki J, Hashimoto M, Yamaguchi S, Itoh Y, Yoshimoto I, Matsumoto T, Imazato S (2015) Fabrication of biomimetic bone tissue using mesenchymal stem cell-derived thee-dimensional constructs incorporating endothelial cells. PLoS One 10:e0129266

Sellaro TL, Ranade A, Faulk DM, Mccabe GP, Dorko K, Badylak SF, Strom SC (2010) Maintenance of human hepatocyte function in vitro by liver-derived extracellular matrix gels. Tissue Eng Part A 16:1075–1082

Sousa BR, Parreira RC, Fonseca EA, Amaya MJ, Tonelli FM, Lacerda SM, Lalwani P, Santos AK, Gomes KN, Ulrich H, Kihara AH, Resende RR (2014) Human adult stem cells from diverse origins: an overview from multiparametric immunophenotyping to clinical applications. Cytometry A 85:43–77

Srivastava A, Bapat M, Ranade S, Srinivasan V, Murugan P, Manjunath S, Thamaraikannan P, Abraham S (2010) Multiple injections of in vitro expanded autologous bone marrow stem cells for cervical level spinal cord injury—a case report. J Stem Cells Regen Med 6:175–176

Strem BM, Hicok KC, Zhu M, Wulur I, Alfonso Z, Scheiber RE, Fraser JK, Hedrick MH (2005) Multipotential differentiation of adipose tissue-derived stem cells. Keio J Med 54:132–141

Tavakol S, Aligholi H, Gorji A, Eshaghabadi A, Hoveizi E, Tavakol B, Rezayat SM, Ai J (2014) Thermogel nanofiber induces human endometrial-derived stromal cells to neural differentiation: in vitro and in vivo studies in rat. J Biomed Mater Res A 102:4590–4597

Terai S, Ishikawa T, Omori K, Aoyama K, Marumoto Y, Urata Y, Yokoyama Y, Uchida K, Yamasaki T, Fujii Y, Okita K, Sakaida I (2006) Improved liver function in patients with liver cirrhosis after autologous bone marrow cell infusion therapy. Stem Cells 24:2292–2298

Thomson JA, Itskovitz-Eldor J, Shapiro SS, Waknitz MA, Swiergiel JJ, Marshall VS, Jones JM (1998) Embryonic stem cell lines derived from human blastocysts. Science 282:1145–1147

Till JE, Mc CE (1961) A direct measurement of the radiation sensitivity of normal mouse bone marrow cells. Radiat Res 14:213–222

Tonelli FMP, Santos AK, Silva SL, Gomes KN, Ladeira LO, Resende RR (2012a) Stem cells and calcium signaling. In: Islam MS (ed) Advances in experimental medicine and biology, 1st edn. Springer, New York

Tonelli FMP, Santos AK, Gomes KN, Lorencon E, Guatimosim S, Ladeira LO, Resende RR (2012b) Carbon nanotube interaction with extracellular matrix proteins producing scaffolds for tissue engineering. Int J Nanomed 7:4511–4529

Tuan RS, Boland G, Tuli R (2003) Adult mesenchymal stem cells and cell-based tissue engineering. Arthitis Res Ther 5:32–45

Valenti MT, Mori A, Malerba G, Dalle Carbonare L (2015) Mesenchymal stem cells: a new diagnostic tool? World J Stem Cells 7:789–792

Wakitani S, Nawata M, Tensho K, Okabe T, Machida H, Ohgushi H (2007) Repair of articular cartilage defects in the patello-femoral joint with autologous bone marrow mesenchymal cell transplantation: the case reports involving nine defects in five knees. J Tissue Eng Regen Med 1:74–79

Wang T, Feng Z-Q, Leach MK, Wu J, Jiang Q (2013) Nanoporous fibers of type-I collagen coated poly(l-lactic acid) for enhancing primary hepatocyte growth and function. J Mater Chem B 1:339–346

Zhang X, Dong J (2015) Direct comparison of different coating matrix on the hepatic differentiation from adipose-derived stem cells. Biochem Biophys Res Commun 456:938–944

Zhang SC, Wernig M, Duncan ID, Brustle O, Thomson JA (2001) *In vitro* differentiation of transplantable neural precursors from human embryonic stem cells. Nat Biotechnol 19:1129–1133

Zhang Y, He Y, Bharadwaj S, Hammam N, Carnagey K, Myers R, Atala A, Van Dyke M (2009) Tissue-specific extracellular matrix coatings for the promotion of cell proliferation and maintenance of cell phenotype. Biomaterials 30:4021–4028

Zuk PA, Zhu M, Mizuno H, Huang J, Futrell JW, Katz AJ, Benhaim P, Lorenz HP, Hedrick MH (2001) Multilineage cells from human adipose tissue: implications for cell-based therapies. Tissue Eng 7:211–228

# Chapter 20
# Scaffolds for Embryonic Stem Cell Growth and Differentiation

**Ana Teresa Semeano, Talita Glaser, Henning Ulrich, and Denise Freitas Siqueira Petri**

**Abstract** This chapter is organized in order to help the reader on choosing materials for scaffolds production, and to serve as a guide for scaffolds preparation and characterization for embryonic stem cell growth and differentiation. First, different methods for the preparation and characterization of scaffolds are briefly described, highlighting critical parameters. Then, we focus on how to evaluate scaffold suitability, such as its stability in culture media and toxicity as well as suitable protocols for sterilization. Qualitative and quantitative analysis of stem cell growth, viability and differentiation by immunofluorescence microscopy and flow cytometry are also presented.

**Keywords** Scaffolds • Polymers • Material properties • Stem cell • Differentiation • Tissue engineering

## 20.1 Introduction

Scaffolds for stem cell growth and differentiation are porous biocompatible materials with high chemical and mechanical stability. Their primary function is promoting cell adhesion and proliferation. Typically, cells are seeded on scaffolds in the absence or in the presence of growth factors with the possibility of mechanical, physical (electric or magnetic field) or chemical (specific molecules) proliferation stimulation. Cells are cultured on scaffolds *in vitro* to create implantable tissues or, scaffolds are implanted *in vivo* to induce tissue regeneration within the organism. After implantation, scaffolds should biodegrade at the same rate that the tissue regenerates, not causing any inflammatory process. 2D or 3D scaffolds might be applied for (1) cell growth, where

A.T. Semeano • T. Glaser • H. Ulrich
Department of Biochemistry, Institute of Chemistry, University of São Paulo, Av. Prof. Lineu Prestes 748, São Paulo, SP 05508-000, Brazil

D.F.S. Petri (✉)
Department of Fundamental Chemistry, Institute of Chemistry, University of São Paulo, Av. Prof. Lineu Prestes 748, São Paulo, SP 05508-000, Brazil
e-mail: dfsp@iq.usp.br


H. Ulrich, P. Davidson Negraes, *Working with Stem Cells*,
DOI 10.1007/978-3-319-30582-0_20

cells attach predominantly to the scaffolds surface, and (2) differentiation, where they diffuse into the matrix and attach to the internal scaffold structures. To visualize attached cells, 2D scaffolds are more convenient. However, 3D scaffolds offer an environment closer to the physiological one, the process can be automatized and the functional activity of (neuronal) cells can be better synchonized than in 2D scaffolds (Bosi et al. 2015). A plethora of commercial 3D scaffolds is available; their advantages and disadvantages are described elsewhere (Comley 2010).

How to make scaffolds for cell growth? There are many possible procedures to create scaffolds in the laboratory; some of them dispense expensive means. For instance, hydrogels of polymers, such as chitosan, alginate, starch and xanthan, are prepared by dissolving the polymer preferably in water, followed by crosslinking with a nontoxic crosslinker agent, such as $Ca^{2+}$ ions, in the case of alginate, genipin or citric acid, in the case of polysaccharides. Glutaraldehyde or epichlorohydrin should be avoided as crosslinkers due to their toxicity. Polymeric electrospun matts can be produced from polymer solutions, emulsions or melts. Polymers' molecular characteristics, solubility/melting properties (viscosity, conductivity, surface tension), flow rate, applied voltage, distance between capillary and collector are some of the many experimental parameters that control the physical characteristics (porosity, uniformity, fiber diameter etc.) of the final matt. Cast protein (laminin, gelatin, fibronectin, vibronectin) film is a traditional method to easily produce 2D scaffolds. Generally, an aqueous protein solution (0.1–0.2 % m/v) is poured onto the plastic dish or tissue culture flask so that the bottom is fully covered, then the solution excess is removed leaving only a thin fluid film; the dish or flask is put inside a laminar flow hood for total evaporation of water. New trends and products continuously appear in this challenging area. Just to mention two of them, decellularization and 3D printer. Decellularization and solubilization of extracellular matrix (ECM) by acid and enzymatic treatments, followed by embedding in a polymeric gel allow creating 3D cell-printed gels (Pati et al. 2014). 3D bioprinting for the generation of tissues and artificial organs has already become reality, although print resolution and speed are parameters that require improvements (Murphy and Atala 2014). Figure 20.1 shows typical structures of scaffolds prepared by some of the methods cited above.

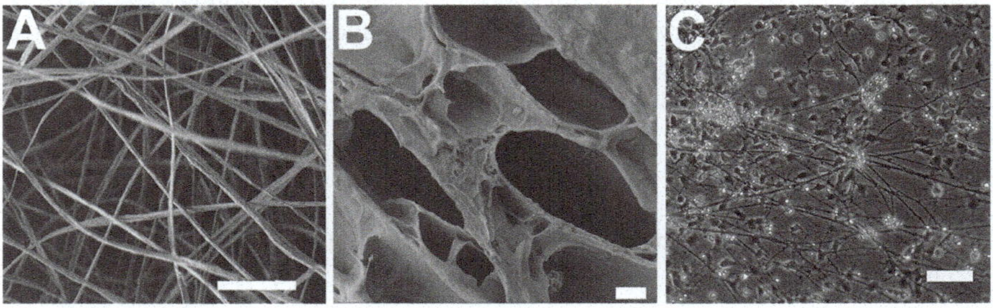


**Fig. 20.1** Examples of typical polymeric structures of scaffolds for cell culture. (**a**) PLLA electrospun mat for fibroblasts (Casarano et al. 2009), scale bar = 10 μm; (**b**) Xanthan network for neuronal cells (Glaser et al. 2015), scale bar = 2 μm; (**c**) Gelatin with magnetic particles for neural precursor cell differentiation. Scale bar = 100 μm

Ceramics, synthetic polymers and natural polymers, or combination of them are materials often used as scaffolds. Nowadays, a large number of commercial scaffolds are available (Dhandayuthapani et al. 2011; Cheng 2008; Place et al. 2009); the most commons are collagen, laminin, hyaluronic acid, alginate, poly(glycolic acid) (PGA), poly(L-lactic acid) (PLLA), poly(lactic-*co*-glycolyc acid) (PLGA) and poly(ethylene glycol) (PEG). Many of these polymers are biodegradable and provide controlled release of growth factors or drugs for regeneration therapy. On the other hand, non-biodegradable polymers are biologically stable, providing a long-lasting support as bone cements.

Regardless of the type of material used in the scaffold, distinct physical and chemical properties are required to make it suitable for cell adhesion, proliferation and differentiation (Fig. 20.2). Surface energy, porosity, surface patterning, mechanical properties and anisotropic topography are relevant parameters that should be considered upon choosing the scaffold material. The ***surface energy*** of the scaffold is important because it affects the adsorption of proteins, such as integrins, which mediate cell attachment to the scaffold (Sniadecki et al. 2006). Materials with low surface energy are hydrophobic, and proteins frequently undergo denaturation upon adsorbing on them. Hydrophilic materials have high surface energy. Cell surface protein binding on them occurs due to H bonding or electrostatic interactions. Fibronectin and collagen are extracellular proteins which provide a suitable layer for cell growth. Glass slides coated with fibronectin and collagen present surface energy values of ~40 and ~48 $mJ/m^2$, respectively (Harnett et al. 2007). Scaffolds with surface energy close to these values are expected to provide adequate environment for cell proliferation. The surface energy can be determined by measuring the contact angle of droplets of test liquids with different polarities (water and diiodomethane, for instance) deposited on the material surface (Kosaka et al. 2009; Yuan and Lee 2013).

Scaffolds should present high ***porosity*** to allow diffusion of nutrients to the cells and the diffusion of waste products from the scaffold. The optimal pore size depends on the cell type and on the tissue to be regenerated. For instance, some published reports indicate pores of 5 μm for neovascularization (Brauker et al. 1995), 30–50 μm for fibroblast (Bueno et al. 2013) or neuronal cells (Glaser et al. 2015), and 100–325 μm for osteoblasts (Karageorgiou and Kaplan 2005; Murphy et al. 2010). Specific surface decreases with pore size increase. Thus, small pores might be beneficial during the initial stage of cell adhesion, but ultimately large pores favor cells migration into the pores, where cells can interact with specific ligands or proteins. The porosity of scaffolds can be determined by different methods. Adsorption isotherms of gas (BET isotherms) yield pore size distribution and surface area; commercial equipment with well-stablished conditions ($N_2$ and 77 K) exists for this application (Sing et al. 1985). However, this method is not always suitable for soft materials, because the pressure applied during the measurement might cause pore collapse. Mercury porosimetry is a capillary penetration technique used to predict the pore volume intruded/extruded corresponding to a certain capillary pressure. In mercury porosimetry, relatively higher capillary pressures are used due to high surface tension and contact angle of the non-wetting mercury. The technique is

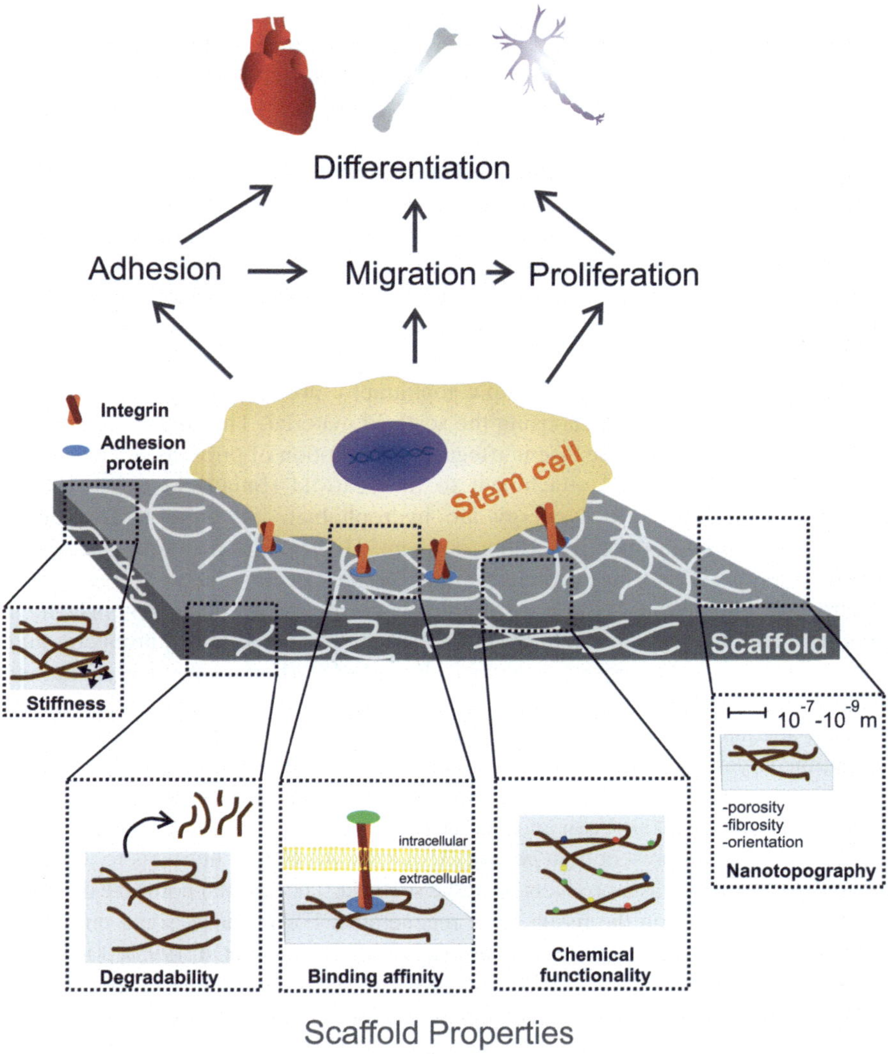


**Fig. 20.2** Inherent material properties. Chemical composition, stiffness, porosity and surface anisotropy of scaffolds define their interactions with key proteins, such as integrins, responsible for cell attachment. Growth and differentiation are stimulated by intrinsic properties of the scaffold and by molecules with biological functionality that are embedded in it

limited to pores with diameters smaller than 200 μm and the high pressure might collapse the pores of soft scaffolds (Gueven and Hicsasmaz 2013). The capillary rise method is widely used for the determination of mean pore radius, contact angle and free surface energy of scaffolds. The linear increase of the square mass (or capillary height) of wetting liquid as a function of time is directly proportional to the mean pore radius (Washburn's equation) (Grundke et al. 1996). The scaffolds should be packed in the sample holder in a reproducible way in order to obtain precise results.

The pore size and size distribution can be estimated through scanning electron microscopy (SEM) image analysis. Scaffolds should be carefully cryofractured or sliced in order to expose the cross-section with minimal structure deformation, dried and coated with a thin (~2 nm) layer of a conducting material (Au, Pt or carbon) prior to the SEM approach. Image analyses can be performed with the help of a software (ImageJ free software, for instance). One should note that the pore size determination from different techniques does not necessarily converge to similar values because techniques might underlie different principles, and the sample might be probed under different ways (wet, dry, 2D or 3D).

Cell adhesion on rough surfaces, resembling native ECM, is favored relative to flat surfaces. The quantitative analysis of ***surface roughness*** can be performed by profilometry or atomic force microscopy (AFM) (Poon and Bhushan 1995). The vertical deviations of the roughness profile from the mean line can be treated with basic statistics; the arithmetic ($R_a$) and root mean square (RMS) averages are frequently calculated from the experimental data to quantify surface roughness

$$R_a = \frac{1}{n}\sum_{i=1}^{n}\left|y_i\right|$$

$$RMS = \sqrt{\frac{1}{n}\sum_{i=1}^{n}y_i^2}$$

where $n$ is the number of data, $y_i$ is the difference between the height at the point $i$ and the mean height of the profile.

***Chemical surface micropatterning*** helps driving the position for cell proliferation on isotropic islands or along anisotropic lines, whereas ***surface nanopatterning*** regulates collective actions, such as adhesion and differentiation. Chemical micropatterning generates cytophilic and cytophobic regions, so that after random seeding the cells specifically adhere to the cytophilic spots. Fibronectin is used as "ink for polydimethylsiloxane-based microstamps" to produce cytophilic regions. Alternatively, octadecyltrichlorosilane can be stamped to create cytophobic spots (Kane et al. 1999; Lehnert et al. 2004). Nanoscale dip-pen lithography is employed to create biomolecular patterns on scales under of 100 nm (Ginger et al. 2004). ***Anisotropic topographic patterning*** favors cell alignment. Cell orientation increases with increasing groove depth of microscale ridges and grooves, while it decreases with increasing groove width (Walboomers et al. 1999). On the other hand, orientation promoted by width and depth of nanogrooves orientation seems to be cell-type dependent (Lim and Donahue 2007).

***Mechanical properties*** of scaffolds are important because they also affect mass transport into and out of the cells. The elastic modulus or Young's modulus (E in Pa) can be calculated from the stress-strain or compression-strain curves; the higher is the E value, the stiffer is the material. Samples dimension and tests conditions are well defined in the ASTM (2010). Scaffolds of similar chemical composition but with different stiffness tend to modulate specific differentiation lineages. For instance, murine mesenchymal progenitor cells on poly-ε-caprolactone-based matts with moduli of 7.1 MPa or 30.6 MPa stimulate chondrogenesis or osteogenesis, respectively (Nam et al. 2011)

## 20.2 Preliminary Tests to Check Scaffold Suitability for Cell Growth and Differentiation

After material characterization, the ***mechanical stability*** of the polymer should be tested in suitable conditions for its functionality, monitoring physical properties of interest. Culture medium resistance test is essential to adjust scaffolds production to its purpose, where adequate properties of scaffold must be kept in medium immersions for cell proliferation and differentiation, saving time and resources whenever possible to perform this assay.

Secondly, choose appropriate ***sterilization method*** is required. Sterilization methods can affect physical and chemical properties of scaffolds. Therefore, it is crucial to assess the most suitable method for the material at hand, so that effective doses can be used with minimal effects on its structure and function, allowing maintenance of scaffold bioactivity. Sterilization is defined as a process intended to remove or destroy all viable forms of microbial life, including bacterial spores, in order to achieve an acceptable level of sterilization (Block 2001). Several effective methods of sterilization are recognized by the U.S. Food and Drug Administration (FDA), including dry heat, moist heat (autoclave), ethylene oxide, ionizing radiation, and liquid chemical sterilants (Freytes and Badylak 2006). Each of these methods involves a different mechanism of action for killing microbes and, therefore, the physical and chemical effects of these methods upon naturally occurring molecules will differ. Steam sterilization (or autoclaving) at high temperature and pressure is a common and simple sterilization process without toxic residues that removes all contaminants from scaffolds (Müller et al. 2006). Ethanol washes are also used as disinfecting agent but often aggressive for some polymers. Alternatively, ethylene oxide gas sterilization has high penetration capability, excellent efficiency and compatibility for degradable and porous scaffolds. Comparing the effects of ethylene oxide and $\gamma$-irradiation sterilization e.g. for tyrosine-derived polycarbonates and PLLA indicates that ethylene oxide induces less structural damage to polymers (Hooper et al. 1997). Decrease of molecular weight and tensile strength of biodegradable polyurethanes was also observed for sterilization methods by $\gamma$-irradiation (Gorna and Gogolewski 2003); however, this process is safe and effective at treating single-use medical devices.

### *20.2.1 Culture Medium Resistance Test*

1. Place the piece of polymer with appropriated geometry into Petri dishes or plates;
2. Fill the plate with suitable culture medium in excess, to not limit the swelling of the polymer;
3. Incubate at 37 °C;

4. Get the sample after 48 h, or several samples over time; *observe changes*;
5. Measure a critical property for a particular scaffold, such as stiffness (or any other previously described).

### 20.2.2 Sterilization Process

**Materials**

- Petri dish.
- Aluminum foil.
- Ethanol 70 % (v/v).
- Phosphate-buffered saline (PBS) 10 mM (see ***Note 1***).
- Self-sealing sterilization pouches.

***Processes***

**Autoclaving**

1. Place scaffolds on a glass Petri dish, avoiding overlaps between them;
2. Wrap up the Petri dish with aluminum foil;
3. Set autoclave at 120 °C and 2.150 bar for 20 min;
4. Dry in an oven at 37 °C overnight and store in a water-free environment.

**Ethanol Treatment**

1. Incubate scaffolds in ethanol 70 % (v/v) for 5 min at 4 °C;
2. Wash three times in 1× PBS;
3. Dry in an oven at 37 °C overnight and store in a water-free environment.

**Ethylene Oxide Gas Treatment**

1. Pack scaffolds in self-sealing sterilization pouches;
2. Expose samples for 3 h to a 100 % ethylene oxide atmosphere at 55 °C;
3. Air-expose the samples for 12 h.

**Gamma Irradiation**

1. Expose scaffolds to γ-irradiation with a dose of 25 kGy;
2. Apply to continuous and batch type gamma γ-irradiators using the radionuclides $^{60}$Co and $^{137}$Cs, and to irradiators using a beam from an electron or X-ray generator (ISO 11137 1995).

**Sterility Determination**

1. Incubate scaffolds in antibiotic-free culture medium at 37 °C;
2. Check samples daily for any changes in color or turbidity, which would denote contamination;
3. Confirm the sterility of the scaffold by the absence of microorganisms through the microscopic observation.

## 20.3 Cell Growth Curves and Cell Viability

In order to be applied for stem cell therapies, organic or synthetic scaffolds should promote cell adhesion (see ***Note 2***), growth/proliferation, and should not promote cell death (toxicity) (Drury and Mooney 2003). A low cost and simple method to explore these characteristics is the cell growth curve analysis, which can be used with all stem cell types. In this chapter, we describe this method using embryonic stem cells as a reference because they proliferate at high rates (Ruiz et al. 2011). A slightly modification of the method allows determination of cell viability, as detailed below.

### *20.3.1 Cell Growth Curve*

Cell growth curves consist of three phases, which are the cell adhesion and reentering of the cell cycle phase (Lag), exponential growth phase (Log), and the decreased growth phase (plateau) due to depletion of culture medium (Mather and Roberts 1998). Figure 20.3 shows an example of a growth curve obtained for embryonic stem cells. This method allows comparing between different scaffolds or conditions by acquisition of parameters such as population-doubling time, saturation density and lag time.

**Materials**

- Stem Cells. ES-E14TG2a feeder free (ATCC CRL-1821).
- Tissue culture dishes or flasks pre-coated with 0.1 % (m/v) gelatin from porcine skin (Sigma-Aldrich, Cat. No. 9000-70-8).
- 1× PBS (see ***Note 1***).
- <u>Growth medium:</u> Dulbecco's Modified Eagle Medium High glucose (DMEM High GIBCO, Cat. No. 11965-126) supplemented with 15 % ESC-qualified Fetal Bovine Serum (GIBCO, Cat. No. 16000-044), 103 U/mL LIF (ESGRO-LIF Merck-Millipore, Cat. No. ESG1106), 100 mM sodium pyruvate (Sigma-

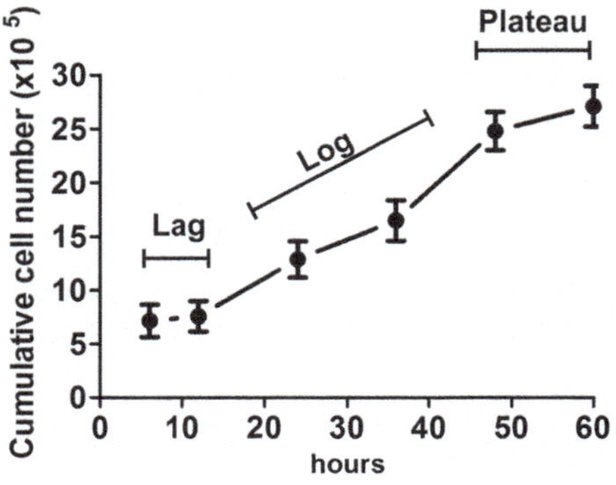


**Fig. 20.3** Cell growth phases. Cell adhesion and reentering the cell cycle phase (Lag), exponential growth phase (Log), and the decreased growth phase (plateau)

Aldrich, Cat. No. 113-24-6), 100 mM L-glutamine (GIBCO, Cat. No. 25030-081), 0.1 mM β-mercaptoethanol (GIBCO, Cat. No. 21985-023), 1 % non-essential aminoacids (GIBCO, Cat. No. 11140-050).
- TrypLE Express Enzyme (1×) (GIBCO, Cat. No. 12604-054).

**Procedure**

1. Wash cells twice with 1× PBS;
2. Add trypsin to the tissue culture flask, 1 mL/T25 $cm^2$ and wait for 2–5 min;
3. Collect the cells in a 15 mL conical tube and centrifuge cells at $1000 \times g$ for 5 min;
4. Resuspend the pellet in 5 ml medium and count the cells using a hemocytometer on an inverted phase microscope. Usually $1 \times 10^6$ cells/mL are acquired;
5. Dilute the cell suspension in order to have an appropriate amount of medium and cells to achieve a seeding density of $2 \times 10^3$ cells/$cm^2$ of surface area. Cell seeding densities vary by cell line;
6. Mix well and seed 2× 35 $cm^2$ dishes for each time point using the appropriate amount of diluted cell suspension;
7. Incubate cells at 37 °C and 5 % $CO_2$;
8. Count the duplicate dishes every 12 h during 4 days;
9. Plot the results on a log-linear scale. The population-doubling time can be determined by identifying a cell number along the exponential phase of the curve, tracing the curve until that number has doubled, and calculating the time between the two (Fig. 20.3).

## *20.3.2 Cell Viability*

Live cells select the compounds that can pass through the cell membrane (Louis and Siegel 2011). Trypan blue is a dye which does not penetrate into live cells; it only stains dead cells that are observed as blue-colored under the microscope. Since live cells are excluded from staining, this procedure is also described as a dye exclusion method (Louis and Siegel 2011). The traditional cell counting method uses a hemocytometer to count the cells and assess their viability. The method cannot distinguish between necrotic and apoptotic cells (Louis and Siegel 2011).

**Materials**

- Trypan Blue Solution (Sigma-Aldrich, 72-57-1), 0.4 % (w/v) in PBS.
- Ethanol, 70 % (v/v).

**Procedure**

1. Prepare a suspension of approximately $1 \times 10^6$ cells/mL,
2. Make sure that the sample is well homogenized. Make a 1:1 mixture of the cell suspension and the 0.4 % (w/v) trypan blue solution. Gently mix and let rest for 1–2 min at room temperature;

3. Wash the hemocytometer and the coverslip with 70 % (v/v) ethanol, allow them to air dry;
4. Apply 10 μL of cell suspension to the edge of the chamber and count the cells

   i. Count both viable (unstained) and non-viable (stained) cells;
   ii. Take the average and multiply by $1 \times 10^4$ to obtain the number of cells per mL in the sample placed in the hemocytometer;
   iii. Multiply by two to take into account the 1:1 dilution of the sample in the trypan blue;
   iv. Multiply by any dilutions in the original sample preparation of the cell suspension.

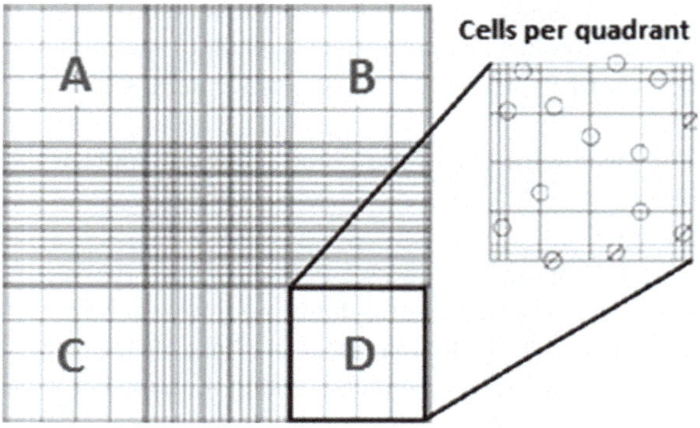


$$\text{number of viable or unviable cells} = \frac{(A+B+C+D)}{4} \times 10^4 \times 2 \times \text{sample dilution}$$

5. The percentage of unstained cells represents the percentage of viable cells in the suspension.

$$\%\,\text{Viable cells} = \frac{\text{Number of viable cells}}{\text{Total number of cells}} \times 100$$

## 20.4 Cell Differentiation Analysis

Cell differentiation can be influenced by signals from the interface of cell/scaffold, since cells perceive the intrinsic properties of substrates when interacting with these materials. The cell differentiation process is usually time consuming and requires appropriated culture media and growth factors to guide the cell to its specific fate. The period of differentiation and requirements for scaffolds, culture medium and growth factors may vary, as these need to be compatible with the respective type of cells, both undifferentiated and differentiated.

The process of differentiation and acquisition of final cell fate is monitored by immunostaining techniques. Qualitative methods such as immunofluorescence microscopy, and quantitative methods, such as image and flow cytometry, are described below.

**Materials**

- Coverslips coated with polyethyleneimine (Sigma-Alrich, Cat. No. 9002-98-6) or poly-L-lysine (Sigma-Alrich, Cat. No. 25988-63-0) for 1 h at room temperature as control. Rinse coverslips well with sterile $H_2O$ (three times, 5 min each). Allow coverslips to completely air dry and sterilize them under UV light for at least 4 h.
- 1× PBS.
- 4 % (m/v) paraformaldehyde (PFA) (Sigma-Alrich, Cat. No. 30525-89-4).
- TrypLE™ Express Enzyme (1×) (GIBCO, Cat. No. 12504-054).
- Blocking solution: 5 % (v/v) serum (see ***Note 3***) and 0.1 % (m/v) triton X-100 (Sigma-Alrich, Cat. No. 9002-93-1) in PBS.
- Digitonin (Sigma-Aldrich, 11024-24-1) or saponin (Sigma-Aldrich, 8047-15-2).
- Bovine serum Albumin (BSA) (Sigma-Aldrich, Cat. No. 9048-46-8).
- Glycine (GIBCO, Cat. No. 15527-013).
- Tween-20 (Sigma-Aldrich, Cat. No. 9005-64-5).
- Hoechst (Sigma-Aldrich, Cat. No. 23491-45-4) or DAPI (Merck Millipore, Cat. No. 1246530100).

### *20.4.1 Direct and Indirect Confocal Immunofluorescence*

To be applied in cell therapy, a scaffold should either maintain or improve cell differentiation. Therefore, to test cell differentiation, some characteristics of the cell should be analyzed such as morphology and expression of markers for specific cell types. For this purpose, we can use an immunofluorescence assay (Bhattacharyya et al. 2010).

**Procedure**

1. Grow cells on glass coverslips or use the scaffolds previously prepared;
2. Rinse briefly in 1× PBS;
3. Fix cells using 4 % (m/v) PFA in PBS pH 7.4 for 10–30 min at room temperature. Cells should be washed three times with ice-cold PBS. Alternatively, incubate the cells in 100 % methanol (chilled at −20 °C) at room temperature for 5 min;
4. Incubate the samples for 10 min with 1× PBS containing 0.1 % (m/v) Triton X-100 (see ***Note 4***) (or 100 μM digitonin or 0.5 % (m/v) saponin);
5. Wash cells in 1× PBS three times for 5 min;
6. Incubate cells with blocking solution for 30 min, or with 1 % (m/v) gelatin or 1 % (m/v) BSA, 22.52 mg/ml glycine in 1× PBS + 0.1 % (v/v) Tween 20;

7. Next, incubate cells with select primary antibodies in 1 % (v/v) serum in 1× PBS+0.1 % (v/v) Triton X-100 or Tween 20 in a humidified chamber for 1 h at room temperature, or overnight at 4 °C;
8. Wash cells three times in 1× PBS, 10 min each wash;
9. Incubate cells with appropriate secondary antibodies in 1 % (v/v) serum for 1 h at room temperature, in the dark;
10. Wash three times with 1× PBS, 10 min each, in the dark (see ***Note 5***);
11. For counterstaining DNA (nucleus), incubate cells with 0.5 μg/mL Hoechst or DAPI for 5 min;
12. Rinse cells with 1× PBS;
13. Drop mounting medium on the coverslip and seal it with nail polish to prevent drying and/or movement under microscope. Store in dark at −20 °C or 4 °C.

## *20.4.2 Flow Cytometry*

Flow cytometry allows measuring and analyzing physical properties of single cells such as size, granularity and relative fluorescence intensity. Light scattering measured in the forward-scattered channel (FSC) determines the cell size, while signals detected by the side-scattered channel (SSC) are proportional to cell granularity or internal complexity. The correlation between FSC and SSC measurements can be used to distinguish cell types in a heterogeneous population. Fluorochome-labeled antibodies can be used to identify antigenic surface markers and more than one fluorochome can be used simultaneously (see ***Note 6***). The fluorescence pattern of each subpopulation, combined with the FSC and SSC data can allow identifying which cells are present in a sample and count their relative percentages (Weaver 2000). Different parameters are analyzed simultaneously for a single cell and can be displayed as a histogram, a two or a three-dimensional plot. Data can also be subdivided by gating upon specific populations. After reaching the desired differentiation stage, cells are individualized, fixed and labeled as described in the protocol written below.

**Procedure**

1. Aspirate the culture medium and wash cells once in 1× PBS;
2. Add 1 ml of trypsin solution, incubate at 37 °C for 3 min and add 3 mL of culture medium with serum to block trypsin action;
3. Spin cells for 1 min at 1000×*g* and resuspend in 1 mL of culture medium. Gently pipette up and down to achieve a homogeneous single cell suspension. It is necessary to completely dissociate suspension or adherent cells into individual cells;
4. After 30 min (time for recovery of the cellular membrane), spin cells and resuspend in 100 μl of 1× PBS, splitting cells (according to the number of antibodies that will be analyzed);

5. Fix cells by adding 1 mL of 4 % (m/v) PFA, mix well and incubate for 15 min at room temperature;
6. Filter cell with a 40 μm strainer;
7. Pellet cells (1000 × *g* for 5 min), decant and resuspend in 1× PBS;
8. Store at 4 °C or proceed to the analysis;
9. *For immunostaining*, spin cells for 5 min at 1000 × *g* and wash twice with 20 volumes of blocking solution;
10. Count cells and adjust the concentration of cell suspension to at least $5 \times 10^6$ cells/mL. Cells should have more than 90 % of viability, as determined by trypan blue exclusion;
11. Determine the number of samples necessary for the experiment and aliquot 100 μL of cell suspension into each sample tube ($1 \times 10^5$ to $1 \times 10^6$ cells/sample, preferably the higher value, if possible);
12. Keep the cells in blocking solution for 30 min at room temperature;
13. Add selected primary antibodies at the appropriate final concentration (see ***Note 7***) in blocking solution and incubate for 30 min at room temperature;
14. Wash twice with 20 volumes of blocking solution;
15. If using directly conjugated antibody, proceed to *step 18*;
16. Add the appropriate secondary antibodies in block solution at the recommended concentration for 30 min at room temperature, in the dark;
17. Repeat *steps 3–6;*
18. Resuspend stained cells in 500 μL of PBS;
19. Proceed to flow cytometry analysis (see ***Note 8***).

**Analysis**

In flow cytometry data analysis, we can define a subset of data through a gate. This gate is a numerical or graphical boundary that allows determining the physical characteristics of particles which will be included in further analysis. An important principle of flow cytometry is to selectively visualize the cells of interest while eliminating results from unwanted particles e.g. dead cells and debris (Fig. 20.4). Mixed populations of cells are represented in typical graphs of SSC vs. FSC. In this type of analysis, we can differentiate cells from each other and from cell debris through their physical properties. Criteria to gating should be applied with the greatest possible accuracy (Rossini et al. 2008).

The argon laser with emission at 488 nm is the most commonly used in flow cytometry because its capability to excite several fluorochomes in the visible light. More than one fluorochome can be used if the wavelengths of their emission peaks do not present major overlaps. For example, the combination of Alexa Fluor 488 and Alexa Fluor 647 satisfies this condition once the peak emission wavelength is 520 nm for Alexa Fluor 488, and 670 nm for Alexa Fluor 647. Therefore, each signal is detected by a separate detector (Fig. 20.4). Distinct laser wavelengths and emission filters can be used to increase the possibilities of combinations without overlap. The choice of fluorochomes should be aided by consulting tables or simulators (BD Biosciences 2015).

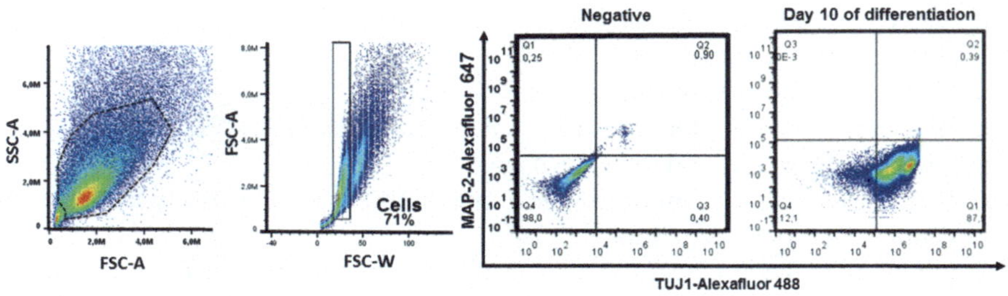


**Fig. 20.4** Gating flow cytometry data to identifying cell types associated with surface markers. Representative flow cytometry data of embryonic stem cells differentiated into neurons and immunostained for protein markers: TUJ1 (early neuron) and MAP-2 (mature neuron). The first plot (*left*) shows debris exclusion (FSC vs. SSC); second plot shows doublet exclusion (FSC-Area vs. FSC-Width); third plot shows negative staining control and; the last plot (*right*) shows expression of TUJ1 and MAP-2 after 10 days of differentiation

## *20.4.3 Real-Time Polymerase Chain Reaction (qRT-PCR)*

Quantitative RT- PCR (qRT-PCR) is a method used to detect relative or absolute gene expression levels. Reactions make use of fluorescence-labelling of amplification products to detect the threshold cycle (Ct) during PCR, when the level of fluorescence is larger than the background signal and is in the linear portion of the amplified curve. The most popular dye is SYBR Green, that intercalates with double-stranded DNA and then fluoresces. The qRT-PCR machine detects the fluorescence and the software calculates Ct values from the intensity of the fluorescence. PCR machines take readings of the amount of double stranded DNA in each well at each cycle and give the critical threshold (Ct) value, which represents the quantization of the product. Since Ct is a relative value, every experiment needs an internal control, usually an endogenous gene, whose expression levels do not vary under the chosen experimental conditions (Bustin et al. 2009). Real-time PCR uses cDNA samples, primers specific for the genes of interest (design primers to ideally function at about 60 °C) and SYBR Green mix, which includes SYBR Green dye, Taq Polymerase, ROX, and dNTP all-in-one. SYBR Green kits are available from many companies and are easy to use, following manufacturers' instruction (e.g. Applied Biosystems or Thermo Fisher Scientific).

***RNA extraction***

Total cell RNA can be extract by using the RNeasy Kit from Qiagen or the TRIZOL reagent from Life Technologies, following the manufacturers' instruction.

***cDNA synthesis***

The cDNA is the DNA strand complementary to the RNA, which is synthetized by using an enzyme capable of reverse transcription, and oligo dT, which is a primer that anneals to the poly-A tail of the mRNA. Nowadays, there are many options of kits for this purpose.

**Table 20.1** Preparation of PCR master mix

| Material | Volume for 1 reaction (μl) | Final concentration |
|---|---|---|
| 2× Reaction mix | 12.5 | 1× |
| Forward primer | 0.5 | 1 μM |
| Reverse primer | 0.5 | 1 μM |
| Rnase-free water | To 25 | – |
| Template | 1–5 | <500 ng/reaction |

**Real time Polymerase chain reaction (qRT-PCR)**

## Materials

- PCR master mix:

  2× Reaction mix (e.g. Applied Biosystems or Thermo Fisher Scientific).
  50 μM forward primer.
  50 μM reverse primer.
  RNase-free water (q.s.p.).
  cDNA.
  1.5 mL microcentrifuge tubes for preparing master mix.
  Thin-walled PCR tubes.
  Real-time PCR thermal cycler.

  ***Procedure***

  The following protocol describes how to conduct a real-time PCR amplification using SYBR Green (Fraga et al. 2014). The determination of amplification efficiency is important for accurate analysis of the PCR results.

1. Put the components described above on ice;
2. Prepare at least five dilutions of a serial dilution of the previously prepared cDNA;
3. Prepare a PCR master mix in 1.5 mL microcentrifuge tubes (see Table 20.1);
4. Distribute equal aliquots of PCT master mix into each thin-walled PCR tube;
5. Add the diluted template to each reaction tube as needed; *remember to include a negative control that does not receive any template*;
6. Place the tubes in the thermal cycler and start the program: 95 °C for 15 min; 40× 95 °C for 30 s, 45 °C to 65 °C for 30 s (depending on primer melting temperature), 72 °C for 30 s; Melting curve analysis;
7. Run the PCR reaction to determine Ct values for each dilution;
8. Plot the Ct values vs. the logarithm of the concentration or the copy number of the template;
9. The line drawn through these points should have a high $R^2$ value (>0.99) and is an indicative of the goodness-of-fit of the line through the data points;
10. Use the slope of the line to determine the efficiency of the PCR reaction by using

$$E = 10^{(-1/slope)} - 1$$

Where $E$ is the efficiency of the reaction and *slope* refers to the slope of the plot of Ct value vs. the log of the input template amount. A slope between −3.6 and −3.1 corresponds to an efficiency between 90 and 110 % (e.g. E=0.9−1.1).

11. Do it for each primer pair.

<u>Analysis</u>

Basically, the data analysis of a control vs. treatment type of experiment uses ΔΔCT method (Livak and Schmittgen 2001), and always require data normalization in order to obtain relative expression values for the gene of interest.

## 20.5 Notes

1. **PBS solution** can be prepared 1× to use or 10× to stock. <u>PBS 1× composition:</u> 137 mM NaCl, 2.7 mM KCl, 10 mM $Na_2HPO_4$ and 1.8 mM $KH_2PO_4$ in $H_2O$. After dissolutions of the reagents, adjust the pH 7.4 and sterilize by autoclaving or by filter sterilization. Store PBS at room temperature.
2. For optimal cell adherence, seeded scaffolds soaked in culture medium should be put into an incubator for at least 1 h. Subsequently, medium is carefully poured around the samples until they are completely covered.
3. In order to **block unspecific binding** of the antibodies, a blocking step is needed. For this purpose, cells should be incubated with serum from the species from which the secondary antibodies were raised. Each antibody needs a different blocking solution. Please, check the datasheet recommendation.
4. It is very important to permeabilize the cells if the target protein is expressed intracellularly, so the antibody can bind to the target protein. Triton X-100 is the most popular detergent for **permeabilization**. However, it is not appropriate for membrane-associated antigens since it destroys membranes.
5. **Multicolor staining** (*optional step*): In order to be able to examine the co-distribution of two (or more) different antigens in the same sample, a double immunofluorescence procedure can be carried out. Please, ensure that you have primary antibodies for different species and the corresponding secondary antibodies. For example, rabbit antibody against antigen A, mouse antibody against antigen B. Alternatively, you can use directly conjugated primary antibodies conjugated to different fluorophores.
6. Emission peaks of fluorochomes used for multiple-staining should be at wavelengths separated as possible in order to avoid **overlapping** of detected signals.
7. Primary and secondary **antibodies must be titrated** on known positive and negative cell populations prior to use.
8. Negative **controls** are required both for untreated (as background auto-fluorescence controls) and for cells treated with primary or secondary antibodies alone

to determine rates of unspecific binding. Compensation controls for distinct fluorochomes are needed to confirm antibody binding and to remove spectral overlaps during flow cytometry measurements.

**Acknowledgments** DFSP and HU acknowledge grant support by Fundação de Amparo à Pesquisa do Estado de São Paulo (FAPESP) and Conselho Nacional de Desenvolvimento Científico e Tecnológico (CNPq), Brazil. ATS's and TG's doctoral theses received support by fellowships from Coordenação de Aperfeiçoamento de Pessoal de Nível Superior (CAPES) and FAPESP, respectively.

## References

ASTM E111–04 (2010) Standard test method for Young's Modulus, Tangent Modulus, and Chord Modulus

Bhattacharyya D, Hammond AT, Glick BS (2010) High-quality immunofluorescence of cultured cells. Methods Mol Biol 619:403–410

BD Biosciences (2015) http://www.bdbiosciences.com/us/applications/research/multicolor-flow/m/745795/resourcestools?cc=US. Accessed 9 Sept 2015

Block SS (2001) Disinfection, sterilization, and preservation, 5th edn. Lippincott Williams & Wilkins, London

Bosi S, Rauti R, Laisham J, Turco A, Lonardoni D, Nieus T, Prato M, Scaini D, Ballerini L (2015) From 2D to 3D: novel nanostructured scaffolds to investigate signalling in reconstructed neuronal networks. Sci Rep 5:9562

Brauker JH, Carr-Brendel VE, Martinson LA, Crudele J, Johnston WD, Johnson RC (1995) Neovascularization of synthetic membranes directed by membrane micro architecture. J Biomed Mater Res 29:1517–1527

Bueno VB, Silva AM, Barbosa LRS, Catalani LH, Teixeira-Neto E, Cornejo DR, Petri DFS (2013) Hybrid composites of xanthan and magnetic nanoparticles for cellular uptake. Chem Commun 49:9911–9913

Bustin SA, Benes V, Garson JA, Hellemans J, Huggett J, Kubista M, Mueller R, Nolan T, Pfaffl MW, Shipley GL, Vandesompele J, Wittwer CT (2009) The MIQE guidelines: minimum information for publication of quantitative real-time PCR experiments. Clin Chem 55:611–622

Casarano R, Bentini R Bueno VB, Iacovella T, Monteiro FBF, Iha FAS, Campa A, Petri DFS, Jaffe M, Catalani LH (2009) Enhanced fibroblast adhesion and proliferation on electrospun fibers obtained from poly(isosorbide succinate-b-L-lactide) block copolymers Polymer 50:6218–6227

Cheng KE (2008) Three dimensional polymer scaffolds for high throughput cell-based assay systems. University of Georgia, Dissertation

Comley J (2010) 3D cell culture easier said than done! Drug Discov World Sum 2010 25–41

Dhandayuthapani B, Yoshida Y, Maekawa T, Kumar DS (2011) Polymeric scaffolds in tissue engineering application: a review. Int J Polym Sci 2011:1–19

Drury JL, Mooney DJ (2003) Hydrogels for tissue engineering: scaffold design variables and applications. Biomaterials 24:4337–4351

Fraga D, Meulia T, Fenster S (2014) Real-time PCR. In: Current protocols essential laboratory techniques 10:10.3:10.3.1–10.3.40

Freytes D, Badylak S (2006) Sterilization of biologic scaffold materials. In: Encyclopedia of medical devices and instrumentation, pp 273–282

Ginger DS, Zhang H, Mirkin CA (2004) The evolution of dip-pen nanolithography. Angew Chem Int Ed 43:30–45

Glaser T, Bueno VB, Cornejo DR, Petri DFS, Ulrich H (2015) Neuronal adhesion, proliferation and differentiation of embryonic stem cells on hybrid scaffolds made of xanthan and magnetite nanoparticles. Biomed Mat 10:045002

Gorna K, Gogolewski S (2003) The effect of gamma radiation on molecular stability and mechanical properties of biodegradable polyurethanes for medical applications. Polym Degrad Stab 79:465–474

Grundke K, Bogumil T, Gietzelt T, Jacobasch HJ, Kwok DY, Neumann AW (1996) Wetting measurements on smooth, rough and porous solid surfaces. Progr Colloid Polym Sci 101:58–68

Gueven A, Hicsasmaz Z (2013) Pore and structure in food. Springer briefs in food, health and nutrition, Doi: 10.1007/978-1-4614-7354-1-2

Harnett EM, Alderman J, Wood T (2007) The surface energy of various biomaterials coated with adhesion molecules used in cell culture. Colloids Surf B Biointerfaces 55:90–97

Hooper KA, Cox JD, Kohn J (1997) Comparison of the effect of ethylene oxide and g-irradiation on selected tyrosine-derived polycarbonates and poly(L-lactic acid). J Appl Polym Sci 63:1499–1510

International Standard Organization (ISO) (1995) Sterilization of healthcare products—requirements for validation and routine control-radiation sterilization, ISO 11137, Geneva

Kane RS, Takayama S, Ostuni E, Ingber DE, Whitesides GM (1999) Patterning proteins and cells using soft lithography. Biomaterials 20:2363–2376

Karageorgiou V, Kaplan D (2005) Porosity of 3D biomaterial scaffolds and osteogenesis. Biomaterials 26:5474–5491

Kosaka PM, Amim J Jr, Saito RSN, Petri DFS (2009) Thermodynamics of cellulose ester surfaces. In: Roman M (ed) Model cellulosic surfaces, vol 1019, ACS symposium series. American Chemical Society, Washington, DC, pp 223–241

Lehnert D, Wehle-Haller B, David C, Weiland U, Ballestrem C, Imhof BA, Bastmeyer M (2004) Cell behaviour on micropatterned substrata: limits of extracellular matrix geometry for spreading and adhesion. J Cell Sci 117:41–52

Lim JY, Donahue HJ (2007) Cell sensing and response to micro- and nanostructured surfaces produced by chemical and topographic patterning. Tissue Eng 13:1879–1891

Livak KJ, Schmittgen TD (2001) Analysis of relative gene expression data using real-time quantitative PCR and the 2(-Delta Delta C(T)) Method. Methods 25:402–408

Louis KS, Siegel AC (2011) Cell viability analysis using trypan blue: manual and automated methods. In: Mammalian cell viability, 1st edn. Humana Press, New York, pp 7–12

Mather JP, Roberts PE (1998) An introduction to cell and tissue culture—theory and technique. Plenum Press, New York

Müller FA, Müller L, Hofmann I, Greil P, Wenzel MM, Staudenmaier R (2006) Cellulose-based scaffold materials for cartilage tissue engineering. Biomaterials 27:3955–3963

Murphy SV, Atala A (2014) 3D bioprinting of tissues and organs. Nat Biotechnol 32:773–785

Murphy CM, Haugh MG, O'Brien FJ (2010) The effect of mean pore size on cell attachment, proliferation and migration in collagen glycosaminoglycan scaffolds for tissue engineering. Biomaterials 31:461–466

Nam J, Johnson J, Lannutti JJ, Agarwal S (2011) Modulation of embryonic mesenchymal progenitor cell differentiation via control over pure mechanical modulus in electrospun nanofibers. Acta Biomater 7:1516–1524

Pati F, Jang J, Ha D-H, Kim SW, Rhie J-W, Shim J-H, Kim D-H, Cho D-W (2014) Printing three-dimensional tissue analogues with decellularized extracellular matrix bioink. Nat Commun 5:3935

Place ES, Evans ND, Stevens MM (2009) Complexity in biomaterials for tissue engineering. Nat Mater 8:457–470

Poon CY, Bhushan B (1995) Comparison of surface roughness measurements by stylus profiler, AFM and non-contact optical profiler. Wear 190:76–88

Rossini AJ, Wan JY, Meur NL (2008) Using RFlowCyt. http://bioconductor.wustl.edu/bioc/vignettes/rflowcyt/inst/doc/rflowcyt.pdf. Accessed 10 Sept 2015

Ruiz S, Panopoulos AD, Herrerias A, Bissig KD, Lutz M, Berggren WT, Verma IM, Belmonte JC (2011) A high proliferation rate is required for cell reprogramming and maintenance of human embryonic stem cell identity. Curr Biol 21:45–52
Sing KSW, Everett DH, Haul RAW, Moscou L, Pierotti RA, Rouquerol J, Siemieniewska T (1985) Reporting physisorption data for gas/solid systems with special reference to the determination of surface area and porosity. Pure Appl Chem 57:603–619
Sniadecki NJ, Desai RA, Ruiz SA, Chen CS (2006) Nanotechnology for cell-substrate interactions. Rev Ann Biomed Eng 34:59–74
Walboomers XF, Monaghan W, Curtis AS, Jansen JA (1999) Attachment of fibroblasts on smooth and microgrooved polystyrene. J Biomed Mater Res 46:212–220
Weaver JL (2000) Introduction to flow cytometry. Methods 21(3):199–201
Yuan Y, Lee TR (2013) Chapter 1, contact angle and wetting properties. In: Bracco G, Holst B (eds) Surface science techniques, Springer series in surface sciences, vol 51. Springer, Heidelberg, Berlin, Doi 10.1007/978-3-642-34243-1_1

# Index




H. Ulrich, P. Davidson Negraes, *Working with Stem Cells*,
DOI 10.1007/978-3-319-30582-0

The manufacturer's authorised representative in the EU is Springer
Nature Customer Service Centre GmbH, Europaplatz 3, 69115 Heidelberg,
Germany. If you have any concerns regarding our products, please
contact ProductSafety@springernature.com

Printed and bound by CPI Group (UK) Ltd, Croydon, CR0 4YY
30/09/2026
02225657-0010